# Simulation Foundations, Methods and Applications

More information about this series at http://www.springer.com/series/10128

John Sokolowski · Umut Durak ·
Navonil Mustafee · Andreas Tolk
Editors

# Summer of Simulation

## 50 Years of Seminal Computer Simulation Research

Springer

*Editors*
John Sokolowski
Department of Modeling, Simulation
and Visualization Engineering
Old Dominion University
Norfolk, VA, USA

Umut Durak
Institute of Flight Systems
German Aerospace
Center (DLR)
Braunschweig, Germany

Navonil Mustafee
Business School
University of Exeter
Exeter, UK

Andreas Tolk
The MITRE Corporation
Hampton, VA, USA

ISSN 2195-2817 ISSN 2195-2825 (electronic)
Simulation Foundations, Methods and Applications
ISBN 978-3-030-17163-6 ISBN 978-3-030-17164-3 (eBook)
https://doi.org/10.1007/978-3-030-17164-3



This Springer imprint is published by the registered company Springer Nature Switzerland AG
The registered company address is: Gewerbestrasse 11, 6330 Cham, Switzerland

*To Marsha, Amy and Whitney, the women in my life.*

John Sokolowski

*To Rabia and Can Levin. The miracles in my life.*

Umut Durak

*To my family and friends. Thank you for your support and understanding.*

Best wishes
Navonil Mustafee

*To all my SCS colleagues and friends who shaped our Society and Profession over the last years, and to my wife Andrea, who had to share me way too often over this time.*

All the best
Andreas Tolk

# Preface

## Introduction

The year 2018 marks the 50th anniversary of Society for Modeling and Simulation International (SCS) Summer Simulation Conference. This conference began in 1969 when Bob Brennan of IBM organized a conference on the application of continuous-system simulation languages. Since that beginning, the conference has grown into one of the two major conferences sponsored by SCS each year. To mark this historical event, we offer this book as a reflection on five decades of contributions to the discipline of modeling and simulation so that all can understand its historical significance.

To accomplish our goal of showcasing 50 years of work, we have organized this book into four major sections. The first section, an introduction, provides a historical perspective of these five decades and describes some of the major themes and ideas that evolved from this body of work. The next section draws from major contributions from SCS Fellows over the years. The title of Fellow is the highest honor bestowed on an SCS member for his or her significant contribution to both SCS and the field of modeling and simulation. The third section takes a detailed look at the conference itself and showcases two of the most influential papers presented. Finally, the fourth section highlights recent works that cover topics across several significant areas of modeling and simulation and represent the best papers for each of the 4 recent years' conferences.

## Historical Perspective

To understand the evolution of this historical event one must have an appreciation for the broader effort that got modeling and simulation (M&S) to where it is today. The catalyst behind the field of modeling and simulation is rooted in the development of analog and digital computer technology. This technology formed the foundation for using modeling and simulation as a tool for design, testing, and evaluation of real-world systems. Using this new technology for the advancement of modeling and simulation started in 1952 when John McLeod, an aerospace engineer, gathered a group of individuals to form an association to increase the effectiveness and expand the application of modeling and simulation. This organization, called the Simulation Council, produced monthly newsletters describing various M&S activities and advancements. The number of subscribers to this newsletter grew consistently, attesting to its relevance to the field. The number of councils also grew to cover various regions of the United States. In 1957, to better coordinate the effort across these various groups, the various council leaders formed a corporation called Simulation Councils, Inc. This corporation is now known as the Society for Modeling and Simulation International (SCS). The monthly newsletters also evolved into what is now the journal *Simulation*, which first appeared in 1963. To facilitate the gathering of SCS members, two conferences were organized by the SCS leadership. They were the Summer Simulation Conference and the Winter Simulation Conference. SCS continues to manage and oversee the summer conference and is one of the four sponsoring groups for the winter conference. It is the Summer Simulation Conference that is the subject of this contributed work.

For these efforts to have survived and thrived for so many years is a testament to the continued relevance of M&S to all areas of life.

## Chapter Summaries

To provide an appreciation for the chapters that follow, we provide a brief summary of the contents of each. They trace the evolution of this conference from its beginnings to current day and showcase some of the important contributions the conference has made to the field of M&S.

Chapter 1 is a panel contribution that took place during the 50th Anniversary of the *Summer Computer Simulation Conference (SCSC)*. The distinguished panel was chaired by Umut Durak and included several academics, researchers, and practitioners. Following the panel contributions at the *Golden Jubilee Panel*, the chapter begins with a brief history of SCSC and how it evolved to be a multi conference. It then discusses disease modeling, simulation software, simulation standards and reuse, M&S and AI, simulation for the future, and simulation in the age of Big Data. The chapter ends with an emphasis on the ubiquitous nature of simulation and how it is here to stay!

The next seven chapters are from the SCS Fellows. In Chap. 2, Bernard Zeigler takes a historical perspective to illustrate the interplay of abstraction, formalization, and implementation in advancing the theory and practice of M&S. The chapter then discusses the potential for the automation of some M&S activities, raising the question as to whether artificial agents/modelers/simulationists can carry out these activities without the need for human intervention.

Chapter 3 presents work on high-speed, low-cost simulation and collaboration that has spanned over five decades. SCS Fellow Roy Crosbie dedicates this chapter to the memory of his former colleagues John Hay (University of Salford) and Richard (Dick) Bednar (California State University, Chico). The chapter presents a personal reflection of the changes that have taken place in the field of computing and the simulation of power electronic systems.

In Chap. 4, SCS Fellow Tuncer Ören reflects on his contribution to the field of M&S over the past 50 years. The chapter is written in the style of a professional autobiography. It includes references to almost 50 first-authored publications and will serve as a reference point for scholars interested in learning about Ören's work, including his *Simulation Body of Knowledge* initiative.

The author of Chap. 5 is SCS Fellow Andreas Tolk. Tolk reviews existing work on CAS, and draws from the philosophy of science and the concepts of emergence, to argue that significant epistemological constraints exist, which limit the usefulness of computer simulation for research in complex adaptive systems.

SCS Fellow Lin Zhang contribution (Chap. 6) is co-authored with colleagues Fei Wang and Feng Li (not SCS Fellows) from the School of Automation Science and Electrical Engineering, Beihang University, China. The paper is on Cloud-based simulation and M&S as a Service (MSaaS). It presents the architecture of a cloud simulation platform and key technologies like virtualization, resource discovery, composition, and scheduling for cloud-based simulation.

Chapter 7 of the book includes the contribution of SCS Fellow Gabriel A. Wainer. The chapter is written in a tutorial style, and presents different modeling methods (e.g., Petri Nets, Finite State Machines, Bond Graphs, Modelica) and their translation into discrete-event systems specifications using the DEVS formalism.

Chapter 8 presents a profiling study of research publications by Navonil Mustafee and Korina Katsaliaki. The objective of this study was to give an overall picture of research published in the SCSC proceedings from 2005 to 2017. The dataset for the analysis consisted of 911 papers. The chapter presents seven different analysis, including, metrics associated with authorship and affiliations, sources of funding, analysis of keywords, and citation analysis.

Chapter 9 is from the author of the *SummerSim 2015 best paper award*—Saurabh Mittal. The paper is on M&S in complex systems engineering. It presents a case of Synthetic Emergence, which is defined as emergence engineering in artificial environments, and provides the pre-, post-conditions and the engineering process to achieve synthetic emergence. It also presents the Simulation Experience Approach (SEA) framework to engineering Live, Virtual and Constructive (LVC) simulation environments.

Chapter 10 is on pedestrian modeling using spatial discrete-event M&S. The authors of this chapter (Gabriel Wainer and Ala'a Al-Habashna) present a Cell-DEVS-based approach for M&S of crowds. The paper includes a review of related work, which is categorized under different modeling approaches, for example, fluid dynamics, social-forces model, agent-based models, and CA models). The proposed Cell-DEVS-based approach is verified by employing case studies related to building and fire evacuation.

Chapter 11 is from the authors of the *SummerSim 2017 best paper award*—Donald J. Berndt, David Boogers, Saurav Chakraborty, and James McCart. This chapter is an extended contribution to their original paper with the same title, "Using agent-based modeling to assess liquidity mismatch in open-end bond funds." The chapter presents an agent-based model of the US corporate bond model. The model is used to simulate realistic market dynamics such as the effect of increasing mutual fund market shares under certain assumptions. The model includes several agent-types (both buyer-side and seller-side agent) and assesses results from agent behaviors in response to external factors such as an increase in interest rates.

The final paper of this book chapter (Chap. 12) is also on agent-based simulation. The authors Rishi Bubna, Jayasree Raveendran, Suman Kumar, Mayuri Duggirala, and Mukul Malik, use ABS to model the likely impact of demonetization on a cash-intensive economy. The likely motivation for this research comes from 2016 demonetization drive that was undertaken by the Indian Government, and the resultant model is able to depict the demonetization outcomes in India. The study further contributes to literature through its use of grounded theories from behavioral sciences literature.

Norfolk, USA — John Sokolowski
Braunschweig, Germany — Umut Durak
Exeter, UK — Navonil Mustafee
Hampton, USA — Andreas Tolk

# Contents

# Contributors

**Ala'a Al-Habashna** Department of Systems and Computer Engineering, Carleton University, Ottawa, ON, Canada

**Jacob Barhak** Independent Researcher, Austin, TX, USA

**Donald J. Berndt** Muma College of Business, University of South Florida, Tampa, FL, USA

**David Boogers** Muma College of Business, University of South Florida, Tampa, FL, USA

**Rishi Bubna** Tata Research Development and Design Center, Pune, Maharashtra, India

**Saurav Chakraborty** Muma College of Business, University of South Florida, Tampa, FL, USA

**Roy Crosbie** California State University, Chico, CA, USA

**Andrea D'Ambrogio** Università Di Roma "Tor Vergata", Rome, Italy

**Saikou Diallo** ODU VMASC, Suffolk, VA, USA

**Mayuri Duggirala** Tata Research Development and Design Center, Pune, Maharashtra, India

**Umut Durak** German Aerospace Center (DLR), Brunswick, Germany

**Ralph Coolidge Huntsinger** California State University, Chico, CA, USA

**Korina Katsaliaki** School of Economics, Business Administration and Legal Studies, International Hellenic University, Thessaloniki, Greece

**Suman Kumar** Tata Research Development and Design Center, Pune, Maharashtra, India

**Feng Li** School of Automation Science and Electrical Engineering, Beihang University, Beijing, China

**Mukul Malik** Tata Research Development and Design Center, Pune, Maharashtra, India

**James McCart** Muma College of Business, University of South Florida, Tampa, FL, USA

**Saurabh Mittal** The MITRE Corporation, VA, Mclean, USA

**Navonil Mustafee** The Centre for Simulation, Analytics and Modelling (CSAM), University of Exeter Business School, Exeter, UK

**Tuncer Ören** Emeritus Professor of Computer Science, School of Electrical Engineering and Computer Science, University of Ottawa, Ottawa, ON, Canada

**M. S. Raunak** Loyola University Maryland, Baltimore, MD, USA

**Jayasree Raveendran** Tata Research Development and Design Center, Pune, Maharashtra, India

**José L. Risco-Martín** Complutense University of Madrid, Madrid, Spain

**Andreas Tolk** The MITRE Corporation, Hampton, VA, USA

**Gabriel A. Wainer** Department of Systems and Computer Engineering, Carleton University, Ottawa, ON, Canada

**Fei Wang** School of Automation Science and Electrical Engineering, Beihang University, Beijing, China

**Gregory Zacharewicz** IMT Mines Alès, Alès Cedex, France

**Bernard P. Zeigler** University of Arizona, Tucson, AZ, USA;
RTSync Corp., Phoenix, AZ, USA

**Lin Zhang** School of Automation Science and Electrical Engineering, Beihang University, Beijing, China

# Chapter 1
# 50 Summers of Computer Simulation

**Umut Durak, Andrea D'Ambrogio, Andreas Tolk, Saikou Diallo, Gregory Zacharewicz, José L. Risco-Martín, Jacob Barhak, Ralph Coolidge Huntsinger and M. S. Raunak**

**Abstract** We are having seasons: summers and winters of many scientific disciplines. Many fields are experiencing hype cycles. Each one of us would remember "AI winter" from the history of Artificial Intelligence. Inflated expectations are followed by disappointment and eventually funding cuts. Renewing the interest takes then years if not decades. The Society for Modeling and Simulation International has achieved outstanding success in the last 50 years to keep Summer Computer Simulation Conference (SCSC) an important event through many seasons of simulation, some of which were more remarkable than others. This chapter summarizes the panel discussion/contributions of the SCSC 2018 about the seasons in computer simulation and the ways to achieve and further prolong summers of computer simulation.

U. Durak
German Aerospace Center (DLR), Brunswick, Germany

A. D'Ambrogio
Università Di Roma "Tor Vergata", Rome, Italy

A. Tolk (✉)
The MITRE Corporation, Hampton, VA 23666, USA
e-mail: andreas.tolk@gmail.com

S. Diallo
ODU VMASC, Suffolk, VA 23435, USA

G. Zacharewicz
IMT Mines Alès, 30319 Alès Cedex, France

J. L. Risco-Martín
Complutense University of Madrid, 28040 Madrid, Spain

J. Barhak
Independent Researcher, Austin, TX, USA

R. C. Huntsinger
California State University, Chico, CA 95929, USA

M. S. Raunak
Loyola University Maryland, Baltimore, MD 21210, USA



J. Sokolowski et al. (eds.), *Summer of Simulation*, Simulation Foundations, Methods and Applications, https://doi.org/10.1007/978-3-030-17164-3_1

## 1.1 Introduction

In 2018, the Summer Computer Simulation Conference (SCSC) marked an important event, as another conference under leadership of the Society for Modeling and Simulation International (SCS); it celebrated its 50th anniversary, its golden jubilee. Having the honor to chair the 50th SCSC, I (*Umut Durak*) used the opportunity to organize a panel discussion in which we discussed the past and the future of SCSC and in this context of simulation.

Frequently scientific disciplines are experiencing hype cycles. Inflated expectations and large resources come together. The following disappointment usually leads to funding cuts. "AI Winter" from the history of Artificial Intelligence (AI) is a famous instance of this pattern. Simulation is not an exception. It is surely true that simulation as a discipline (Tolk and Ören 2017) and simulation-based disciplines (Mittal et al. 2017) had also ups and downs, some of which were more remarkable than others. Notwithstanding, the golden jubilee of SCSC, as its name suggests, is giving us a never ending 50 years of summer feeling.

This chapter summarizes the discussion from the *SCSC Golden Jubilee Panel*. It follows the order of speakers. We start with Ralph Coolidge Huntsinger, an eminent fellow of SCS and the historian of the society. He introduced us the first years of SCSC and how it evolved to a multi-conference, named as SummerSim. Then Gabriel Wainer gave a brief history of the later years with an outlook.

SCSC started 50 years ago as a simulation application conference. The methodology and tools areas were added over the years. I tried to structure the panel discussion in the same way. After the history part the application domains were discussed along with the SCSC. Jacob Barhak presented the seasons of disease modeling. The following section on methodology and tools consisted of a discussions about simulation software from José L. Risco-Martín, standardization and reuse from M S Raunak, and modeling and simulation (M&S) and AI from Andrea D'Ambrogio. The panel ended with the last three speakers focusing on the future of simulation: Gregory Zacharewicz discussed the "future" and simulation, Saikou Diallo promoted simulation for the big problems, and Andreas Tolk emphasized the ubiquitous nature of simulation today and in the future.

## 1.2 The History of Summer Computer Simulation Conference

### 1.2.1 *The First Years*

I (*Ralph Coolidge Huntsinger*) am the only person left alive who has been to all 50 SCSCs. The first one was by Takeshi Utsumi, who is an emeritus of the Columbia University at New York. I met him in 1970 in Denver. Then I got involved in SCSC, presented papers, later being a session chairman or being a program chairman. Eventually in 1979, I was the overall program chair of the SCSC and we had a "big" conference. This was really a "great" fun because SCSC was the prime conference and our focus was simulation applications. Now we have more philosophical ideas about simulation. That time probably more than the half of the attendees were coming from industry. This changed slowly with the increasing interest to the conference from the academia. The conference focus also evolved towards modeling and simulation methodologies and tools. The methodology and tools group eventually got so big that they became a separate conference that is co-located with SCSC. This was the foundation of the SummerSim Multi-Conference. Later, Roy Crosbie started the grand challenges of simulation track, which took a considerable interest from the defense modeling and simulation community. The track still exists in the 50th edition.

### 1.2.2 Establishment and Beyond

My (*Gabriel Wainer*) first SummerSim was in 1999, in Chicago. Coming from Argentina, my first SCS conference in the USA was something impressive, in a historic hotel downtown. Mohammad Obaidat opened the conference, and I was able to listen to top presentations on simulation topics from experts that had been involved in the conference during the first 30 years. I had the honor to meet great researchers like Axel Lehmann, François E. Cellier, Hassan Rajaei, Josep Granda, Hans Vangheluwe, and many others. Methodologies and applications on continuous and discrete-event simulation showed the maturity of a research field that was ready to move towards the next stage. At that time, SummerSim hosted SPECTS, the Symposium on Performance of Computer and Telecommunications Systems, where I presented my research on serialization of Cell-DEVS models (Wainer and Giambiasi 1999). I was able to discuss my research with top experts in the area who gave me insight on how to advance in the field. At that time Cellular Automat0061 (CA) were popular to solve complex problems (Wolfram 1986). CA are discrete-time discrete models described as n-dimensional lattices. The use of discrete time poses constraints in the precision and execution performance of these complex models, and Cell-DEVS solved these issues combining DEVS and CA (Wainer 2009). At that time there was research on how to use multiple processors and on how to prevent inherently parallel models from executing serially.

At that time, there was varied research showing how to transform modeling formalisms. We presented various efforts in this sense in SummerSim; first, numerous CA models (Ameghino and Wainer 2000), Petri Nets (Jacques and Wainer 2002), Finite State Machines (Zheng and Wainer 2003), and later Bond Graphs (D'Abreu and Wainer 2006) and Modelica (Chechiu and Wainer 2005). At this time, DEVS was proven to be the most generic Discrete-Event system Specification (Vangheluwe 2000). Another important fact was the research on modeling of continuous and hybrid systems trying to provide a uniform approach to model hybrid systems, i.e. composed of both continuous and discrete components.

A new idea proposed to use quantization of the state variables to obtain a discrete-event approximation of the continuous system (Kofman and Junco 2001).

At this time, simulation technology also took another direction: web-based simulation. Our first article in SummerSim (Wainer and Chen 2003) showed a mechanism for remote execution of simulation models using web-based technologies. Shortly after, a large number of researchers focused on model standardization, and the High Level Architecture (HLA) standard was in focus (Saghir et al. 2004), spawning hundreds of researchers focusing on how to coordinate distributed simulations on the web. At present, and in the future, simulation will be ubiquitous, and there are numerous examples of simulation software running in the cloud and on mobile devices (Jeffery et al. 2013). Many of the coordination algorithms for distributed simulation were based on a large body of knowledge in the field of parallel discrete-event simulation, in which a number of conservative and optimistic algorithms and variations provided the means to guarantee correct execution and to prevent causality errors (Jafer and Wainer 2011).

Many advanced implementations are now built on Web Services to communicate. Nevertheless, building Service Oriented Architecture (SOA)-based simulations is still complex, as the services usually address the interoperability of simulation engines at a low level of abstraction. In recent years, Grid and Cloud computing introduced new ways of sharing computing power and storage in heterogeneous environments in which resources are virtualized as services consumed on demand (with minimal limitation for resource location). The Representational State Transfer (REST) style can help solving the interoperability limitations, and it makes easy the development of mashups, which can be developed in a shorter period of time The future of this area lies in making interaction of these distributed models more efficient using a plug-and-play model and easier reuse of existing models. We had the chance to see this field evolve in the 2010's, in particular in SummerSim 2007 and 2010, which I helped organizing. I was able to involve Andreas Tolk and Pieter Mosterman in the organization of the conference, and later we involved younger and energetic emerging leaders like José-Luis Risco-Martín, Saurabh Mittal, Saikou Diallo, Gregory Zacharewicz, Abdy Abhari, Shafagh Jafer, Mohammad Moallemi, and Andrea D'Ambrogio, who have shaped the conference to be in good hands for the next 50 years.

Recent years have also seen an evolution of simulation and real-time applications. Real-Time (RT) systems are built as sets of components interacting with their surrounding environment. These are highly reactive systems, in which not only correctness is critical, but also the timing for executing the system tasks. Failing to guarantee that all the computations meet their deadlines could be catastrophic. Most design methods are known to be hard and difficult to apply to large-scale systems, and they do not guarantee error-free systems. In recent years, Modeling and Simulation (M&S) has been used as a practical approach to verification of these systems with reduced costs and risks (Shang and Wainer 2007; Yu and Wainer 2007; Ahmed et al. 2011). These techniques allow testing the systems in a risk-free environment. In particular, formal M&S provides even better results as the software artifacts can be built faster and safer, and the formal models can be used for formal verification. However, M&S techniques often require extra effort to model features of specific target systems, for example, the timing constraints in RT systems.

We thus suggest that, although there have been numerous advances in this field; that the following questions still need to be addressed:

1. How to interface simulation software with Smartphone Application Programming Interfaces (APIs)? How to deal with the inherent performance issues of these devices (power consumption, CPU speed, communication latency)?
2. How to deal with power and communication disruptions?
3. How to enable multi-user collaboration between numerous users participating in a joint experiment?
4. How to integrate different online services and real-time data available in the Cloud?
5. How to include advanced algorithms and methods for combining discrete-event simulation, cloud computing (with web service interfaces) and mobile devices for distributed simulation and collaboration?
6. How to build advanced mashup applications using simulation, sharing and reusing models and experiments?
7. How to use a more abstract approach to deal with these problems? (Instead of dealing with the data and simulation levels, interoperability will be dealt with at the modeling and experimentation levels, improving reuse and providing better ways to mashup models, experiments and other services).
8. How to handle the large amounts of simulation data through instrumentation of scenarios, aggregation policies and dynamic adaptation of the simulation to varying computing conditions based on different policies?
9. How to integrate RT tasks and simulation in a seamless fashion?
10. How to integrate machine learning methods and simulation? (Floyd and Wainer 2010)
11. How to interface simulation models for telecommunications and networking protocols in real-time?

50 more years and we will have Simulation Everywhere!

## 1.3 Summers of Simulation Application

### *1.3.1 Seasons of Disease Modeling*

The burden of disease was a good incentive for modelers to analyze diseases and attempt to forecast their progression. Chronic diseases that bear long term economic effects were targeted. Specifically the burden of heart disease was in focus and continues to be in focus until these days. In this section, I (*Jacob Barhak*) would like to roughly summarize trends and seasons within some disease modeling communities in the last three decades (Table 1).

**Table 1** Seasons of disease modeling

| Season | Modeling Focus |
|---|---|
| 1980s+ | Markov Models |
| 1995–2004 | Risk equations |
| 2000s | Microsimulation & Discrete Event Simulation |
| 2010s | High Performance Computing |
| 2015+ | Model and Data Exchange Standards |

An early famous disease model was created by Weinstein for coronary heart disease (Weinstein et al. 1987). It was a fairly complex model for that time since it included influx of population whereas most disease models use a fixed starting baseline population, even these days. Most models at that time were constrained by computational power and therefore simpler modeling techniques were used.

Markov models were considered the standard and even the state of the art. In those models, cohorts of individuals were modeled and each state counted the number of individuals in that state. Transitions between states were governed by transition probabilities. Correct estimation of those transition probabilities was part of the art of modeling. This kind of modeling was not restricted to chronic diseases and one good example of a Markov disease model came from the mental health perspective (Leff and Dada 1986), a basic explanation for medical prognosis is provided in (Beck and Pauker 1983).

However, Markov models were no longer sufficient since modelers wanted to get additional information such as age and gender into the model. Chronic disease modeling therefore started transitioning into more complicated simulation types that allow incorporation of population parameters.

Risk equations that model populations were extracted from large longitudinal cohort studies. One of the most famous studies was the Framingham study that produced the Framingham model (Wilson et al. 1998; D'Agostino et al. 2008) that represented heart disease complications and was updated multiple times. It is perhaps the most famous disease model to date.

Similar attempts were carried by the United Kingdom Prospective Diabetes Study (UKPDS) group that generated multiple models for different aspects of diabetes. These included risk equations (Stevens et al. 2001; Kothari et al. 2002) and cost effectiveness models (Clarke et al. 2005). Eventually the group assembled the pieces into a more advanced microsimulation model (Clarke et al. 2004; Hayes et al. 2013). Such a model is used to conduct Monte Carlo simulations and required more computing power and provided richer modeling capabilities.

The diabetes modeling community then started comparing these richer models. The Mount Hood challenge was created (Brown et al. 2000; The Mount Hood 4 Modeling Group 2007; Palmer et al. 2013) and diabetes modelers from around the world started meeting on a regular basis to compare and contrast their models. Most interesting were the validation challenges that repeatedly showed the differences between models and the gap of understanding of disease processes.

Regular participants in those challenges included the UKPDS model, IMS model, IHE model, Mikado, and the Michigan Model.[1] The latter model was interesting since work on it started with Markov modeling in mind with emphasis on parameter estimation based on multiple trials (Isaman et al. 2006), yet over the years it evolved into a micro-simulation model with a set of public tools to support it (Barhak et al. 2010).

Another notable participant in those challenges was the Archimedes model (Eddy and Schlessinger 2003; Schlessinger and Eddy 2002). It went past the trend of building models on top of one population set and validated the model against multiple studies. The model was advanced and addressed issues of efficient simulation through discrete event simulation and dedicated code development. It also addressed issues of baseline population generation.

However, despite the efficiency of the model, it was not emphasizing the use of the growing availability of computing power.

The Reference Model[2] (Barhak 2017) that was developed in the last decade took advantage of High Performance Computing (HPC) capabilities and could run simulations on the cloud. The Reference Model was a split from the Michigan Model and its open source set of tools. It took the idea of knowledge accumulation to the next level since the model validates multiple models against multiple populations. Its use of public data sources and open source tools allow simulations that compare models. It uses optimization techniques used in machine learning and evolutionary computation to accumulate knowledge.

It is likely that future simulations will use the availability of computing power and open source code to better accumulate knowledge. Yet the current simulation problem is less about modeling technique or modeling approach. The disease modeling community is now facing a crisis of reproducibility. The 2016 Mount Hood Challenge exposed this when multiple teams of modelers around the world could not reproduce two published models.

Therefore future disease modeling will have to focus on model exchange mechanisms such as the Systems Biology Markup Language (SBML),[3] the initial work has started in this direction (Smith et al. 2016). Moreover, recently disease databases started to gain traction. Two notable examples are ClinicalTrials.Gov (Ide et al. 2016; Zarin et al. 2016) and the Global Burden of Disease (GBD) database.[4] Those databases now feed information into newer disease models. Therefore the modeling focus in the next decade will most probably move towards model and data sharing.

---

[1]Michigan Model for Diabetes. Online: http://diabetesresearch.med.umich.edu/Core_MCDTR_Methods_DM_MMD.php.

[2]The Reference Model for Disease Progression. Online: https://simtk.org/projects/therefmodel/.

[3]SBML.org The Systems Biology Markup Language: http://sbml.org/Main_Page.

[4]26. IMHE——Global Burden of Disease (GBD) online: http://www.healthdata.org/gbd.

## 1.4 Summers of Methodology and Tools

### *1.4.1 Simulation Software: A Historical Perspective and Future Trends*

Simulation is one of the most multifaceted topics present today in both industry and academia. Simulation has been traditionally used as a tool to increase production and capacity. Nowadays, many other aspects are studied in simulation, like analysis, reliability, scalability, verification, validation, human training, etc. To visualize the future of simulation, we must first analyze its historical perspective. Such perspective can be presented from several angles: uses of simulation, simulation languages, simulation environments or application domains. Here I (*José Luis Risco Martín*) offer a brief treatment from the perspective of the simulation software, which somehow captures simulation languages and simulation environments. This section shows the historical perspective of the simulation software during the last 50 years. It also provides insights regarding the future directions of simulation software and simulation paradigms and how the SCSC can be a main witness of this future.

My discussion of the history of simulation software is based on Nance (Nance 1995). I have taken, simplified and adapted his original classification periods into a more "modern" point of view, resulting in the following four periods:

- [1955–1974] First simulation languages
- [1975–1989] First consolidation and regeneration
- [1990–2008] Integrated environments
- [2009–????] Second consolidation and regeneration

**1955–1974 First simulation languages**
Simulation was firstly conducted in FORTRAN and other general programming languages. Obviously, there was no support for simulation specific routines. In 1960, K.D. Rocher and D.G. Owen launched what is considered the first simulation language effort, named the General Simulation Program. Later from 1961 to 1965 several Simulation Programming Languages (SPL) appeared, like the General Purpose Simulation System (GPSS) developed by Geoffrey Gordon at IBM. Philip J. Kiviat began the development of the General Activity Simulation Program (GASP), in 1961. Hary Markowitz provided the major conceptual guidance for SIMSCRIPT in 1963. In Europe other simulation programming languages appeared, like SIMULA or the Control and Simulation Language (CSL).

From 1966 to 1974, the previous tools were upgraded: GPSS was released as GPSS/360 and later as GPSS/NORDEN, SIMSCRIPT evolved to SIMSCRIPT II, GASP to GASP IV, and CSL to ECSL. SIMULA also added some object-oriented programming concepts, considered the initial steps towards the modern object-oriented programming languages.

**1975–1989 First consolidation and regeneration**

During this period, traditional SPLs were adapted to desktop computers, with the era of the microprocessor. GPSS/H was also released in 1977 for specific IBM mainframes, which became the principal version of GPSS in use today. Two major descendants of GASP appeared: the Simulation Software for Alternative Modeling II (SLAMM II) and the SIMulation Analysis (SIMAN), both including multiple modeling perspectives and combined modeling capabilities. SIMAN was the first major simulation language for the IBM PC and designed to run in MS-DOS.

**1990–2008 Integrated environments**

This period is remarkable by the growth of SPLs on the personal computer and the creation of many simulation environments with graphical user interfaces, automatic reports generation, data analyzers, animation and specific visualization tools. Most of these environments attempt to simplify the modeling process avoiding the need to learn a programming syntax. Animations include from schematic-like representations to 2D and 3D approximations to reality.

Some of the most popular integrated environments created over this period include: Arena, AutoMod, Extend, Flexim, Micro Saint, ProModel, Quest, Simul8 or Witness.

**2009–2018 Second consolidation and regeneration**

This period is completely analogue to the first consolidation and regeneration period, but instead of an accommodation of the traditional simulation software to the personal computer, in this case we have had an evolution from multi-processor integrated environments to the cloud simulation or the simulation as a service in cloud infrastructures. The evolution of traditional message passing programming techniques (through MPI for example) to modern cloud programming paradigms has facilitated the appearance of new modeling and simulation paradigms like DEVS/SOA (Mittal 2009). Beyond that, a plethora of new simulation software (or an evolution of the previous integrated environments) has appeared in the last ten years. Some examples we found are: AnyLogic, Arena, AutoMod, Enterprise Dynamics, ExtendSim, FlexSim, GoldSim, GPSS, MS4, Plant Simulation, ProModel, Simcad Pro, SimEvents, Simio, Simul8, VisualSim, WitNess, DESMO-J, Ptolemy II, SimPy, SystemC, etc.

SCSC is 50 years old. It is enough to presume a vast experience. Regarding simulation software, SCSC has seen brilliant papers focused on simulation languages, platforms and tools like:

- J. Leon, C. O. Alford and J. Hammond (1970). DIHYSYS—a hybrid systems simulator.
- T. I. Ören (1971). A basis for the taxonomy of simulation languages.
- G. E. Miles, R. N. Peart, and A. A. B. Pritsker (1976). CROPS: A GASP IV Based Crop Simulation Language.
- R. M. Fujimoto (1985). The SIMON simulation and development system.

- M. Gourgand and P. Kellert (1992). An object-oriented methodology for manufacturing systems modelling.
- O. Balci and R. E. Nance (1998). A taxonomy of layout composition techniques for visual simulation.
- J. Ameghino, E. Glinsky and G. Wainer (2003). Applying Cell-DEVS in Models of Complex Systems.

There is still room for 50 or 500 years of new simulation engines. In a new world full of complex systems of systems and with very exigent time-to-market constraints, SCSC should serve as a conductor vehicle to check the validation of such simulation software, its performance, applicability, scalability, and usefulness to both industry and academia. SCSC must serve as a joint forum to converge to (why not) a unified methodology in the art of modeling and simulation, always from a pure practical point of view.

### *1.4.2 Standardization and Reuse*

In the past 50 years, we have had great success in defining the discipline of modeling and simulation. The discipline has steadily grown and found its presence in many disciplines including, but not limited to, engineering, computer science, operations research, and management science. Amongst many highlights from this time period, we can include:

- Establishing processes and practices for developing effective simulation models and performing useful studies using them.
- Development of general purpose as well as domain specific languages for creating simulation models.
- Development of many commercial and non-commercial simulation frameworks.
- Establishing rigorous and formal approaches to define the simulation models.
- Establishing techniques and processes for verification and validation of simulation models.

The rapid advancement of technology and its use is likely to facilitate the proliferation of using simulation in many more areas over the next 50 years.

At the Golden Jubilee year, one important question that I (*M S Raunak*) would like to ask is: How mature does the discipline of modeling and simulation appear to be? An approach towards judging the maturity of the field would require us to contemplate the following questions:

- Do we have rigorous building blocks for creating models and performing simulation studies with them?
- Do we have fundamental rules to govern the activities of simulation practitioners?

- Are simulation studies getting reproduced regularly for corroborating their results or to identify potential problems?
- Do we have a standardized way of communicating the verification and validation (V&V) performed on a model?
- Are there building blocks that are readily available for reuse in constructing new models?

The answers to many of these questions would still come out to be negative. This is an indication that the field has not yet matured like some of the other natural science and engineering disciplines.

The simulation community needs to continue working on developing standards for simulation modeling approaches, notations, implementation, and experimentation. In the 2016 winter simulation conference, there was a panel discussion on standards related to smart manufacturing systems focusing on data, process, and environmental aspects (Beck et al. 2016). More general purpose standards need to be developed in every area of simulation. One collaboration that could facilitate and fast-track this process is to involve government research agencies such as the National Institute of Standards and Technology (NIST) in the US.

We also need to work on developing the practice of proper use and reporting of validations performed on simulation models presented in published research. A 2014 survey identified the unusually low validation efforts reported in health-care related simulation research (Raunak and Olsen 2014a). This leads to reduced confidence in published results, which in turn, reduces the use of the results in the real world policy-change. In addition to following established standards for modeling and performing verification and validation, a standard way of communicating about them is also needed. A new set of research has shown ways to quantify and communicate about simulation validation (Raunak and Olsen 2014b; Olsen and Raunak 2015). With the adoption of such standard processes and practices, we are going to increase our confidence in our simulation experiments and results. Verification and validation of simulation models can also benefit from new approaches and techniques from other fields such as software testing (Olsen and Raunak 2016).

A key factor of a mature research field is the practice of reusing models, components, frameworks, and reproducing results. The simulation community is still behind in achieving reasonable progress in this regard. There are some application areas such as network simulation or agent based models, where standard modeling tools have facilitated some level of model sharing and reuse. In many other areas, sharing and reuse of simulation models remains a rarity. Factors including closed or classified environments (e.g. military simulation), intellectual property, and the size and complexity of the models have challenged the proliferation of model reuse and result reproduction. There are encouraging signs in the community, however, as they have recognized the need for reproducible research and the challenges surrounding it (Uhrmacher et al. 2016). What is missing in the discussion so far is the establishment of a repository of simulation models and related artifacts for researchers and practitioners to use in their experimentation and analysis.

The software engineering community has greatly benefited from the creation of the Software artifact Infrastructure Repository (SIR)[5], where researchers get access to many different versions of software programs, test suites, bug reports and other software artifacts to perform rigorous controlled experiments. This has immensely facilitated the reuse and reproduction of research results. Developed and maintained through collaborative efforts from multiple institutions and funding support from the National Science Foundation, SIR artifacts have been used by more than 600 universities and research institutes all over the globe and have resulted in at least 700 software analysis and testing related research studies and publications. To facilitate the path to becoming a more mature discipline, a simulation model related repository is extremely essential and will help us leap-frog in the direction of model reuse and result reproduction.

The field of M&S has made great strides in the last 50 years. M&S now permeates many different areas of scientific research. With the exponential growth of technological advancement, especially in the areas of medicine, robotics, autonomous cars, unmanned aerial vehicles, and smart cities, M&S is likely to see another fifty years of intense activity. However, we need to take stock of the places we are lagging behind in terms of maturing as a scientific discipline. More focus on developing and using rigorous standards in all areas of simulation is an important aspect. Encouraging government standard bodies to get more actively involved in this process will benefit us a lot.

We need to put more emphasis on V&V activities in simulation and an effective standard way of talking about it. Our community needs to look into the factors that have contributed to a lack of reuse and reproducible results. Establishing a simulation-artifact infrastructure repository is a very important need of the community and it will help graduate the field to become a more matured discipline. Finally, with the advent of new technologies, confluence of disciplines and the challenges that come with them, our community needs to be open and in the search for discovering and adopting new ideas, methods and processes. With the right focus and a collaborative effort, the next fifty summers of computer simulation are surely going to be exciting!

### *1.4.3 M&S and AI: The Odd Couple*

Celebrating the 50th anniversary of the SCSC is a significant step that witnesses the relevance and longstanding tradition of the computer simulation field, which is now preferably referred to as the M&S field, so to emphasize the role of modeling as the essence of any computer-based simulation effort.

Significant contributions can be found that report about the history of M&S in terms of different generations of simulation software, such as in (Nance 1995).

[5]The software-artifact infrastructure repository. http://sir.unl.edu/portal/index.php.

What I (*Andrea D'Ambrogio*) would like to focus on is instead the thin (sometimes thick) thread that connects the lifelines of M&S and AI, by specifically looking at some cases in the past 50 years in which M&S has provided support to and/or has been influenced by AI technologies, as well as by giving a look at how these technologies could have an impact on the next generation of M&S.

A clear example of M&S and AI crossing their paths is dated back to the 1980's, when expert systems brought a renewed enthusiasm in the AI field after one of the so-called "AI winters", i.e., periods of reduced funding and interest that have been experienced as a direct consequence of over-inflated and unmet expectations. Expert systems were seen as systems capable of simulating the knowledge and the analytical capabilities of human beings, and simulators were essentially seen as knowledge-based expert systems using a combination of symbolic reasoning and data processing.

The proceedings of SCSC editions held in that period provide several contributions that refer both to the use of M&S to improve the prediction properties of expert systems (see, e.g., Vansteenkiste 1985) and to so-called "expert simulation systems", or systems that result from the combination of expert systems and conventional simulation technology, so to solve complex simulation problems [3]. In the same period, specific events were organized that focused on the combination of M&S and AI (see, e.g., Xindong 1990 and Uttamsingh and Wildberger 1989). The unmet hype around expert systems led to another winter of AI, in the first half of the 1990s, with a consequential dropped interest by M&S researchers and practitioners (Gupta and Biegel 1990).

In the last two decades AI revamped, mostly due to the incredibly vast amount of storage and processing resources, which led to the introduction of innovative techniques, specifically those under the umbrella of machine learning and, recently, deep learning. Analogously, a renewed interest has been observed in the M&S community, with contributions focused on the tremendous potential resulting from the synergy of M&S, big data and deep learning for the next generation of M&S (Tolk 2015).

It's not known if another winter of AI will be observed sooner or later, as skeptics predict in response to some well-known failures of deep learning applications in relevant domains, such as autonomous driving. What is known is that M&S didn't experience similar "winters" in its history, as also witnessed by the 50 "summers" of simulation success stories reported in SCSC proceedings. M&S is no more seen as a variant of the experimental method, but as a novel way of doing science, thus being recognized as the third pillar of science, alongside the traditional theory and experiment pillars.

The lesson we can learn from the paths crossing AI and M&S is that M&S should not chase AI technology only to properly integrate the latest advances into M&S efforts, but rather exploit the potential behind such advances as the driver of foundational innovations that would allow M&S to work out problems that are harder or impossible to solve.

In this respect, what I expect for the next generation of M&S is something that goes far beyond the integration of recent and future technology advances and the availability of almost unlimited amounts of data and resources.

What is ever sought and deserves to be addressed is closing the "reality gap", which refers to the difficulty of transferring simulated experience into the real world. Significant efforts in the computational biomedicine field, which aim to build the "virtual human" through the modeling and simulation of all aspects of the human body, from the genomic level down to the whole human (Lumley and Pringle 2017), are not intended to produce successful results only by exploiting the most advanced high-performance computing facilities.

To approach this and similar ambitious objectives, the role of modeling approaches for properly representing the observed reality will be essential, with the simulation aspects dealt with by increasingly powerful model transformation and execution platforms (Bocciarelli and D'Ambrogio 2016). The ability to use available data to properly build, map and orchestrate models at various levels of abstraction will be key to nurture an effective and successful M&S development in the next years.

## 1.5 Upcoming Summers

### *1.5.1 Is Simulation An Option for the Future?*

Maybe my (*Gregory Zacharewicz*) speculation developed in this session about the future of simulation can start by asking ourselves the following general question: Do we have the choice? Is simulation an option for the future? The world is changing quickly, due to human progress, new technologies appear and help humans being more performant in all domains but at the same time human activity has a damaging influence on the environment. It appears that simulation will have to be kept at the center for the analysis of past phenomena and anticipation of the future. It will be a lever for decision making process support, giving clues and answers to anticipate the potential changes (desired and undesired) that are already appearing in the world. From my point of view, one central question that is now in all minds is: Will the world of tomorrow be still livable enough for humans or will the future generations have to spend, in the better case, their lifetime in "artificial bubbles" of simulated worlds? So simulation is now demanded to anticipate this future and convince people that it is urgent to change. But if we fail, simulation might still also be there as the only way to participate and to interact with our lost environment, where a virtual world will progressively replace all the ecosystems that will be broken down.

This pessimistic hypothesis of an open-air world that will be almost not livable (dystopian world), was still very unlikely only 10–20 years ago. But one century of human impacts accumulated on the environment has rushed effects within a short period of 10 years demonstrating that the situation is critical. The climate changes are warning us more frequently than ever about the potential ruin of our environment. It has been now developed in numerous scientific visions that the future will be difficult (Schiermeier 2018). Scientists predict that the conditions for life will be potentially broken within the next 100–200 years. For instance, microbiologist Frank Fenner postulated in 2010[6] that “humans will probably be extinct within 100 years, because of overpopulation, environmental destruction, and climate change”. Others such as (Nolan et al. 2018) report that biodiversity will be drastically affected by pollution emission. Rather optimistic scenarios are now not very common in the scientific field.

In the domain of literature and arts, on the one hand, fiction scenarios were providing utopia visions of a future with a mostly positive development. Alexandra Whittington, forecasting consultant at Fast Future[7], enumerates that scientists, including the late Stephen Hawking, already warned that we have only 100 years of life left on earth. However, she is part of the more optimistic branch that thinks that we still may have a desirable, functional and safe ecosystem for future generations if we start reacting now. On the other hand, real life with a broken environment in the next 200 years is depicted in several opinion talks such as the one intoned by astrophysician Aurélien Barrau at the climax festival[8] and also in popular books and movies such as Mad Max and the recent Ready Player One from Spielberg. To convince the people that do not consider this situation as critical, simulation can help to run, verify and validate warning scenarios.

Simulation can anticipate accurately this future; it has to be a tool for avoiding irreversible situations.

In addition to this crucial use of simulation in a close future, several domains will be calling simulation to give support and answers to human life.

The industrial domain, for instance under the keyword Industry 4.0, is full of perspectives only reachable thanks to simulation. We, by the use of simulation, want to glue many different things (technologies, software, concepts, …) together but not only gluing them but also making them interoperable. Nowadays, a huge amount of information exists, but the challenge is to link and give it a meaning that can be shared among the different potential stakeholders and connecting them in a big simulation world (Zacharewicz et al. 2017b).

Semantics of things in the simulation world will be another challenge. It is clear that in simulation, big data technologies can be utilized to deal with huge amounts of data. But will we still be able to understand the meaning of the data produced by simulation? It has to be correctly captured to profit from simulation. AI and semantics are already helping for matching of concepts. But in the future, it will go further by not only proposing matching of concepts, for instance to couple different simulation concepts, but also to create a new information corpus, new concepts,

---

[6]https://www.independent.co.uk/environment/global-warming-climate-change-plastic-extinction-open-letter-bradley-cooper-juliette-binoche-a8522411.html.

[7]http://fastfuturepublishing.com/main/about/our-team/.

[8]https://youtu.be/tArqczVLImc.

also starting from detail going by aggregation to more general views and vice versa, thanks to the capacity to create information based on observed and documented similar situations.

I believe also that model driven approaches are part of the future simulations (Zacharewicz et al. 2017a). These approaches can lead the transformation from concepts and human understanding to implementation of models and simulation. Thanks to AI approaches the transformation will be able to be self-driven and to deduce missing information that is today a barrier when transforming concepts to executable models.

Computation power will permit massive replication of simulations, allowing us to repeat, train and prepare better before action in real life. The leitmotiv will be to never give up with testing and anticipating situations before reaching the almost zero default solution. As assumed by NASA, defeat is not an option.

With the emergency to survive in a hostile world, computers will be focused to calculate the continuous evolution of the environment. Developing simulations that will warn us when anticipating important issues is maybe one the most urgent priorities to tackle.

Immersive environments will have the potential to train people, they will be everywhere. The use of augmented reality will prevent or reduce environment-destructive training.

To open the conclusion and end with a more optimistic discussion, we can consider that all the artwork since the appearance of the human on earth are models and simulations of the world or of a desired world. It is our responsibility to keep designing and planning what we want for the future. Simulation is only providing a support but our future is still in our physical hands.

### *1.5.2 Big Theory and Big Simulation*

The ubiquitous presence of simulation technology is undeniable. Simulations are so pervasive and embedded within our lives that they are invisible to everyday people. The journey to where we are has been long, arduous and full of fits and starts. As a discipline, the M&S journey is in lockstep with that of Computing Sciences and Systems Sciences. This symbiotic relationship is so ingrained in the collective psyche of M&S practitioners that we tend to think of modeling as developing a system and simulation as only computer simulation. The SCSC embodies this worldview of M&S as a practice rooted in engineering and measured by the utility it delivers to mankind. In this section, I (*Saikou Diallo*) argue that while M&S has been successful in many application domains in terms of providing tools and solutions in the past fifty years, our biggest challenges in the next fifty years lie in our ability to tackle societal problems such as universal access to science and technology for people across all spectrums (sight, hearing, mental and physical), large scale migrations, child sex trafficking and other big problems that affect human beings at the local, national and global scale. In other words, we need methods for modeling and simulating humans and societies with the level of complexity necessary to allow us to represent and study problems that humans care most about. We discuss necessary advances in M&S theories, frameworks and tools necessary to achieve this goal in the next fifty years.

The idea of using M&S to study societal problems is not new. However, there is no integrated theory that allows us to build artificial humans in a way that reflects their emotional, cognitive and affective states in a way that is consistent with accepted theories in psychology and cognitive sciences. Similarly, we do not have a consistent way to develop artificial societies at the scale of real human societies (billions of people and objects) where the mechanism of social dynamics between people and groups reflects accepted theories in social sciences and the humanities. This type of comprehensive big theories can only be achieved in transdisciplinary teams of equals where the only concern is a meaningful blending of theories and methods from all disciplines with a shared understanding of each other's epistemological constraints. From an M&S standpoint, the contributions to big theories are in the areas of:

- Collaborative Model: Big theory requires a collaborative environment where engineers, social scientists, humanists and computer scientists can come together to construct a universe. Since this universe is shared, it has to be understandable by people from all disciplines including how to use it as means to investigate questions of interest but also understand and even empathize with the simulation. For simulation engineers, it means additional training in elicitation, soft system methodology and even design thinking have to take place in order to make simulations that are more appealing to a wider audience.
- Computable Models: Big theories have a narrative, mathematical and logical component. While it is possible to derive a computational model from big theories, it might not be possible to implement it using one framework, tool or paradigm. Early attempts at big theory implementation have shown that a multi-stage, multi-simulation or multi-paradigm approach was best suited to reflect the theory. In addition, a "computation only" approach might be limiting, which means a virtual and live component might be necessary to account for non-computable aspects of big theory.
- Verification and Validation: For large scale societies with complex cognitive processes, how do we guarantee in a reasonable amount of time that a simulation model is correct, i.e. that it is a correct representation of the model? Our current approaches are mostly informal and will not scale in light of the number of processes involved. This observation points us towards semi-automation which means that we have to be able to decide which parts of the verification process are best to automate. Consequently, formal theories of verification that can be implemented in tools or useful model transformation techniques need to be developed to deal with the size and complexity of the problem space we are dealing with.
- Experimentation and Analysis: Large scale societies have the potential to generate large scale (even "big") data. Current techniques for analysis and experimentation are inadequate to successfully convey the leading causes of change

in such simulations. As a result, we need new techniques that immerse and engage the observers such that they can achieve the same level of insight as they currently do. Visual analytics combined with virtual and augmented reality need to be further investigated as a potential way to deliver useful and reliable insight from a simulation study of millions if not billions of people.
- Universal Access: The idea of providing access to all users across all spectrums is an important component for future simulation design. Currently, we rely heavily on human computer interface design principles rooted in task-based applications. Principles of aesthetics, inclusion, multi-sensory feedback and presence are not always taken into account when designing simulation interfaces.

Ultimately, the goal is to have several artificial societies operating around the world. These societies should be open and accessible for all investigators. Ideally, the implementation of these societies will be validated by independent teams of researchers, and alternative artificial societies implementing competing theories will be available. Within these societies, researchers will be able to study societal challenges in a safer environment and should be able to look at alternatives and compare potential policies within and across societies. The engineering work that is required to achieve such artificial societies demands collaboration from engineering, sciences, and the humanities. Results and lessons learned will affect how we design, present and analyze simulations in the future. It has the potential to affect the way we design M&S curricula by putting more emphasis on effective modeling and communication across disciplines and problem domains. In the next fifty years, if we are successful, M&S can contribute to the important debate on the future of humanity that started in the advent of the internet and social media.

### *1.5.3 Ubiquity of Simulation*

Today, simulation is literally everywhere, although it may not always be called simulation. In my plenary presentation for this 50th SCSC, I (*Andreas Tolk*) show the close relation of modeling and simulation and computational sciences (Tolk 2018). In this book celebrating the 50th anniversary of the summer simulation, the question of simulation of complex adaptive systems is addressed in detail (Tolk 2019). The broader picture is presented in a new book on simulation-based disciplines (Mittal et al. 2017); this guide deals with engineering and architecture, natural sciences, and social science and management applications of simulation methods. The application of simulation continues to thrive. In parallel to these application driven activities, the work on understanding modeling and simulation as a discipline continues as well. In the recent book on the profession of modeling and simulation (Tolk and Ören 2017), the various chapters deal with ethics, education, vocation, societies, and economic questions, providing an overview of current activities.

The success of modeling and simulation is furthermore shown by the many conference anniversaries we have been able to celebrate: the Annual Simulation Symposium, the Interservice/Industry Training, Simulation, and Education Conference, and the Winter Simulation Conference all celebrated 50 years of supporting our community. Simulation continuously pushes the boundaries of what could only be done theoretically yesterday to what we can accomplish with practical tools supporting the researchers and scientists today.

But despite all these success stories, modeling and simulation did not make it into the main stream of scientific success stories. The reason for this is that even many simulation experts see modeling and simulation mainly as a *computational tool* that helps to make better decisions, which can be technical or managerial in nature. Other sciences recognize the power of simulation, but as a supporting method that extends the discipline that is supported, not a discipline in itself. As a result, insights in such applied domains are hardly generalized and shared. Even more important, known validity and applicability limits and constraints are not shared either. Too often, simulation stays in the shadow of the supported discipline with the focus on providing a specific solution. Instead the general supporting methods and the theory from which such methods can be derived should be at the center of the simulationist's attention, allowing to develop a general *simulation theory* to drive *simulation thinking*. In a recent study, Chen and Crilly (2016) evaluated commonality of issues between practitioners in the fields of synthetic biology and swarm robotics and showed that these practitioners shared more complexity related issues between each other than they did with colleagues in their original domains. Nonetheless, the sharing of information and reuse of solutions was hindered by the different terms and concepts used to describe them within their home domains. The application of a cross-domain framework allowed not only to identify shared issues, but also to align available solutions. This is a typical problem for simulation practitioners supporting different domains and disciplines as well: they are divided by the language and methods of the supported field, and they do not have a language to share the knowledge on their own.

Describing the need to think as a simulationist, the obvious similarity to systems theory and systems thinking—as presented by Arnold and Wade (Arnold and Wade 2015)—is intentional. They propose the following definition:

> *Systems thinking is a set of synergistic analytic skills used to improve the capability of identifying and understanding systems, predicting their behaviors, and devising modifications to them in order to produce desired effects. These skills work together as a system.*
>
> (Arnold and Wade 2015, p. 675)

Simulation thinking must also be such a set of synergistic analytic skills, utilizing the various modeling paradigms regarding modeling methodologies—such as discrete event systems, system dynamics, and agent based approaches—and model types—such as ordinary differential equations, process algebra, and temporal logic, as explained in detail in Fishwick (Fishwick 2007). The development of a simulation theory will allow for a consistent definition of hybrid simulation in support of various application domains (Mustafee et al. 2017). While systems thinking focuses on identifying and understanding systems, simulation thinking focuses on predicting the behavior, as simulation allows the numerical evaluation of the dynamic behavior of systems by generating quasi-empirical data, as long as a valid simulation is used. As such, systems and simulation thinking are mutual supportive, whereby simulation thinking provides additional insights into the simplification and abstraction of systems for the purpose of simulation.

In general, there is no other way to predict but to simulate, no matter what other term we use: smart inter- and extrapolations are simulations, estimates are simulations, etc. Furthermore, scientific work is tightly connected to modeling, as shown in the already mentioned paper by Tolk (2018). How can we as a community and a discipline step into the light, as we are doing fantastic things that deserve more recognition and that should be celebrated as simulation success stories? Here are just a couple of examples: *Robotics* and *autonomous systems* are in high demand, and simulation is a core piece of their planning and control functions. If a robot has to make a decision, it has to evaluate how this decision will influence the situated environment in the foreseeable future, which means it has to simulate. The same is true for *cyber-physical systems*. Mustafiz et al. (2016) observe that *"the engineering of a complex cyber-physical system involves the creation and simulation of hybrid models often encompassing multiple levels of abstraction and combining different formalisms, often not expressible in any single existing formalisms."* This matches easily with the application of our findings as modeling and simulation experts in hybrid simulation in support of cyber-physical systems, which immediately translates into Industry 4.0 and the Internet of Things. One of the most challenging topics of today is to better understand and manage complex systems. When looking through the recommended methods in the complexity primer for systems engineers (Sheard et al. 2015), many simulation methods are enumerated.

The era of modeling and simulation has just begun, and many of the computational application domains are using simulation without being truly aware of it. Simulation is more powerful than data science as it adds causality to correlations. Simulations enable robots to think. Simulation is part of the solution set of the big challenges of today's complex decision environments. We should be proud and active. By pushing the boundaries from what is theoretically possible to what is practically feasible, we are directly contributing to the progress science and research. Like a picture says more than thousand words, a simulation says more than thousand pictures. Like a telescope allows to observe a multitude more than the naked eye, the use of simulation extends the use of mathematics for a multitude of alternative evaluations. We create immersive, virtual worlds enabling a new way to share knowledge and reach people of all educational and disciplinary backgrounds. Simulation is everywhere, and our role is strengthening. The simulation winter may be coming someday, but for now we are still in the early days of a beautiful summer.

## 1.6 Summary

This chapter provides a summary of what has been discussed at the *Summer Computer Simulation Conference Golden Jubilee Panel*. It compiles the topics presented by the eminent members of the SummerSim society in their talks. The chapter does not only give an historical perspective, but also includes bold sentences from controversial viewpoints of panelists who have words to say for today and tomorrow of simulation.

**Acknowledgements** Authors would like to thank Gabriel Wainer (Carleton University) for his contribution to the panel discussion. The *Establishment and Beyond* section is based on his talk.

**Disclaimer** Andreas Tolk's affiliation with The MITRE Corporation is provided for identification purposes only, and is not intended to convey or imply MITRE's concurrence with, or support for, the positions, opinions, or viewpoints expressed by the author. This contribution has been approved for Public Release; Distribution Unlimited. Case Number 17-3081-20.

## References

Ahmed, A., Moallemi, M., Wainer, G., & Mahmoud, S. (2011). VCELL: A 3D real-time visual simulation in support of combat. In *Proceedings of the 2011 ACM/SCS Summer Computer Simulation Conference*. The Hague, The Netherlands.

Ameghino, J., and Wainer, G. (2000). Application of the Cell-DEVS paradigm using N-CD ++. In *Proceedings of the 32nd SCS Summer Computer Simulation Conference*. Vancouver, Canada.

Arnold, R. D., & Wade, J. P. (2015). A definition of systems thinking: A systems approach. *Procedia Computer Science, 44*, 669–678.

Barhak, J., Isaman, D. J. M., Ye, W., & Lee, D. (2010). Chronic disease modeling and simulation software. *Journal of Biomedical Informatics, 43*(5), 791–799.

Barhak, J. (2017). The reference model models clinicaltrials.gov. In SummerSim. Bellevue, WA.

Beck, J.R.,& Pauker, S.G. (2016) The Markov process in medical prognosis. *Medical decision making*, 3(4).

Bock, C. E., Shao, G., Lyons, K. W., Ak, R., Morris, K. C., & Johansson, B. (2016). Standards supporting simulations of smart manufacturing systems. In *Proceedings of the Winter Simulation Conferences*. Washington, D.C.

Brown, J.B., Palmer, A.J., Bisgaard, P., Chan, W., Pedula, K., & Russell, A. (2000). The Mt. Hood challenge: Cross-testing two diabetes simulation models. *Diabetes Research and Clinical Practice* 50 (3): 57–64.

Bocciarelli, P., & D'Ambrogio, A. (2016). Modeling-to-simulation: transformation approaches to boost automation in modeling & simulation. In *Proceedings of the 2016 Summer Computer Simulation Conference* (SCSC'16). Montreal, QC, Canada.

Chechiu, L., & Wainer, G. (2005). Experimental results on the use of Modelica/CD ++. In *Proceedings of the 2005 SCS Summer Computer Simulation Conference* (Student Workshop). Philadelphia, PA.

Chen, C.-C., & Crilly, N. (2016). Describing complex design practices with a cross-domain framework: Learning from Synthetic Biology and Swarm Robotics. *Research in Engineering Design, 27*(3), 291–305.

Clarke, P. M., Gray, A. M., Briggs, A., Stevens, R. J., Matthews, D. R., & Holman, R. R. (2005). Cost-utility analyses of intensive blood glucose and tight blood pressure control in type 2 diabetes (UKPDS 72). *Diabetologia, 48*(5), 868–877.

Clarke, P. M., Gray, A. M., Briggs, A., Farmer, A. J., Fenn, P., Stevens, R. J., et al. (2004). A model to estimate the lifetime health outcomes of patients with type 2 diabetes: The United Kingdom Prospective Diabetes Study (UKPDS) Outcomes Model (UKPDS 68). *Diabetologia, 47*(10), 1747–1759.

D'Abreu, M., & Wainer, G. (2006). A bond-graph mapping mechanism for M/CD++. In *Proceedings of the 2006 SCS Summer Computer Simulation Conference*. Calgary, AB, Canada.

D'Agostino, R. B., Vasan, R. S., Pencina, M. J., Wolf, P. A., Cobain, M., Massaro, J. M., et al. (2008). General cardiovascular risk profile for use in primary care: The Framingham heart study. *Circulation, 117*(6), 743–753.

Eddy, D. M., & Schlessinger, L. (2003). Validation of the Archimedes diabetes model. *Diabetes Care, 26*(11), 3102–3110.

Fishwick, P. A. (Ed.). (2007). *Handbook of Dynamic System Modeling*. Boca Raton, Florida: Chapman & Hall/CRC Press, Taylor and Francis Group.

Floyd, M., & Wainer, G. (2010). Creation of DEVS models using imitation learning. In *Proceedings of the 2010 ACM/SCS Summer Computer Simulation Conference*. Ottawa, ON, Canada.

Gupta, U. G., & Biegel, J. (1990). Expert systems can fail. In *Proceedings of the 1990 Summer Computer Simulation Conference* (SCSC'90). Calgary, Alberta, Canada.

Hayes, A. J., Leal, J., Gray, A. M., Holman, R. R., & Clarke, P. M. (2013). UKPDS outcomes model 2: A new version of a model to simulate lifetime health outcomes of patients with type 2 diabetes mellitus using data from the 30 year United Kingdom Prospective Diabetes Study: UKPDS 82. *Diabetologia, 56*(9), 1925–1933.

Ide, N. C., Loane, R. F., & Demner-Fushman, D. (2016). Essie: A concept-based search engine for structured biomedical text. *Journal of the American Medical Informatics Association, 14*(3), 253–263.

Isaman, D. J. M., Herman, W. H., & Brown, M. B. (2006). A discrete-state and discrete-time model using indirect estimates. *Statistics in Medicine, 25,* 1035–1049.

Jacques, C., & Wainer, G. (2002). Using the CD ++ DEVS toolkit to develop Petri nets. In *Proceedings of the 2002 SCS Summer Computer Simulation Conference*. San Diego, CA.

Jafer, S., & Wainer, G. (2011) Conservative synchronization methods for parallel DEVS and Cell-DEVS. In *Proceedings of the 2011 ACM/SCS Summer Computer Simulation Conference*. The Hague, The Netherlands.

Jeffery, A., Panke, J., Eacket, N., & Wainer, G. (2013). Mobile simulation with applications for serious gaming. In *Proceedings of the 2013 SCS/ACM Summer Computer Simulation Conference* (SCSC'13). Toronto, ON, Canada.

Kofman, E., & Junco, S. (2001). Quantized state systems. A DEVS approach for continuous system simulation. *Transactions of the SCS, 18*(3), 123–132.

Kothari, V., Stevens, R. J., Adler, A. I., Stratton, I. M., Manley, S. E., Neil, H. A., et al. (2002). Risk of stroke in type 2 diabetes estimated by the UK Prospective Diabetes Study risk engine (UKPDS 60). *Stroke, 33,* 1776–1781.

Leff, H. S., Dada, M., & Graves, S. C. (1986). An LP planning model for a mental health community support system. *Management Science, 32*(2), 139–155.

Lumley, E., & Pringle, G. J. (2017). CompBioMed: A new centre of excellence in computational biomedicine. *EPCC, News,* 81.

Mittal, S., Risco-Martín, J. L., & Zeigler, B. P. (2009). DEVS/SOA: A cross-platform framework for net-centric modeling and simulation in DEVS unified process. *Simulation, 85*(7), 419–450.

Mittal, S., Durak, U., & Ören, T. (Eds.). (2017). *Guide to simulation-based disciplines: Advancing our computational future*. Cham, Switzerland: Springer International Publishing AG.

Mustafee, N., Brailsford, S., Djanatliev, A., Eldabi, T., Kunc, M., & Tolk, A. (2017). Purpose and benefits of hybrid simulation: contributing to the convergence of its definition. In *Proceedings of the Winter Simulation Conference* (pp. 1631–1645). Las Vegas, NV.

Mustafiz, S., Gomes, C., Barroca, B., & Vangheluwe, H. (2016). Modular design of hybrid languages by explicit modeling of semantic adaptation. In *Proceedings of the Symposium on Theory of Modeling & Simulation* (pp. 29–36). Pasadena, CA.

Nance, R. E. (1995). Simulation programming languages: an abridged history. In *Winter Simulation Conference*. Arlington, VA.

Nolan, C., Overpeck, J. T., Allen, J. R., Anderson, P. M., Betancourt, J. L., Binney, H. A., et al. (2018). Past and future global transformation of terrestrial ecosystems under climate change. *Science, 361*(6405), 920–923.

Olsen, M., & Raunak, M. (2015). A method for quantified confidence of DEVS validation. In *Proceedings of SpringSim TMS/DEVS*. Alexandria, VA.

Olsen, M., & Raunak, M. (2016). Metamorphic validation for agent-based simulation models. In *Proceedings of the Summer Computer Simulation Conference (SCSC '16)*. Montreal, Canada.

Palmer, A.J., and The Mount Hood 5 Modeling Group. (2013). Computer modeling of diabetes and its complications: A report on the fifth mount hood challenge meeting. *Value in Health, 16* (4), 670–685.

Raunak, M., & Olsen, M. (2014a). A survey of validation in health care simulation studies. In *Proceedings of the 2014 Winter Simulation Conference* (WSC '14) (pp. 4089–4090). Savannah, GA.

Raunak, M., & Olsen, M. (2014b). Quantifying validation of discrete event models. In *Proceedings of the Winter Simulation Conference* (WSC '14). Savannah, GA.

Saghir, A., Pearce, T., & Wainer, G. (2004). Modeling computer hardware platforms using DEVS and HLA simulation. In *Proceedings of the 2004 Summer Computer Simulation Conference* (SCSC'04). San Jose, CA.

Schiermeier, Q. (2018). Droughts, heatwaves and floods: How to tell when climate change is to blame. *Nature, 560*(7716), 20–22.

Schlessinger, L., & Eddy, D. M. (2002). Archimedes: A new model for simulating health care systems: The mathematical formulation. *Journal of Biomedical Informatics, 35,* 37–50.

Shang, H., & Wainer, G. (2007). A flexible dynamic structure DEVS algorithm towards real-time systems. In *Proceedings of the 2007 ACM/SCS Summer Computer Simulation Conference*. San Diego, CA.

Sheard, S., Cook, S., Honour, E., Hybertson, D., Krupa, J., McEver, J., McKinney, D., Ondrus, P., Ryan, A., Scheurer, R., & Singer, J. (2015). A complexity primer for systems engineers. INCOSE Complex Systems Working Group White Paper. Retrieved from March 2018 http://www.incose.org/docs/default-source/ProductsPublications/a-complexity-primer-for-systems-engineers.pdf.

Smith, L., Swat, M. J., & Barhak, J. (2016). Sharing formats for disease models. In *Proceedings of the 2016 Summer Computer Simulation Conference* (SCSC'16). Montreal, Canada.

Stevens, R., Kothari, V., Adler, A., & Stratton, I. (2001). The UKPDS risk engine: a model for the risk of coronary heart disease in type II diabetes (UKPDS 56). *Clinical Science, 101,* 671–679.

The Mount Hood 4 Modeling Group. (2007). Computer modeling of diabetes and its complications, a report on the fourth mount hood challenge meeting. *Diabetes Care, 30,* 1638–1646.

Tolk, A. (2015). The next generation of modeling & simulation: Integrating big data and deep learning. In *Proceedings of the 2015 Summer Computer Simulation Conference* (SCSC'15). Chicago, IL, USA.

Tolk, A., & Ören, T. (Eds.). (2017). *The profession of modeling and simulation: discipline, ethics, education, vocation, societies, and economics*. John Wiley & Sons.

Tolk, A. (2018). Simulation and modeling as the essence of computational science. In *Proceedings of the 50th Summer Computer Simulation Conference*. Bordeaux, France.

Tolk, A. (2019). What are Computer Simulations of Complex Adaptive Systems Good for? In J. Sokolowski et al (Eds.) Summer of simulation—50 years of seminal computing research. Cham, Switzerland: Springer International Publishing AG, in preparation.

Uhrmacher, A. M., Brailsford, S., Liu, J., Rabe, M., & Tolk, A. (2016). Panel—Reproducible research in discrete event simulation—A must or rather a maybe?. In *Winter Simulation Conference* (WSC) (pp. 1301–1315). Washington, D.C.

Uttamsingh, R., & Wildberger, A. M. (1989). In *Proceedings of the SCS Multi-Conference on Advances in AI and Simulation*. Tampa, FL.

Vangheluwe, H. (2000). DEVS as a common denominator for multi-formalism hybrid systems modelling. In *Proceedings of IEEE International Symposium on Computer-Aided Control System Design* (CACSD) (pp. 129–134). Anchorage, AK.

Vansteenkiste, G. C. (1985). New challenges in system simulation. In *Proceedings of the 1985 Summer Computer Simulation Conference* (SCSC'85). Chicago, IL.

Wainer, G., & Giambiasi, N. (1999). Avoiding serialization in timed Cell-DEVS models. In *Proceedings of the International Symposium on Performance Evaluation of Computer and Telecommunication Systems* (SPECTS). Chicago, IL.

Wainer, G. A. (2009). Discrete-event modeling and simulation: A practitioner's approach. Boca Raton, FL, USA: CRC Press.

Wainer, G., & Chen, W. (2003). Remote execution and 3D visualization of Cell-DEVS model. In *Proceedings of the 2003 SCS Summer Computer Simulation Conference*. Montreal, Canada.

Weinstein, M. C., Coxson, P. G., Williams, L. W., Pass, T. M., Stason, W. B., & Goldman, L. (1987). Forecasting coronary heart disease incidence, mortality, and cost: The Coronary Heart Disease Policy Model. *Journal of Public Health, 77*(11), 1417–1426.

Wilson, P. W. F., D'Agostino, R. B., Levy, D., Belanger, A. M., Silbershatz, H., & Kannel, W. B. (1998). Prediction of coronary heart disease using risk factor categories. *Circulation, 97,* 1837–1847.

Wolfram, S. (1986). Theory and applications of cellular automata. In Advanced Series on Complex Systems (Vol. 1). World Scientific, Singapore.

Xindong, W. (1990). On expert simulation systems. In *Proceedings of the 1990 Summer Computer Simulation Conference* (SCSC'90). Calgary, Canada.

Yu, J., and Wainer, G. (2007). eCD ++: An engine for executing DEVS models in embedded plat-forms. In *Proceedings of the 2007 ACM/SCS Summer Computer Simulation Conference*. San Diego, CA.

Zacharewicz, G., Diallo, S. Y., Ducq, Y., Agostinho, C., Jardim-Gonçalves, R., Bazoun, H., et al. (2017a). Model-based approaches for interoperability of next generation enterprise information systems: State of the art and future challenges. *Information Systems and e-Business Management, 15,* 229–256.

Zacharewicz, G., Pirayesh-Neghab, A., Seregni, M., Ducq, Y., Doumeingts, G. (2017b). Simulation-based enterprise management. In S. Mittal, U. Durak, T. Ören (Eds.), *Guide to simulation-based disciplines. Simulation Foundations, Methods and Applications* (pp. 261–289). Cham: Springer.

Zarin, D. A., Tse, T., Williams, R. J., & Carr, S. (2016). Trial reporting in clinicaltrials.gov—The final rule. *New England Journal of Medicine, 375,* 1998–2004.

Zheng, T., & Wainer, G. (2003). Implementing finite state machines using the CD++ toolkit. In *Proceedings of the 2003 SCS Summer Computer Simulation Conference*. Montreal, Canada.

# Chapter 2
# How Abstraction, Formalization and Implementation Drive the Next Stage in Modeling and Simulation

**Bernard P. Zeigler**

**Abstract** The progression of abstraction, formalization, and implementation have played a critical role in advancing the theory and practice of modeling and simulation. In this chapter, we first review the historical record to illustrate how this progression characterizes the pattern of development of both the precursors, and the essence, of discrete event simulation. Then we review Judea Pearl's Ladder of Causation to put it into correspondence with the levels of specification of systems theory. This provides the basis for discussion of potential automation of some of the activities of the M&S enterprise with the help of causal inference methods from artificial intelligence.

## 2.1 Introduction

With the 50th anniversary of the summer simulation conference at hand it's time to take a historical look back at the evolution and development of modeling and simulation (M&S). Are there patterns that emerge that might reveal how we might better understand and influence its future course? Evolutionary biologists argue whether natural evolution is inherently contingent or whether it has a predictable trajectories. Historians accept the fundamental non-determinism of human history but seek lessons to learn that assume history can repeat itself in essential ways. In this chapter, we suggest a pattern that gives form to important landmarks in the evolution of computational technology in general that may apply to certain aspects of discrete event M&S in particular. We identify the sequence or progression of abstraction, formalization, and implementation that appear to have played a critical role in advancing the theory and practice of modeling and simulation. We review the historical record to illustrate how this progression characterizes the pattern of

B. P. Zeigler (✉)
University of Arizona, Tucson, AZ 20854, USA
e-mail: zeigler@rtsync.com

B. P. Zeigler
RTSync Corp., Phoenix, AZ 85226, USA

 
J. Sokolowski et al. (eds.), *Summer of Simulation*, Simulation Foundations, Methods and Applications, https://doi.org/10.1007/978-3-030-17164-3_2

development of precursors (such as microelectronic hardware and computational software), as well as the core, of discrete event simulation (including the Discrete Event Systems Specification (DEVS) formalism). Then we review Judea Pearl's Ladder of Causation to put it into correspondence with the levels of specification of systems theory. This provides the basis for discussion of potential automation of some of the activities of the M&S enterprise with the help of causal inference methods from artificial intelligence.

## 2.2 Abstraction, Formalization, and Implementation

The interplay of abstraction and concreteness have played a critical role in advancing the theory and practice of modeling and simulation. Indeed, this interplay and the associated interplay of the generic and the specific also impel more broadly, the development of knowledge. Abstraction is defined as the drawing out of a general quality or characteristic, apart from concrete realities, specific objects, or actual instances. To formalize an idea or abstraction is to develop and state rules in symbolic form that characterize the abstraction. Important advances in human knowledge have been made through the progression from an abstraction to its formalization and subsequently to implementation in reality as illustrated in Fig. 2.1.

An abstraction focuses on an aspect of reality and, almost by definition, greatly reduces the complexity of the reality being considered. Subsequent formalization makes it easier to work out implications of the abstraction and implement them in reality. Implementation can be considered as providing a concrete realization of the abstraction (often called "reduction to concrete form"). Such implementation reintroduces "messy" real world complexity into the situation and may stimulate another round of abstraction to address the emerging problems.

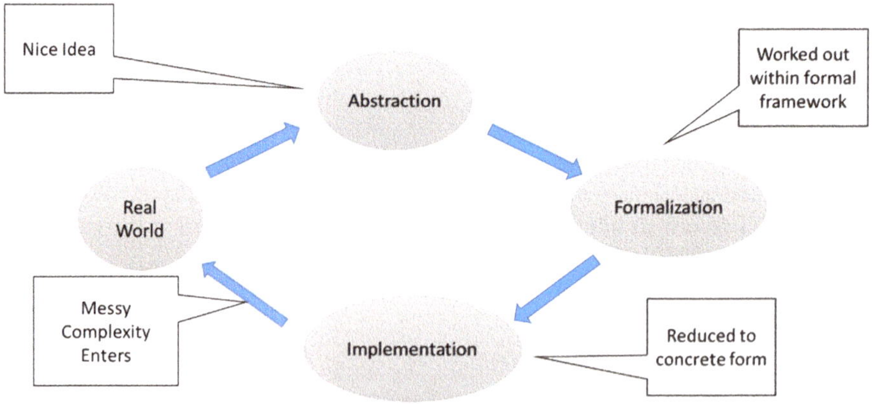


**Fig. 2.1** Progression of abstraction to implementation

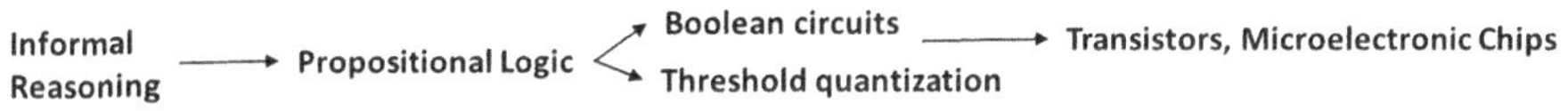


**Fig. 2.2** Instance of the progression of Fig. 2.1 for Boolean logic

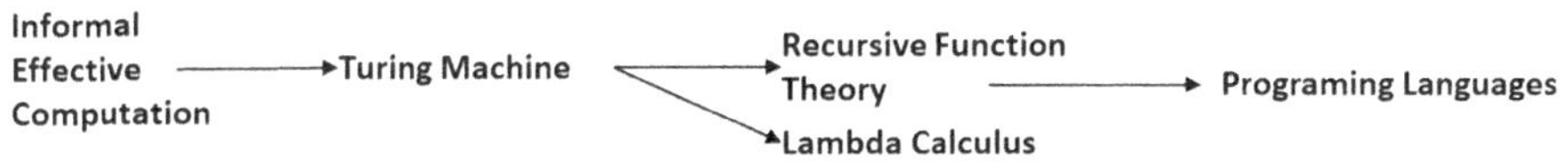


**Fig. 2.3** Instance of the progression of Fig. 2.1 for effective computation

A few instances of the abstraction progression of Fig. 2.1 may help to understand this idea. One can see the eventual breakthrough innovation in computer fabrication technology as stemming from the abstraction of propositional logic derived from common sense informal reasoning (Fig. 2.2). "An Investigation of the Laws of Thought on Which are Founded the Mathematical Theories of Logic and Probabilities" was authored by George Boole in 1854. Subsequently the very simple rules of Boolean algebra and the representation of 0's and 1's through thresholding of continuous valued voltages enabled the design of digital computers and eventual mass volume production of Very Large Scale Integrated Circuits. None of this could have happened had the benefit of computation based on 0's and 1's not been realized by engineers a century after Boole's abstraction (Fig. 2.2).

A second example of the progression in Fig. 2.1 is that from the invention of the fundamental model of computation by Turing to today's powerful programming languages and software systems (Fig. 2.3). Turing Machines gave a formal basis for understanding the heretofore informal concepts of algorithmic computation. This abstraction led to recursive function theory and the Lambda Calculus (among other frameworks for computation) that led to the explosive growth of software. Hinsen (2017) discussed the current problematic state of software sharability and whether return to the simplicity of the Turing Machine could ameliorate the situation.

The abstraction inherent in the concept of "discrete event" brings these ideas closer to modeling and simulation. K.D. Tocher appears to be the first to conceive of discrete events as being the right abstraction to characterize the models underlying the event-oriented simulation techniques that he and others were adopting in the mid1950s. Although there was clearly a similarity in purpose across the United Steel steel plants that Tocher was attempting capture with the limited computational facilities of the era, the various technologies, equipment and layouts would have to be taken into account for any accurate modelling to be possible. Nevertheless, to have any hope of meeting the challenging complexity, Tocher had to conceive a framework that would address the steel plant problem more generically and exploit the commonality of purpose. According to Hollocks (2008), Tocher's core idea conceived of a manufacturing system as consisting of individual components, or 'machines', progressing as time unfolds through 'states' that change only at discrete 'events'. As in Fig. 2.4, Tocher's abstraction of discrete events, states, and components was formalized in the Discrete Event System Specification (DEVS) and led

**Fig. 2.4** Instance of the progression of Fig. 2.1 for evolution of DEVS

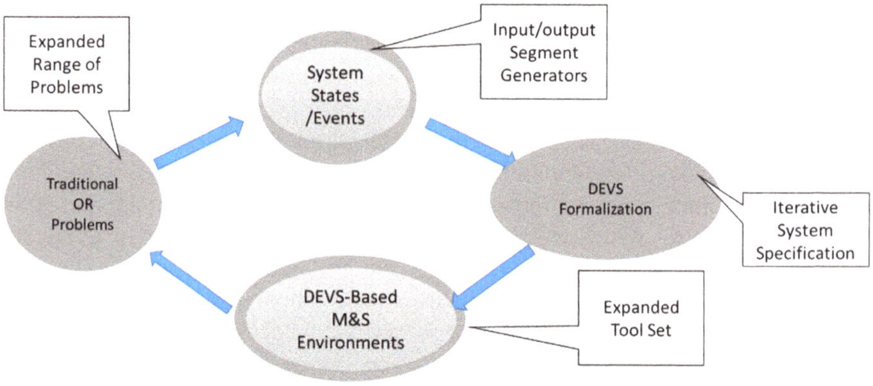


**Fig. 2.5** Second spiral of progression of abstraction to implementation

to implementations in simulation environments based on DEVS (see Zeigler and Sarjoughian 2017 for discussion of DEVS-based environments.[1]).

As in spiral software system development, the progression in Fig. 2.1 can spawn a second (or more) round of abstraction, formalization, and implementation. Figure 2.5 illustrates how this idea applies to the concept of iterative system specification (Zeigler et al. 2018). The maturity of DEVS-based environments lead to demands for application of DEVS to an expanded range of problems beyond the operations research and early applications at which the discrete event simulation were aimed (Fig. 2.5).

The expanded range of problems is characterized by problems that require multiple disciplines to solve. Tolk (2017) distinguishes between interdisciplinary, multidisciplinary and transdisciplinary approaches to such problems. Although only transdisciplinary solutions can result in completely consistent representations of

[1]See (Nance 1996) for a comprehensive history of Simulation Programing Languages (SPL) that fills out the evolution of SPL's in great detail. It is interesting that the author (Richard Nance) notes that a convergence of the languages surveyed to a set of variants of each other perhaps due to tendencies to include features to mimic those that others have introduced (thereby minimizing the risk of loss in completion). However, he also opines that this convergence may be due to lack of firm basis in an underlying modeling semantics for discrete event simulation. In contrast the development of the DEVS family of environments proceeded from a firm basis in system theory-based model semantics (Oren and Zeigler 2012).

truth, the concurrent use of different paradigms within multi and interdisciplinary environments must also be supported in practice. Powell and Mustafee (2016) distinguish between hybrid simulation and hybrid M&S. Hybrid simulation is the use of multiple M&S techniques in the model implementation, while hybrid M&S refers to the application of methods and techniques from different disciplines to one or more stages of a simulation development. DEVS is one of the key formalisms (along with UML) that they include within their recommended framework for hybrid studies (which typically include hybrid simulation) due to its general systems basis.

## 2.3 Ladder of Causation Versus Systems Specification Levels

One potential extension of M&S methodology to address the extended range of problems is to increase its ability to build new models in a more bottom up fashion leveraging emerging Big Data methods. Pearl (2018) paints a picture of recent science as grounded in (big) data and reluctant to resort to assumption of underlying causes in deriving new knowledge. Asserting that data mining can reveal associations (correlations), but not causal links between variables (which one causes the other), he argues that humans are distinct from other creatures in that they employ causal thinking as a critical faculty in everyday life. Pearl's research is primarily directed at enabling artificial intelligence (AI) programs to employ causal reasoning at near human levels.

The progression from abstraction to implementation discussed earlier can help understand the progress Pearl is trying to make toward achieving such AI capability.

Figure 2.6 offers an interpretation of Pearl's research as a progression involving causal reasoning (phenomenon), to the ladder of causation (abstraction) and the causal directed acyclic graph (DAG) (formalization) to end in an effective causal inference engine (implementation). These elements and their roles in the sequence are now briefly explained.

The ladder of causation has 3 rungs ascending from working with associations at Rung 1 to working with causes at Rung 2 to working with causal models at Rung 3. As described in Table 2.1, Rung 1 is where statistics (and Big Data) stop. Rung 2 requires adding causal directions (e.g., the direction: "A causes B" rather than

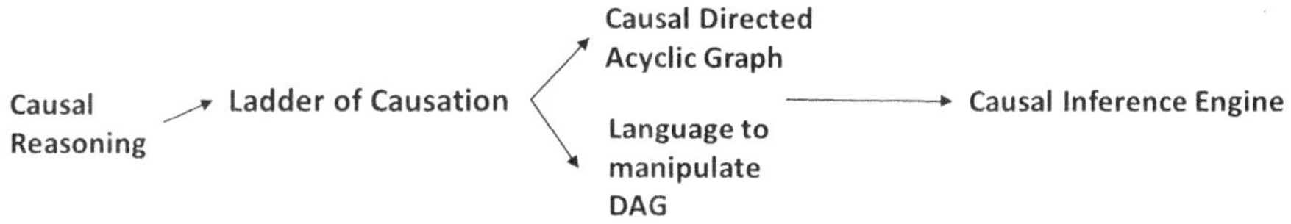


**Fig. 2.6** Instance of the progression of Fig. 2.1 for evolution of causal inference enginies

**Table 2.1** Ladder of causation (Pearl 2018)

| Rung | Name | Activity | Questions | Examples |
|---|---|---|---|---|
| 1 | Association | Seeing, observing | How are the variables related? | How would seeing X change my belief in Y? What does a symptom tell me about a disease? What does a survey tell us about the election results? |
| 2 | Intervention | Doing, intervening | What would Y be if I do X? How can I make Y happen? | If I take aspirin, will my headache be cured? What if we ban cigarettes? |
| 3 | Counterfactuals | Imagining, retrospection, understanding | Was it X that caused Y? What if X had not occurred? What if I had acted differently? | Was it the aspirin that stopped my headache? Would Kennedv be alive if Oswald had not killed him? What if I had not smoked for the last 2 years? |

"B causes A") to association paths to enable predictions to be made for "what if" questions. Rung 3 supports counterfactuals, i.e., making changes in model structures developed at Rung 2.

Pearl introduced the abstraction of a DAG to represent the network of associations that form when addressing questions of the type in Table 2.1. The DAG is an abstraction in that its links represent only the existence (not the content) of the numerical or symbolic relationships among variables. He also developed a language (or calculus) to manipulate the DAG which includes rules for manipulating the "Do" operator, a concept that only starts to be available at Rung 2. This operator represents the act of setting a variable to a value or otherwise constraining the values it can assume and asking how the constraint propagates through the DAG. The essence of the abstraction is that much of the reasoning required to trace the propagation of such constraint can be done on the DAG alone without knowledge of the particulars of the functional relationships that its edges signify.

Counterfactuals make their introduction at the third rung where questions are asked that are not supported by the data obtained and analyzed at the lower rungs. An observation that "A and B were observed and that A caused B" does not enable one to answer what would have happened if A had not happened (c.f., the question about Kennedy in Table 2.1). Pearl's insight in this regard is that the DAG itself must be manipulated in fundamental ways to make the required inference. For example, in regard to Kennedy, this requires (counter-to-fact) disarming Oswald and restraining any propagation of influence on him; then asking what are the effects of other factors on Kennedy (such as those of growing old) that would impact his being alive today.

We return to discuss the causal inference engine in more detail in the context of implementation of an artificial agent capable of executing M&S activities.

**Table 2.2** System specification hierarchy and epistemological levels

| Level | Specification name | What we know at this level |
|---|---|---|
| 0 | Observation frame | How to stimulate the system with inputs; what variables to measure and how to observe them over a time base; |
| 1 | I/O behavior | Time-indexed data collected from a source system; consists of input/output pairs |
| 2 | I/O function | Knowledge of initial state; given an initial state, every input stimulus produces a unique output |
| 3 | State transition | How states are affected by inputs; given a state and an input what is the state after the input stimulus is over; what output event is generated by a state |
| 4 | Coupled component | Components and how they are coupled together. The components can be specified at lower levels or can even be structure systems themselves—leading to hierarchical structure |

Pearl's ladder of causation and its supporting operations can be correlated with the levels of knowledge (epistemological levels) about a system as summarized in Table 2.2 from Theory of Modeling and Simulation (Zeigler et al. 2018).

Climbing up Pearl's ladder corresponds to climbing up the levels of system specification in Table 2.2. Indeed, to ascend from one level to the next higher one requires adding more information about a system, i.e., having more knowledge about the internals of the system. On the other hand, simulation corresponds to going from the top level (having a model that represents knowledge of how the system works) to generating its behavior (getting predictive data that can be compared to data that can be subsequently gathered to validate the model). So modeling and simulation can be thought of as cycling between climbing up (modeling), and coming down (simulation), the levels—hopefully converging on validated knowledge.

Although the same concept of ascending in epistemological levels is common to both stratifications, the correspondence between the rungs and levels is less well-defined and therefore instructive. Rung 1 corresponds roughly to Levels 0 and 1. Rung 2 to Level 2 and Rung 3 to Level 4. However, the major difference is that general systems theory underlies the stratification in M&S while not so in the ladder of causation. Thus such concepts as time base, causality, and concept of state, and coupling of component systems support the methods employed in the former but not in the latter.

These differences stem primarily from the objectives motivating the disciplines of statistical inference (now Data Science) and M&S. Statistical inference methodology is largely directed at inferring correlations from (now voluminous) raw data while M&S is largely concerned with model-based support of design and engineering decisions. However, as emphasized by Pearl, making meaningful inferences from raw data is severely limited (Rung 1) and needs to be supported by an extended set of concepts (Rungs 2 and 3). The theory of M&S (Zeigler et al. 2018) asserts that mathematical systems theory is the proper foundation for formulating the model-based activities of M&S. Indeed, as suggested by Table 2.1, the theory, summarize in the levels of system specification, can provide the extensions needed to enlarge the scope of application of Data Science. Meanwhile, M&S can benefit from the elaborated methodology developed in Data Science.

## 2.4 Synergy Between Data Science and Simulation Modeling

The potential for fruitful synergy between the Data Science and M&S methodologies was recently explicated by Kim et al. (2017) and is summarized in Table 2.3 based on their Table 2.1.

In this table we have already discussed the relation of the methodologies to system knowledge. Also that simulation modeling, in contrast to data modeling, views state representation as one of the critical levels of such knowledge. In fact, we can consider that memoryless systems (see Appendix 1 for details) as the restricted subclass of systems underlying conventional data modeling. Further, as discussed, the basic approach to data modeling is limited to climbing up the ladder

**Table 2.3** Comparison of the data modeling and the simulation modeling approaches (based on Table 2.1 of Kim et al. 2017)

| Dimensions | Data modeling | Simulation modeling |
|---|---|---|
| Model representation and system knowledge | Associational relationship between variables; definitional knowledge about system required (only System Specification Level 0) | Cause-effect relationship between variables; System knowledge required (System Specification Levels 1 and above) |
| Dynamics of the model | Memoryless (stateless) model | State transition and coupled component system specifications |
| Modeling methodology | Data mining<br>Machine learning<br>Causal inference | Assumptions made about system structure expressed in model structure based on first principles, physical laws, previously derived causal relationships, etc. |
| Analysis levels | Descriptive → Predictive (No decision on control policy) | Predictive → Prescriptive → Cognitive (Control policy)(Planning) |
| Condition for valid prediction | System structure remains the same before and after training (Rung 2) | Model Validation |
| Simulation time (to make a decision) | Very short—real time | Relatively long |
| Anomaly (non-existing system) | Not applicable | Applicable (as in rare event or new design) |

of causation or system specification levels, while that of simulation models can include motion in either up or down directions. As Kim et al. (2017) point out, the static nature of the memoryless system and the reliance on data acquired under particular conditions, tend to limit the capability of derived data models to replicate the data used in training or predict hitherto unseen data gathered under the same circumstances. This simplified structure enables fast enough computation for use in real-time. On the other hand, since the simulation-modeling approach is based on high resolution structural knowledge, it enables prediction of the future under a different conditions than those employed in training (calibration, or parameter tuning). This predictive capability supports interventions on the system such as involved in planning, management, control, and engineering. Moreover, such capability underlies therapy (humans) and repair (machines) which deal with abnormal, rarely seen, or otherwise novel system states as (Level 3) or system configurations (Level 4)..

Kim et al. (2017) provide a case study in which a greenhouse environmental control system is designed using a conventional proportional-gain controller with the difference that the conditions-based control law is derived by fitting a neural network to observational data. In the next section we close with a view on how the next spiral in the abstraction-formalization-implementation cycle of Fig. 2.5 can lead to fundamental synthesis of Data Science and M&S.

### *2.4.1 Automating the Activities of M&S*

Zeigler et al. (2009) depict the various capabilities that would be involved in automating all, or parts, of the M&S enterprise, raising the question of whether some or all of M&S activities can be executed by an autonomous artificial modeler/ simulationist. In other words, can an artificial agent carry out all the separate functions identified in the M&S Framework (Zeigler et al. 2018) as well as the high level management of these functions that is currently under exclusively human control.

At least for the initial part of its life, such a modeling agent would need to work on a “first order” assumption about its environment, namely, that it can exploit only semantics-free properties [39]. Regularities, such as periodic behaviors, and stimulus-response associations, are one source of such semantics-free properties. To the extent that such behaviors can be produced by memoryless systems, they could be discovered by Data Science methods, as in Table 2.2.

Figure 2.7 sketches a proposed progression from abstraction to implementation toward an agent capable of automating the activities involved in M&S. To unify the underlying abstractions, the Ladder of Causation and associated concepts would be formulated within the M&S Framework and its systems theory foundation. This would set the basis for the formalization of the causal graph and its associated functional manipulations within the hierarchy of systems specifications (as suggested in the Appendix). Also needed would be the morphisms that relate system specifications at the same and different levels (Zeigler et al. 2018) allowing formal capture of the equivalence relations that are preserved by transformation rules employed to simplify causal graphs to equivalent counterparts with desired measurement properties.

Pearl (2018) formulated a scheme for an artificial agent that could work with the data relationships based on the Ladder of Causation. As illustrated in Fig. 2.7, which is adapted and extended from Fig. 2.1 of Pearl (2018), the agent is provided a query and must ultimately come up with an answer. The agent is supposed to be able to assess whether the query can be answered with the available knowledge and if so, to formulate a statistical inference procedure to estimate the answer based on available data. This agent is proposed as the kernel for the realization of an agent capable of executing M&S activities, the implementation stage of Fig. 2.7, as we discuss next.

In the proposed extended formulation, knowledge consists of a set of models (model base or library) that have been developed to address earlier queries. If the existing models are not responsive to the query, the agent develops a new (causal) model by imposing assumptions and combining/extending existing models. This loop iterates until an adequate model is developed. As a byproduct, implications of the developed causal model are tested to validate the model. In the extension, the answer can consist of a prediction in response to a "what-if" question on management or engineering of system with given parameter settings or more generally the model generated from the pruning of a System Entity Structure within a particular Experimental Frame (Zeigler and Sarjouhian 2017). A long-horizon approach emphasizing cumulative growth of knowledge in the form of models and data should be adopted. This would be to develop, and keep expanding, a library of reusable models at different levels of abstraction for the types of problems that will arise. This means that rapid modification/composition can be done with simpler more general models that can give quick and dirty "back-of-the-envelope" estimates. While more complex and accurate simulations can be done as users' requests become more penetrating.

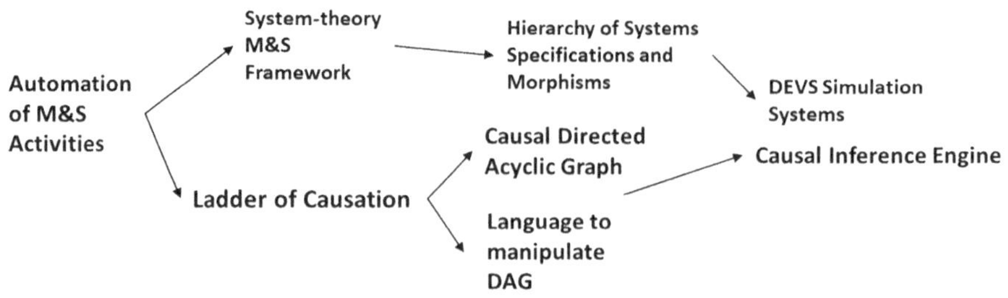


**Fig. 2.7** Merging M&S and Causal Inference progressions

## 2.5 Summary and Future Work

This chapter reviewed the historical record to illustrate how the progression from abstraction to formalization and implementation characterizes the pattern of development of the hardware and software precursors of discrete event simulation as well as the system-theoretically-based DEVS formalism and its simulation realization. Then we reviewed Judea Pearl's Ladder of Causation to put it into correspondence with the levels of specification of systems theory. This provided the basis for discussion of potential automation of some of the activities of the M&S enterprise with the help of causal inference methods from artificial intelligence as expressed in Fig. 2.8.

While the functions required in Fig. 2.8 are beyond the current state of the art, they imply a follow-on challenge: can an agent with the capabilities implied by Fig. 2.8 ask queries about, and develop models of, itself? Such endomorphic modeling capability (Zeigler 1996), is an advanced feature that such an agent must have if it is to fully emulate human M&S capability. Since this capacity implies an infinite regress in which models contain models of themselves without end, it can only be had to a limited extent. The degree to which an endomorphic agent can construct and employ models of its own "mind" as well of the 'minds" of other agents depends on both logical factors (inherent self-reference), objectives (the motivation for such model development), and intelligence (ability to create abstractions and to marshal the necessary computational resources). The enigma of such endomorphic agents provides extreme challenges to research in AI and M&S, as well as related disciplines such as cognitive science and philosophy.

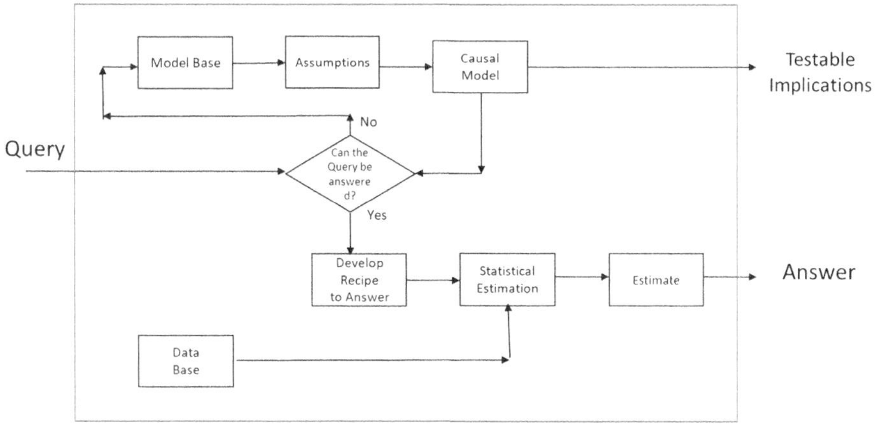


**Fig. 2.8** Adapted Fig. 2.1 from Pearl 2018

**Fig. 2.9** Memoryless system

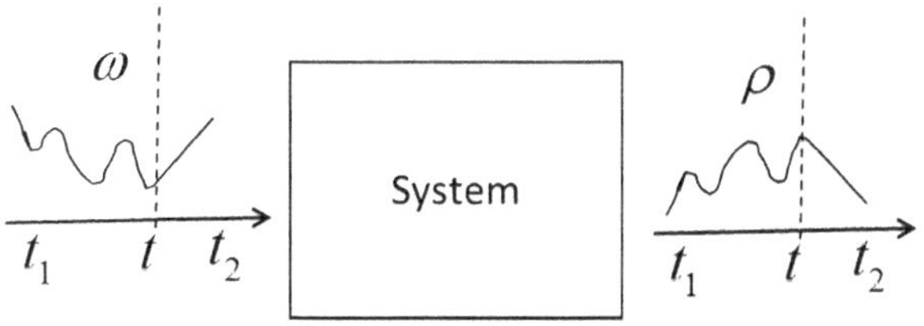


**Table 2.4** Hierarchy of system specifications for memoryless systems

| Level | Specification name | Memoryless system |
|---|---|---|
| 0 | Observation frame | Same as general |
| 1 | I/O behavior | Time-indexed data collected from a source system; consists of input/output pairs $(\omega, \rho)$ where $\rho(t) = \lambda(\omega(t))$, i.e., the output at time t is purely a function of the input at that time |
| 2 | I/O function | Effectively there is only one state, so that the I/O Behavior and I/O Function form a single mapping from input segments to output segments |
| 3 | State transition | The single state "transitions" to itself |
| 4 | Coupled component | Memoryless systems can be coupled together and result in a memoryless system under conditions that the coupling is feedback free (the underlying digraph is a DAG) |

## Appendix 1: Memoryless Systems

A memoryless system, illustrated in Fig. 2.9, is a special case of a system for which the output segment value at any time $t$ is a function of the input segment at that time, i.e., for all $t, \rho(t) = \lambda(\omega(t))$.

Table 2.4 shows how the hierarchy of system specifications gets specialized in the case of memoryless systems.

## References

Hinsen, K. (2017). A dream of simplicity: Scientific computing on turing machines. *Computing in Science and Engineering, 19*(3), 78.

Hollocks, B. W. (2008). Intelligence, innovation and integrity—KD Tocher and the dawn of simulation. *Journal of Simulation, 2*(3), 128–137.

Kim, B. S., Kang, B. G., Choi, S. H., & Kim, T. G. (2017). Data modeling versus simulation modeling in the big data era: Case study of a greenhouse control system. *Simulation-Transactions of the Society for Modeling and Simulation International, 93*(7), 579–594. 10.1.177/0037549717692866.

Nance. R. E. (1996). A history of discrete event simulation programming languages. In T. J. Bergin, Jr. & R. G. Gibson, Jr. (Eds.), *History of programming languages—II*, New York, NY, USA: ACM (pp. 369–427). https://doi.org/10.1145/234286.1057822.

Ören, T. I., & Zeigler, B. P. (2012). System theoretic foundations of modeling and simulation: A historic perspective and the legacy of A Wayne Wymore. *Simulation, 88*(9), 10331046.

Pearl, J., & Mackenzie, D. (2018, May 15). *The book of why: The new science of cause and effect* (1st ed.). Basic Books.

Powell, J. H., & Mustafee, N. (2016). Widening requirements capture with soft methods: An investigation of hybrid M&S studies in healthcare. *Journal of the Operational Research Society.*

Tolk A. (2017). Multidisciplinary, interdisciplinary and transdisciplinary federations in support of new medical simulation concepts: Harmonics for the music of life. In M. Traore (Ed.), *Computational Frameworks* (pp. 47–59).

Zeigler, B. P., & Sarjoughian, H. S. (2017). *Guide to modeling and simulation of systems of systems*. Heidelberg, Germany: Springer.

Zeigler, B. P., Muzy, A., & Kofman, E. (2018). *Theory of modeling and simulation: Integrating discrete event and continuous complex dynamic systems*. NY: Academic Press.

Zeigler, B., Muzy, A., & Yilmaz, L. (2009). Artificial intelligence in modeling and simulation. In R. A. Meyers (Ed.), *Encyclopedia of Complexity and System Science*, Editor-in-Chief. Heidelberg, Germany: Springer.

# Chapter 3
# High-Speed, Low-Cost Simulation of Power Electronic Systems—A 50+ Year Collaboration

**Roy Crosbie**

**Abstract** Ship's electric power systems use many power converters to generate the ac or dc supplies needed by a wide variety of loads. The switching frequencies used in these converters continue to increase, reducing unwanted harmonics and the size and weight of transformers and other components. Higher switching speeds pose problems for simulations of these systems particularly where real-time execution is necessary. The development of low-cost, high-speed simulatiosn of power electronic systems has been a goal of research at California State University, Chico for over 20 years. This research was the result of a collaboration that began in the UK over 50 years ago.

## 3.1 Introduction

This is the story of a simulation partnership that spans more than 50 years. It also tracks many of the changes that have taken place in the field of computing in general and specifically in computer simulation during that period. These changes are, of course, covered in much more detail in the many papers that have been presented at and recorded in the Proceedings of the 50 Summer Computer Simulation Conferences that this volume celebrates.

## 3.2 Early Days

I arrived at the Royal College of Advanced Technology, Salford (soon to become the University of Salford) in the summer of 1963. On arrival I was advised to find a more senior faculty member to whose research I might be able to contribute.

This chapter is dedicated to the memory of former colleagues John Hay of the University of Salford and Richard (Dick) Bednar of CSU, Chico.

R. Crosbie (✉)
California State University, Chico, CA, USA
e-mail: rcrosbie@csuchico.edu

 
J. Sokolowski et al. (eds.), *Summer of Simulation*, Simulation Foundations, Methods and Applications, https://doi.org/10.1007/978-3-030-17164-3_3

My main interest was strongly focused on finding a project that required analog computing support. My final year undergraduate project at the University of Liverpool in 1956/7 had involved the construction of a 4-amplifier analog computer to solve 2nd-order linear ordinary differential equations. This prepared me for a PhD project, also at Liverpool, in which I built an analog computer to study a nuclear reactor problem. This experience led me to Dr Narain (Nari) Hingorani and the field of high-voltage direct-current (HVDC) power systems which was, at that time, a rapidly developing technology used to transmit bulk electrical power economically over large distances or under water. Nari had arrived at Salford shortly before I did and was in the process of setting up an HVDC laboratory in the basement of the building that housed the Department of Electrical Engineering (appropriately called the Maxwell building) across the street from a former home of physicist James Joule where I was to lecture occasionally. Nari had completed his PhD at the University of Manchester Institute of Technology (UMIST). His thesis was the basis of the first comprehensive text on HVDC systems (Adamson and Hingorani 1960), co-authored with his PhD supervisor Professor Colin Adamson. As he pursued the procurement of the electrical equipment needed to set up the lab, he was also interested in establishing a simulation facility to support the planned research. He saw my analog computer experience as a natural fit and agreed that we should work together.

My first task was to find some way of procuring an analog computer. The University did not have a digital computer at that time and had just one small analog computer housed in another department. Building an analog computer from scratch as I had done in Liverpool was an option, but that would be time consuming and a home-built system would lack the advanced features of available commercial systems (such as a removable patch panel to allow a rapid change of problems). Analog computers are based on high-gain direct-coupled amplifiers with precision feedback components that configure them to perform addition, inversion or integration with respect to time on one or more input voltages (Korn and Korn 1952; Howe 1961; Karplus and Soroka 1959; McLeod 1968). Additional components can produce output voltages that are the product of two input voltages (multipliers) or non-linear functions of the input voltage (function generators). The computer was programmed by making appropriate connections with patch cords between the individual components, setting parameters and initial conditions on potentiometers (usually with a helical winding capable of ten turns to improve precision) and starting the simulation by switching from an initial condition configuration to a compute configuration with all elements interconnected. At that time the components were based on vacuum tubes. The values of variables represented by the output voltages of the analog computer components were measured in volts against a reference voltage, usually in the range +/− 100 volts set by a precise and highly stabilized reference power supply. Readings using a digital voltmeter offered 4-digit precision (but rarely 0.1% accuracy).

Change was, however, on the way. The replacement of vacuum tubes by transistors was changing the design of a lot of electronic equipment and the first solid-state analog computers, based on a 10-volt reference, were becoming available. The leading manufacturer was Electronic Associates Inc (EAI) based in

Princeton NJ which had a UK subsidiary, Electronic Associates Ltd (EAL) in Burgess Hill, Sussex. They were prepared to give a generous educational discount for a basic configuration of their TR48 10-volt system. The problem was that I had access to a very small capital budget but a more generous amount for consumables. Fortunately, I was able to persuade the university to place an order for a TR48 cabinet with power supply as a capital item, and the individual plug-in computing units as consumables.

With the TR48 installed one of the first problems I was asked by Nari to address was a calculation of a ship's magnetic compass deflection caused by an undersea HVDC power cable such as the one under the English Channel connecting the power systems of France and England. It was a simple problem, easy to implement on the analog computer and it led to my first professional publication as lead author (Crosbie and Hingorani 1967).

The analog computer was used on several power system problems including one setup in which it was interfaced to scaled down hardware in a hardware-in-the-loop configuration. In addition to HVDC problems, one study in which the analog computer prove useful was related to an electric arc discharge characteristic being studied by a colleague in the HVDC research team. He was using an analytical solution to the defining differential equation that required the summation of a very slowly converging series and very long execution times on the DEUCE. So much so that he was only allowed to run his program overnight. Rather than developing a program to solve the differential equation on DEUCE he asked me if we could solve it on the analog computer. This proved to be a relatively simple task and solutions became available in seconds rather than hours (Crosbie and Halder 1986).

The analog computer continued to be useful but the main simulation focus turned to the University's newly acquired digital computer. This was an English Electric DEUCE computer which was the production version of the ACE, one of the first stored-program computers designed in the UK. ACE was developed at the UK's National Physical Laboratory (NPL) and was a cut down version of a design by Alan Turing. The DEUCE contained 1450 vacuum tubes and used mercury delay lines for its main memory with a capacity of 1536 bytes supplemented by an 8192-word (32 K) magnetic drum. I/O was provided via a card reader and punch plus optional paper-tape reader and punch.

Use of the DEUCE was very controlled. The computer itself was installed in the only air-conditioned room in the University and was accessible only by a very few senior computer staff. Programs were punched on paper tape using a Teletype Model 33 ASR teleprinter and the language of choice for most users was initially Algol (a Fortran compiler came later). The punched paper tape was deposited in a receptacle outside the computer lab. The next day one returned to pick up the paper tape containing the output produced by the computer (a high-speed line printer was added later). This could then be printed on the teleprinter and reviewed. It was good news if it contained actual numerical results, bad news if it read "compiler error on line NN". Error correction involved cutting out the offending section of tape and carefully splicing in a segment of tape with the corrected punched code. This was then resubmitted, and fingers were firmly crossed until the next day. Even so, the

DEUCE proved very effective in providing results for numerous publications (Hingorani et al. 1968a, b; Hingorani and Chadwick 1986; Hay and Hingorani 1970) describing our research, with much of the work being performed by the late Dr. John Hay and other PhD students. John Hay was at that time a very talented PhD student and later faculty member and valued colleague, who was the primary developer of the ISIM, ISIS, and ESL simulation languages (Hay et al. 1974; Hay and Griffin 1980; Hay 1981, 1975).

## 3.3 A Time of Change

I did not realize at the time how fortunate I was to work with Nari. Already recognized internationally as an expert in HVDC technology, he was just at the beginning of an outstanding career as a distinguished and much-honored power systems expert. His research at Salford lasted just 4 years, produced several publications (Hingorani et al. 1966, 1968a, b; Crosbie and Hingorani 1967; Hingorani and Chadwick 1986; Hay and Hingorani 1970)—mostly on digital computer simulations of HVDC components and systems—and ended when he moved to the USA for an appointment as Senior Scientist with the Bonneville Power Administration (BPA) where he worked on and commissioned the Pacific DC Intertie which was to deliver over 3000 MW of power from the Pacific North West to the Los Angeles area over an 864-mile, +/−1000 kV DC link. Although alternating current is normally used to distribute and deliver electric power, it is less expensive to use direct current to transmit large amounts of power over long distances or under water. The DC link between France and England was referred to earlier and there are numerous other examples including those between the North and South Islands of New Zealand, between Japanese islands and an overland link in Russia.

In 1974 Nari moved to the Electric Power Research Institute (EPRI) in Palo Alto, California. In his time there he received recognition as the developer of important new technologies—FACTS (Flexible AC Transmission Systems) and Custom Power. After 20 years at EPRI, the last 5 as Vice President of the Electric Systems Division, he retired to set up his own consulting business. He is a member of the National Academy of Engineering. He received a DSc degree from the University of Manchester and the Lamme medal from the IEEE. The IEEE Power Engineering Society awarded him its Uno Lamme award and introduced FACTS and Custom Power awards named after him. In 2006 he received the Franklin Institute's Bower Award and Prize for Achievement in Science.

With the departure of Nari to the US, research activities at Salford turned to a more general simulation theme including collaboration with department colleagues John Hay and John Pearce (at that time a Ph.D. student) in the development of simulation languages. This was a time when minicomputers were gaining popularity and we were impressed with the ease of using BASIC on the newly acquired PDP8. This prompted us to consider the development of a similar interpretive language for simulation in which one could produce solutions to differential equations directly and generate a solution similarly to the way BASIC produced solutions to algebraic equations. The first result was ISIS (Interpretive Simulation System) (Hay et al. 1974; Hay and Griffin 1980; Hay 1981). Later, with the advent of microcomputers, this was followed by ISIM (Interpretive Simulation for Microcomputers) Hay (1975). This led to the formation of a company, ISIM Simulation Ltd., to market these products. We also established a simulation laboratory at Salford with a new hybrid computer linking the PDP8 to a new analog system and the first MS degree in Computer Simulation in the UK was developed. One of the topics of interest was the accurate handling of discontinuities in continuous simulations which resulted in a paper presentation at the 1974 Summer Computer Simulation Conference in Houston (Crosbie and Hay 1974; Hay 1975) and had relevance to later research described below. The continuing work on simulating discontinuous systems led to an invitation to participate in a NATO Advanced Science Institute meeting and publication (Crosbie and Hay 1977). There was great interest at that time in specifications for improved simulation languages and we participated in the debate in 1981 by developing and distributing a report "Outline proposals for a new standard for continuous-system simulation languages (CSSL 81)". This included new features for accurately handling discontinuities and support for a segmented program structure that allowed multi-rate and distributed simulations. This led to a contract from the European Space Technology Centre (ESTEC) at Noordwijk in the Netherlands (part of the European Space Agency (ESA)) to develop a language (ESL) based on our specification (Hay 1978; Hay and Crosbie 1982). This work was still in progress when I left Salford. The language continued to be developed first by John Hay and then by John Pearce and is still in use today.

## 3.4 Society for Computer Simulation

My early involvement with SCS in the UK has been recorded elsewhere (Crosbie 1984). It led to an invitation to join the SCS Executive Committee which allowed me to attend the SCSC in Montreal in 1973 for my first SCSC and XCOM meeting. Through SCS I made many contacts in the US which served me well when 10 years later I had to make a serious decision about my future career.

## 3.5 California State University, Chico

In 1983, at a time of deep cuts in UK University funding imposed by the Thatcher government, I accepted voluntary severance from Salford and moved to California State University, Chico; first in the Computer Science department, then in the newly established Computer Engineering department, and finally in the combined Department of Electrical and Computer Engineering. It was a welcome change of environment. One of my senior Salford colleagues, who had responsibility for our area of research, had described computer simulation *as "an* ad hoc *technique that is sometimes useful to control engineers"* which did not bode well for future support. Fortunately, this was not an attitude that greeted me in the US. Within days of my arrival in Chico, I was invited by SCS Conference Chair Stew Schlesinger of the Aerospace Corporation (later SCS President) to accept the position of Program Chair of the 1984 Summer Computer Simulation Conference in Reno, Nevada with Norm Pobanz of the Bechtel Corporation (later Treasurer and President of SCS) as General Chair. Reno turned out to be the most successful SCSC to date with an attendance of over 500 and a healthy financial surplus. This prompted a return of the SCSC to Reno in 1986 in which I served as General Chair and a UK expat colleague then at CSU Chico, Paul Luker, was Program Chair. This was all part of a strong relationship between Chico's Computer Science Department and SCS at that time led by Dr. Ralph Huntsinger (now Society Historian) and Department Chair Dr. Orlando Madrigal. The initiative for the McLeod Institute of Simulation Sciences (MISS), named in honor of SCS Founder John McLeod, was led by Ralph and the first MISS center was established in Chico in the 1980s. Ralph became SCS President in 1986 and I followed in 1988 producing joking references to "The Chico Mafia".

## 3.6 Partnership Renewed

Meanwhile, following his time at BPA and EPRI, Nari retired to start his own consulting business in the development of power electronics and devices, and the application of power electronics to transmission, distribution, industrial power, and marine power systems. I had maintained intermittent contact with him and in 1998 he contacted me asking if I could help with a simulation problem. At that time the accepted standard for real-time power system simulations was to use a 50-μS time step, adequate for representing systems with a 60-Hz fundamental frequency and a few lower harmonics. With the increasing use of power electronic techniques which required rapid switching of converters Nari was prompted to investigate the possibility of developing real-time simulations of power systems with time steps of 20 μS or less. He was working with a Japanese power company and they had initially produced a simulation of a simple power system consisting of an a.c. generator, transmission line and load. Initial tests were not executed in real-time. The three components of the simulation all gave satisfactory results when executed individually with a 20-μS step using a simple numerical integration algorithm, but in combination the simulation became unstable. Nari asked me if I could explain why. The answer was not difficult to find and involved a familiar problem with the integration algorithms used to solve the model differential equations. Combining the individual model components into a simulation of the complete system introduced eigenvalues outside the stability region of the integration algorithm being used. This prompted Nari to fund an initial study at Chico State to investigate techniques that might deliver real-time power system simulations with time steps of

20 μS or less. A small group of researchers was established at Chico that included departmental colleagues John Zenor and Richard Bednar both of whom became vital to the progress of the research in subsequent years. John is a very talented software and computer systems engineer who led the development of the software and its architecture and interfacing issues. The late and much missed Richard (Dick) Bednar was an outstanding controls and mathematics expert, who's meticulous work and reports were admired by all who worked with him. An 18-month study in 1999/2000 used a single digital signal processor (DSP) and produced a real-time simulation of a simple 3-phase converter model executing with a time step within the 20-μS maximum.

## 3.7 ONR Support

This success prompted Nari to encourage us to submit a proposal on the development of high-speed simulation techniques to the US Office of Naval Research (ONR) given the Navy's interest in the design of power systems for its future all-electric warship. Propulsion systems for ships have traditionally used diesel engines or gas turbines driving shafts that turn the ships propellers. They also drive ac generators to serve the ship's electrical loads. More recently ships have been using electric motors for propulsion. Modern cruise ships generally follow this approach and can be classed as all-electric. Using this approach on Navy ships presents additional problems with higher loads, including weapon systems, in a smaller space. The design of an integrated power system (IPS) for future US Navy all-electric warships has been the source of much research including the development of suitable simulation tools and techniques. The general approach involves the conversion of the output of ac generators to a medium voltage (5–10 kV) dc bus that feeds individual loads through dc/ac or dc/dc converters that produce power in the form required by the load. These converters are based on electronic switches that switch at kHz frequencies. In general, the higher the switching frequency, the better the quality of the output and the smaller the converter for a given power output, so the tendency has been to increase switching frequency as advancing technology permits.

The power system for an all-electric warship must supply many different loads and types of load. The electrical plant was previously required to provide only hotel services, air conditioning and instrumentation. An all-electric ship adds propulsion and, in the case of a warship, some weapons systems including pulsed loads for electric guns and aircraft launchers. The total electrical system involves electric power generation, conversion to supply dc busbars, and further conversion at multiple points to feed the different loads.

It follows that simulation of the power system for an all-electric warship involves the simulation of many switched converters. The experience gained earlier in simulating discontinuities at Salford plus the more recent DSP work at Chico State provided a sound basis for further research aimed at developing efficient, cost-effective simulations of ship power systems. Added to this was the willingness of Nari Hingorani to advise the research team. On this basis the US Office of Naval Research (ONR) funded an initial 3-year study, starting March 2001, with a grant of $871 K.

The research has continued and expanded to include additional topics related to the main theme. Nari Hingorani has been a partner throughout the program, normally visiting Chico monthly to advise on the detail of systems to be simulated and analyzing the results of simulations, judging their credibility, and pointing out any unexpected features. During his earlier visits, Nari defined an initial benchmark based on a 6-pulse, back-to-back, converter system (Fig. 3.1). This consisted basically of two 3-phase ac systems connected via a dc link. Each ac system is connected to a transformer via filter components. The two converters each consist of 3 switches that convert ac to dc or dc to ac as the power flow dictates. The pulse-width modulation (PWM) controllers that drive these switches are not shown but are included in the benchmark code.

This benchmark formed the basis of the research for several years. During this time the proposed switching frequency of the power electronic components was increasing, which puts further constraints on the acceptable simulation time steps. With the goal of achieving shorter time steps while addressing larger systems the DSP work was extended to use commercial boards providing a small array of 4 DSPs in addition to supporting different I/O protocols. These boards could be inserted into desk top PCs and this became the standard configuration for much of the later research (Crosbie et al. 1985, 2004; Hay 1986; Crosbie 2007, 2008, 2012). In addition to the need to develop highly efficient code, much work was done on developing and improving algorithms, including integration algorithms, that combined accuracy with speed of execution (Bednar and Crosbie 2009, 2010).

Among the techniques that were used to accelerate execution in appropriate cases was the use of multi-rate simulation (Crosbie and Hingorani 2004; Crosbie et al. 2004; Bednar and Crosbie 2006; Crosbie et al. 2007; Zenor et al. 2007a, b; Word et al. 2007; Crosbie et al. 2009b, c; Pearce et al. 2009; Zenor et al. 2010). The use of different time steps for different parts of a simulation is a well-established

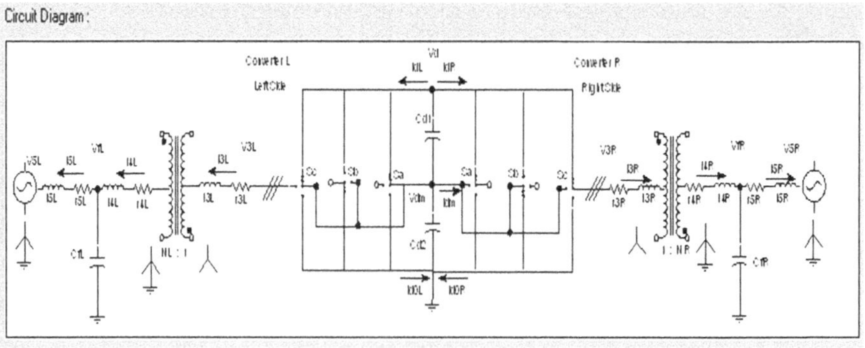


**Fig. 3.1** 6-pulse back-to-back converter system used as a benchmark

and familiar technique that can be applied to systems that can be separated into sub-systems with different dynamics. Slowly changing elements can use longer time steps than those with more rapid dynamics. Separation into slow, medium, fast subsystems can often be done heuristically based on knowledge of the system dynamics.

To demonstrate the capabilities of multi-rate simulation and to provide a test system for evaluating the technique Chico State collaborated with the University of South Carolina (USC) and the University of Glasgow in Scotland in an ONR-funded study to develop a multi-rate simulation of an unmanned underwater vehicle (UUV). The research group at USC, led by Roger Dougal, had developed the virtual test bed (VTB) a simulation tool funded by ONR which was to be used in the joint project. David Murray-Smith and Euan McGookin of the University of Glasgow developed the UUV model to be used in the study, Roger Dougal and his colleagues at USC provided the battery and motor models and added a multi-rate solver to the VTB simulator, and CSU, Chico provided the high-speed converter model and was responsible for the integration of the complete simulation (Bednar and Crosbie 2007). In the simulation the battery and the vessel ran at the low rate (long time-step of 10–100 mS), the motor ran at medium rate (10–100 μS) and the converter at the highest rate (2–5 μS). Figure 3.2 contains an outline of the system, separated into its multi-rate components. A supplementary award supported a contribution from former Salford colleague John Pearce of ISIM Simulation who developed an interface that allows ESL models to be incorporated in VTB simulations (Crosbie et al. 2007).

In addition to developing code for sample problems, Dick Bednar worked on stability analysis of bi-rate systems and on techniques to separate the system automatically into subsystems with differing dynamics using similarity transforms (Crosbie et al. 2004b, c). Dick showed that similarity transforms can be used to separate a linear system or subsystem into independent first-order equations that can be solved totally in parallel. Although the approach is limited to linear

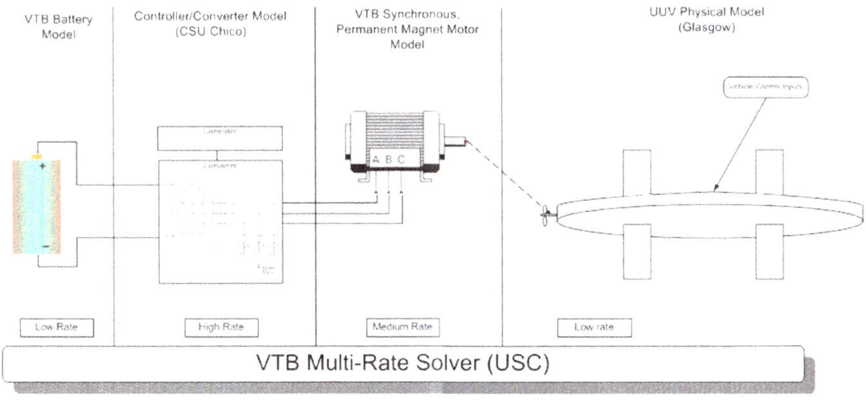


Fig. 3.2 Multi-rate model of unmanned underwater vehicle

(or linearized) systems it is potentially useful in electrical network simulations which often contain large linear subsystems that could use the technique. Another area of investigation covered interfacing issues between devices and in real-time hardware-in-the-loop simulations (Crosbie et al. 2010; Yusta et al. 2011).

By 2007 we were able to report a 2-μS time-step using 4 Analog Devices TS-201 DSPs at the 2007 IEEE Electric Ship Technology Symposium (Crosbie et al. 2007). At this point a careful analysis of the problem and the capability of available and expected future DSPs suggested that no significant further improvement in performance was likely in an acceptable time frame.

This led to the consideration of the field-programmable gate array (FPGA) as an alternative. After some initial studies using fixed-point representation of data it was determined that the use of double-precision floating point did not impose an unacceptable performance penalty. Simulations were converted to FPGA processors and eventually it proved possible to execute the same 6-pulse back-to-back benchmark in 200 nS (Crosbie et al. 2009a; Yusta et al. 2011; Zenor et al. 2011, 2012; Zenor and Kredo 2012; Bednar and Crosbie 2012), an order of magnitude improvement on the DSP performance and two orders better than the original goal.

This result prompted ONR to fund a Technology Transfer project in which CSU, Chico developed a simulator based on 2 FPGA cards within a desk top PC. The FPGAs contained models of two types of electrical machine. Simulink software was included, and the FPGA models were accessible as components in a Simulink simulation (Kredo et al. 2015). This system was delivered to NSWC in 2014 for evaluation. Tests indicated that models based on the FPGA-coded elements executed up to 120 times faster than equivalent Simulink-only models.

Another development was the creation of a simple hardware-in-the-loop simulation in which a simulated controller was interfaced to a small educational generator-motor machine set driving a mechanical load. The equipment was part of a lab setup developed by Dr Ned Mohan of the University of Minnesota who visited Chico to discuss his pioneering work in developing new educational programs in energy systems. It was interesting to learn that on beginning his PhD he was given a copy of the 1966 paper (Hingorani et al. 1966) and asked to reproduce its results, which added another aspect to the continuity of the research.

Recent work (Kredo et al. 2017) has two main themes. A Simulink model provided by the Naval Ship Systems Engineering Station (NAVSSES) in Philadelphia (Fig. 3.3) forms the basis for an investigation of the process of converting a graphical model representation to efficient FPGA code. Dick Bednar has been studying the extraction of the equations that form the math model from the graphical representation. Kurtis Kredo is evaluating available software tools that help in the process of converting a Simulink diagram into FPGA code and is developing additional tools to support the process with the goal of providing a means of converting a Simulink diagram transparently into efficient executable FPGA code—in effect, a Simulink to FPGA compiler.

The NAVSSES model is based on a diesel generator backup system and includes multiple machine models, a transformer, and multiple loads. Figure 3.1 shows the Simulink schematic for the model. The alternator and motor models come from

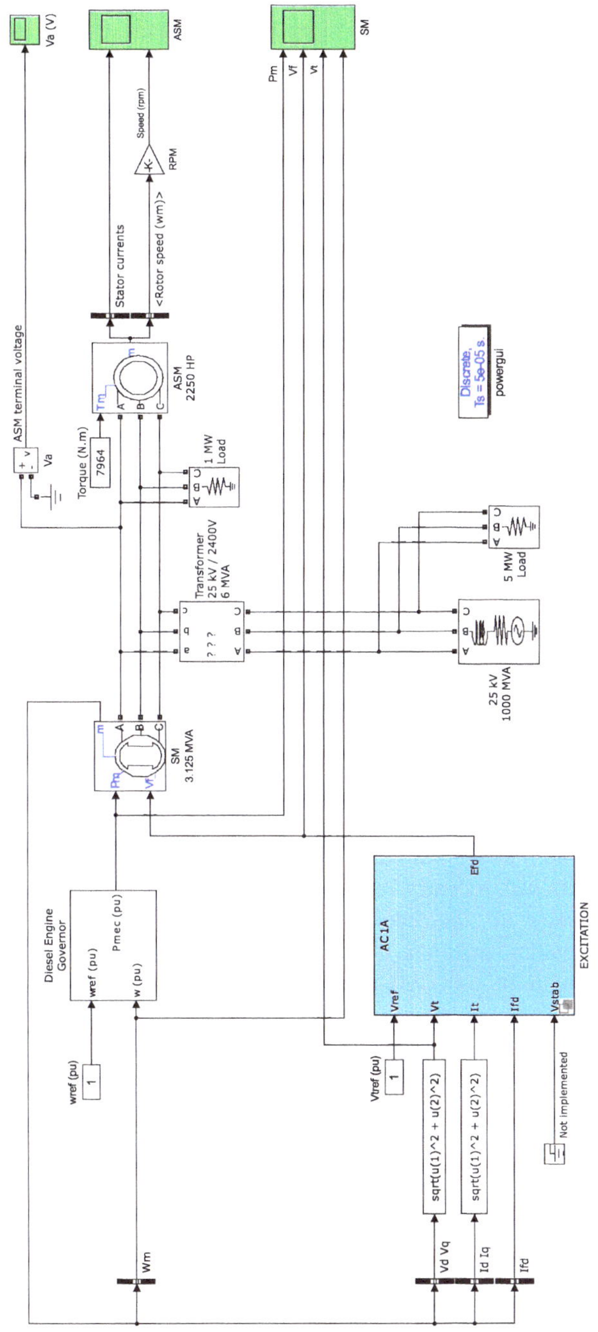


**Fig. 3.3** SimPowerSystems simulation diagram provided by the navy

Simscape Power Systems (SPS library (2018)). Converting this type of model to efficient FPGA code requires the extraction of the model differential equations from the Simulink model, and this has proved to be a very difficult and time-consuming process. Simulink does provide features that support this process, but they fall well short of fully converting the graphical model into the underlying differential equations. In the model in question the mix of basic Simulink and SPS elements introduces additional problems. Full details of the difficulties experienced is a topic for a future publication that will require further study of the reports prepared by Dick Bednar shortly before his untimely death. One important lesson that was reinforced by this study is the necessity of always providing accompanying documentation of graphical models such as these, that include a detailed math model, underlying assumptions, limitations of the domain in which the model is valid, and any available information related to the verification and validation of the model.

## 3.8 Conclusion

The list of references indicates that the research has been regularly reported at SCS and other simulation conferences and in journals, including the Simulation special issue on power electronics (Word et al. 2008). Nari Hingorani has been a co-author in many of these publications. At the time of writing the research group has been strengthened by the arrival of new Chico State faculty Hadil Mustafa, whose interests include embedded systems and computer architecture, and Zahrasadat Alavi, whose interests include control systems, stability analysis and simulation. It is difficult in 4-year schools for new faculty, who have little or no time allocated for research, to develop funded research projects and one of the goals of our program is to provide opportunities for new faculty to develop their research portfolios.

A new project, which addresses the prospect of converter switching frequencies increasing to 500 kHz was recently funded with an award of $1.1M by ONR. This requires a mean interval of 2 μS between switching events and a simulation step size several times smaller, certainly no more than 200 nS and probably considerably less. Nari will again be part of the team, hopefully extending this long running partnership to nearly 60 years. It has continued to exemplify the contribution of simulation to science and engineering as a partnership between domain expertise and simulation expertise that aims to advance scientific knowledge or engineering design.

**Acknowledgements** The extensive and continuing support of the research described in this chapter by the US Office of Naval Research is gratefully acknowledged.

## References

Adamson, C., & Hingorani, N. G. (1960). *High voltage direct current power transmission*. Garraway.

Bednar, R., & Crosbie, R. E. (2006). Applications of power spectral and harmonic distortion techniques to computer simulation parameter selection. In *Proceedings of the Summer Computer Simulation Conference, Society for Modeling and Simulation International, Calgary, Alberta, Canada, July 2006*.

Bednar, R., & Crosbie, R. E. (2007). Stability of multi-rate simulation algorithms. In *Proceedings of the Summer Computer Simulation Conference, San Diego CA, 15–18 July, 2007*, SCS.

Bednar, R., & Crosbie, R. E. (2009). Solution of non-linear differential equations using state transition methods. In *Proceedings of SCS Conference on Grand Challenges in Modeling and Simulation, Istanbul, Turkey, 13–16 July, 2009* (pp. 5–9).

Bednar, R., & Crosbie, R. E. (2010). Multi-rate simulation accuracy improvement for linear systems using hold networks. In *Proceedings of SCS Conference on Grand Challenges in Modeling and Simulation, Ottawa, 12–14 July, 2010* (pp. 261–267) SCS.

Bednar, R., & Crosbie, R. E. (2012). Multi-rate solution of linear differential equations using a similarity transformation. In *Proceedings of SCS Conference on Grand Challenges in Modeling and Simulation, Genoa, Italy, 8–11 July, 2012* (Conference Best Paper Award).

Crosbie, R. E. (1984). Continuous and discontinuous-change models: Concepts for simulation languages. In T. I. Ören, B. P. Zeigler, & M. S. Elzas (Eds.) *Simulation and model-based methodologies: An integrative view* (Vol. 10). NATO ASI Series (Series F: Computer and Systems Sciences). Springer, Heidelberg.

Crosbie, R. E. (2007). High-speed real-time simulation. In *Asia International Conference on Modelling & Simulation (AMS), Prince of Songkla University, Phuket, Thailand, 2007* (pp. 7–13).

Crosbie, R. E. (2008). Advances in high-speed real-time simulation. In *Keynote presentation at UKSIM European Symposium, 8–10 September, 2008* (pp. 2–8). Liverpool, England: IEEE.

Crosbie, R. E. (2012). A professional lifetime with the society for modeling and simulation international—a personal memoir. *Transactions of the Society for Modeling and Simulation International, 88*(9), 1025–1032.

Crosbie, R. E., & Halder, A. K. (1986). Improved representation of arc static characteristic. *Electronics Letters, 4*(16) (Institution of Engineering and Technology, London, England).

Crosbie, R. E., & Hay, J. L. (1974). Digital techniques for the simulation of discontinuities. In *Proceedings of the Summer Computer Simulation Conference*. Houston, Texas.

Crosbie, R. E., & Hay, J. L. (1977). Continuous simulation languages for minicomputers. In *A New Approach Proceedings Summer Computer Simulation Conference Chicago, Illinois* (pp. 22–24).

Crosbie, R. E., & Hingorani, N. G. (1967). Analog computer solution of ship's deviation due to submarine DC cable. *Maths and Computers in Simulation, 9*(2), 81–85 (Elsevier).

Crosbie, R. E., & Hingorani, N. G. (2004) High-speed, real-time simulators for engineering design. In *Presented at the Advanced Engineering Design Conference, University of Glasgow, September 2004*.

Crosbie, R. E., Javey, S., Hay, J. L. & Pearce, J. G. (1985). ESL-a new continuous system simulation language. *Simulation, Transactions of the Society for Modeling and Simulation International, 44*(5), 242–246.

Crosbie, R. E., Zenor, J. J., Bednar, R., Word, D., Hingorani, N., & Ericsen, T. (2004a). High-speed, scalable, real-time simulation using DSP arrays. In *Presented at IEEE/ACM Parallel and Distributed Simulation Workshop*. Kufstein, Austria.

Crosbie R. E., Zenor, J. J., Word, D., & Hingorani, N. G. (2004b). Fast real-time DSP simulations for on-line testing of hardware and software. In *Proceedings of Summer Computer Simulation Conference, Society for Modeling and Simulation International, San Jose, July 2004*.

Crosbie, R. E., Zenor, J. J., Word, D., & Hingorani, N. G. (2004c). Fast real-time DSP simulations for on-line testing of hardware and software. *Modeling & Simulation, 3*(4) (SCS, San Diego).

Crosbie, R. E., Zenor, J. J., Bednar, R., Word, D., & Hingorani, N. G. (2007). Multi-rate simulation techniques for electric ship design. In *2007 IEEE Electric Ship Technologies Symposium, Arlington VA, 21–23 May, 2007*.

Crosbie, R. E., Zenor, J. J., Word, D., Bednar, R., Hingorani, N. G. (2009a). Advances in high-speed real-time multi-rate simulation techniques for ship power systems. In *2009 IEEE Electric Ship Technology Symposium, Baltimore MD, 20 22 April, 2009*.

Crosbie, R. E., Word, D., Zenor, J. J., Bednar, R., & Hingorani, N. G. (2009b). High-speed real-time multi-rate simulation. In *Proceedings of SCS Conference on Grand Challenges in Modeling and Simulation, Istanbul, Turkey, 13–16 July, 2009* (pp. 10–16).

Crosbie, R. E., Zenor, J. J., Bednar, R., Word, D., & Hingorani, N. G. (2009c). Embedding high-speed power electronics simulations in real-time power system simulations. In *Proceedings of 9th IASTED European Conference on Power and Energy Systems, Palma de Mallorca, Spain, 7–9 September, 2009* (pp. 86–90).

Crosbie, R. E., Zenor, J. J., Word, D., Bednar, R., & Hingorani, N. G. (2010). FPGA-based real-time simulation of power electronic systems. In *Proceedings of 7th Eurosim Congress, Prague, Czech Republic, 6–10 September, 2010*.

Crosbie, R., Zenor, J., Bednar, R., Word, D., Kredo II, K., & Hingorani, N. G. (2013). Using FPGAs for low-cost high-performance simulations. In *IEEE Electric Ship Technologies Symposium (ESTS), April 2013*.

English electric DEUCE home page. http://users.tpg.com.au/eedeuce/index.html.

Hay, J. L. (1975). ISIS—an interactive simulation language. In *Proceedings of the United Kingdom Simulation Council Conference on Computer Simulation. Bowness-on-Windermere, England, 6–8 May, 1975*.

Hay, J. L. (1978). Interactive simulation on minicomputers: Part 1-ISIS, a CSSL language. *Simulation, Transactions of the Society for Modeling and Simulation International, 31*(1), 1–7.

Hay, J. L. (1981). A new CSSL standard—an implementation view. In *UKSC Conference on Computer Simulation*. Harrogate, UK.

Hay, J. L. (1986). ESL—advanced simulation language for parallel processors. In M. G. Singh, A. Y. Allidina, & B. K. Daniels (Eds.), *Parallel processing techniques for simulation*. Springer, Boston, MA: Applied Information Technology.

Hay, J. L,. & Crosbie, R. E. (1982). ISIM-A simulation language for microprocessors. In *Proceedings Summer Computer Simulation Conference*. Denver, Colorado.

Hay, J. L., & Griffin, A. W. J. (1980). Simulation of discontinuous systems. In L. Dekker et al. (Ed.), *Simulation of systems '79*. North Holland Pub co.

Hay, J. L., & Hingorani, N. G. (1970). Dynamic simulation of multiconverter HVDC systems by digital computer, part I: Mathematical model and part II: Computer program. *IEEE Transactions on PAS, 89,* 218–228.

Hay, J. L., Crosbie, R. E., & Chaplin, R. I. (1974). Integration subroutines for systems with discontinuities. *Computer Journal, 17*(3).

Hingorani, N. G., & Chadwick, P. (1986). A new constant extinction angle control for AC/DC/AC static converters. *IEEE Transactions on PAS, 3*, 866–872.

Hingorani, N. G., Hay, J. L., & Crosbie, R. E. (1966). Dynamic simulation of HVDC transmission systems on digital computers. *Proceedings of the IEEE, 113*(5), 793–802.

Hingorani, N. G., Crosbie, R. E., & McConnach, J. (1968a). Steady-state simulation of HVDC convertors by digital computers. *Proceedings of the IEEE, 115*(5).

Hingorani, N. G., Kitchin, R. H., Hay, J. L. (1968b). Dynamic simulation of HVDC power transmission systems on digital computer—generalised mesh analysis approach. *IEEE Transactions on PAS, 87*, 989–996.

Howe, R. M. (1961). *Design fundamentals of analog computer components*. Princeton: Van Nostrand.

Karplus, W. J., & Soroka, W. W. (1959). *Analog methods: Computation and simulation* (2nd ed.). New York: McGraw-Hill.

Korn, G. A., & Korn, T. M. (1952). *Electronic analog computers*. New York: McGraw-Hill.

Kredo K., Zenor, J. J., Yusta, J., Moore, D., & Mish, S. (2012). Interfacing techniques for FPGA-based distributed simulation. In *Proceedings of SCS Conference on Grand Challenges in Modeling and Simulation, Genoa, Italy, 8–11 July, 2012.*

Kredo, K., Zenor, J. J., Bednar, R., & Crosbie, R. E. (2015). FPGA-accelerated simulink simulations of electrical machines. In *IEEE Electric Ship Technologies Symposium (ESTS), Arlington, VA, USA, June 2015.*

Kredo, K., Bednar, R., Crosbie, R. E., & Zenor, J. J. (2017). Techniques for accelerating simulations of electric ships. In *IEEE Electric Ship Technologies Symposium (ESTS), Arlington, VA, USA, August 2017.*

McLeod, J. (Ed.). (1968). *Simulation: The dynamic of ideas and systems with computers*. New York: McGraw-Hill.

Pearce, J. G., Crosbie, R. E., Zenor, J. J., Bednar, R., Word, D., Hingorani, N. G. (2009). Developments and applications of multi-rate simulation. In *Proceedings of UKSim 2009, Emmanuel College Cambridge, UK, 25–27 March, 2009* (pp. 129–133).

The Mathworks, Inc., SimscapePowerSystems. Retrieved March 29, 2018, from https://www.mathworks.com/products/simpower.html.

Word, D., Zenor, J. J., Bednar, R., Crosbie, R. E., Hingorani, N. G. (2007). Multi-rate real-time simulation techniques. In *Proceedings of the Summer Computer Simulation Conference, San Diego CA, 15–18 July, 2007*, SCS.

Word, D., Zenor, J. J., Bednar, R., & Hingorani, N. G. (2008). High-speed real-time simulation for power electronic systems. *Simulation: Transactions of the Society for Modeling and Simulation International*, *84*(8/9), 441–456 (R. E. Crosbie (Ed.), Special Issue on Power Electronic Systems).

Yusta, J., Zenor, J. J., & Kredo II, K. (2011). Efficient high-speed ethernet for real-time simulation. In *Proceedings of SCS Conference on Grand Challenges in Modeling and Simulation, The Hague, Netherlands, 27–30 June, 2011* (pp. 137–143).

Zenor, J. J., Word, D., Bednar, R., Crosbie, R. E., & Hingorani, N. G. (2010). Progress towards a low-cost high-speed real-time multi-rate simulator. In *Proceedings of SCS Conference on Grand Challenges in Modeling and Simulation, Ottawa, 12–14 July, 2010* (pp. 247–252) SCS.

Zenor, J. J., & Kredo, K. (2012) Interfacing techniques for FPGA-based distributed simulation. In *Proceedings of SCS Conference on Grand Challenges in Modeling and Simulation, Genoa, Italy, 8–11 July, 2012.*

Zenor, J. J., Bednar, R., Word, D., Hingorani, N. G., & McGookin, E. (2007a). Simulation of an unmanned underwater vehicle (UUV): A multi-rate simulation. In *Proceedings of the Summer Computer Simulation Conference, San Diego CA, 15–18 July, 2007*, SCS.

Zenor J. J., Pearce, J. G., Bednar, R. (2007b). Testing of multi-rate simulations using the esl simulation language. In *Proceedings of the Summer Computer Simulation Conference, San Diego CA, 15–18 July, 2007*, SCS.

Zenor, J. J., Bednar, R., Crosbie, R. E., & Hingorani, N. G. (2011). Efficient real-time simulation of linear differential equations arising from simulation of electronic power systems. In *Proceedings of SCS Conference on Grand Challenges in Modeling and Simulation, The Hague, Netherlands, 27–30 June, 2011* (pp. 144–149).

# Chapter 4
# Over Fifty Years of My Involvement in Simulation

**Tuncer Ören**

> *Those –be it an individual, an institution, or a country– unable to surpass themselves cannot exceed others. Therefore, in achieving progress, what is difficult is to supersede oneself.*
> (*Ören* 1995).

**Abstract** Author's involvement and witnessing the advancements of modeling and simulation over 50 years are highlighted. Some concepts are outlined: inputs, data, quality and failure avoidance in simulation, ethics, machine understanding, synergies of simulation with several disciplines, intelligence and simulation, agent-directed simulation, simulation terminology, modeling and simulation body of knowledge, big picture of simulation, bigger picture of similarity, and some of the aspirations of the author for the future of simulation.

## 4.1 Introduction

Recently, two important proceedings cast light on the past and the evolution of simulation: One is on the 30 years of the European Council of Modeling and Simulation (ECMS) (Al-Begain and Bargiela 2016); the other one goes farther back and documents the "Seminal Research from the 50 Years of Winter Simulation Conferences" (Tolk et al. 2017).

I have been involved in simulation since 1965, had well over 500 publications including over 45 books and proceedings, and have been active in over 500 conferences and seminars in 40 countries. I had several opportunities of documenting historic aspects of my involvements in modeling and simulation. The first occasion was on the 25th anniversary of SCS (Society for Modeling and Simulation International); I wrote "Simulation—as it has been, is and should be (Ören 1977a). Some of my wishful thinking for the future of simulation were:

T. Ören (✉)
Emeritus Professor of Computer Science, School of Electrical Engineering and Computer Science, University of Ottawa, Ottawa, ON, Canada
e-mail: oren@eecs.uOttawa.ca



J. Sokolowski et al. (eds.), *Summer of Simulation*, Simulation Foundations, Methods and Applications, https://doi.org/10.1007/978-3-030-17164-3_4

1. "Simulation models will be comprehensible." Later, during 1980–1996 when I was a member of the Technical Advisory Committee (TAC) of the Atomic Energy of Canada Ltd. (AECL) on the Nuclear Fuel Waste Management Program as the Representative of the Canadian Information Processing Society (CIPS) and Chairman of the System Analysis Subcommittee of TAC, I realized how important it was the comprehensibility of simulation programs by non-programmer scientists.
2. "If a simulation program is to be used several times, a list of questions answerable by the model will be part of the documentation provided to the user." It seems the problem persists.
3. "Advanced modelling concepts will be used to simulate complex phenomena. For example, behaviorally anticipatory models have definite advantages over classical feedback models."
4. "New simulation software implementing advanced concepts in modelling and model manipulation will be part of computer-assisted model building, handling, and documentation systems. Some of the algorithmic model manipulations may be done for consistency checks, decomposition, simplification, coarsening, elaboration, or comparison of models." Currently, it is comforting to see the advent of simulation model engineering (Zhang et al. 2019)

The following year, on another occasion, I wrote: "A Personal View on the Future of Simulation Languages" (Ören 1978b). On the 50th anniversary of the SCS, it was "SCS and Simulation: Fifty Years of Progress" (Ören 2002a). On the 60th anniversary of SCS, Bernie Zeigler and I wrote: "System Theoretic Foundations of Modeling and Simulation: A Historic Perspective and the Legacy of A. Wayne Wymore" (Ören and Zeigler 2012). (My intellectual pedigree which goes back to Gauss is documented in this article.) The same year, I also wrote: "Evolution of the Discontinuity Concept in Modeling and Simulation: From Original Idea to Model Switching, Switchable Understanding, and Beyond" (Ören 2012). A recent article documents the nine evolution aspects of modeling and simulation (Ören et al. 2017).

The kind invitation from the editors of this book gave me an occasion to concentrate on some aspects of the history and advancement of modeling and simulation intertwined with my witnessing; and in some areas, with my involvements. Since, I have contributed –with talented colleagues or alone, to well over 500 publications, including 46 books and proceedings –mostly in modeling and simulation, in this chapter I would like to concentrate on the threads of advancements and on some anecdotes.

## 4.2 The Beginnings

Normally, I don't remember when I heard a word for the first time. The only exception is "simulation." In 1965, while working as a systems engineer at IBM in my native Istanbul, Turkey, I decided to enrol the Ph.D. program at the Technical

University of Istanbul. I asked a theme for doctoral studies to Prof. Faruk Akün, the then head of the Industrial Engineering Department of the Mechanical Engineering Faculty. He suggested "simulation" a concept which intrigued me then and which continues to interest me with increasing zeal after over half a century.

In the mean time, I had applied for a scholarship to a US government agency. Considering that I was an IBM systems engineer, the Agency chose the then Systems Engineering Department (later it became, Department of Systems and Industrial Engineering) of the University of Arizona in Tucson, Arizona. The department was established by Prof. Albert Wayne Wymore, one of the founding fathers of systems engineering. I went to Tucson in 1967, the year Dr. Wymore published his seminal book "A Mathematical Theory of Systems Engineering: The Elements" (Wymore 1967). In 1967, Dr. Wymore was spending his sabbatical in Hawaii. I enrolled in many courses. With my background as mechanical engineer, some of them such as finite state machines appealed to me very much as abstract machines.

My scholarship was for one year. The Agency's handbook had the following note: "It is seldom possible to approve a request to extend the time limit of a quest to extend the time limit of a training program." However, thanks to two favorable brief notes from my professors on annual evaluation forms, my scholarship was extended twice.

The second year Dr. Wymore returned to Tucson –albeit after a surf accident. He kindly accepted to be my thesis advisor. My thesis work was to develop a simulation model specification language based on his systems theory.

At the end of 1970, I satisfied all requirements for my doctoral studies. The thesis: "GEST: General System Theory Implementor, A Combined Digital Simulation Language," was submitted to the University (Ören 1971a). There was a position at the newly established Computer Science Department of the University of Ottawa in the Canadian national capital. I communicated with the founding chair, Prof. Louis G. Birta and joined the department on December 1970.

## 4.3 The Reviewers Were Not Ready

After getting my Ph.D., I thought I was ready to disseminate my views; however, soon I realized that the reviewers were not. Here is a sequence of some initial failures (of the reviewers): Before I left Tucson, I prepared an article on GEST and I submitted it late 1970 to Simulation, the Journal of the Simulation Councils –SCi (Later became SCS—The Society for Modeling and Simulation International). That was the time of high success of CSSL (Continuous System Simulation Language) (SCi Software Committee 1967). The reviewers who were accustomed to CSSL –a simulation *programming* language– and its variants, flatly rejected my article which was the first *model specification language* based on a systems theory. I wrote several letters to SCi to persuade them; to no avail. Finally –after about two dozen letters– John McLeod, the founder of SCi wrote me to announce that the article is not accepted for Simulation. As I clarified elsewhere, I never blamed John for the refusal of my article; I realized, to my dismay, that the referees couldn't appreciate it (Ören 2002a). Three years later, I was an Associate editor of Simulation. Over the years, I was honored several times by SCS as SCS McLeod Founder's Award for Distinguished Service to the profession, SCS Fellow, SCS Hall of Fame, Distinguished Service Award, and Distinguished Professional Achievements Award (SCS-AR). And early 1980s after a NATO Advanced Study Institute in Ottawa, John wrote very favourably:

> This summer (July 26-August 6) Suzy and I participated in the NATO Advanced Study Institute "Simulation and Model-Based Methodologies: An Integrative View" held at the University of Ottawa, Ottawa, Canada; I now know vastly more about modeling and simulation (or the possibilities thereof) than I did before. I also learned that there is a lot more to learn!
>
> Well, the 58 lecturers and participants from 18 countries at the Institute, organized under the direction of Dr. Tuncer Ören of the University of Ottawa's Computer Science Department, are certainly doing their best to remedy that. In fact, the degree of mathematical abstraction (which seems necessary to develop theory) was so advanced that this writer found it difficult at times to relate it to reality.
>
> Dr. Ören was the first speaker. After an appropriate greeting he presented a lecture on 'Computerization of Model-Based Activities: A Paradigm Shift and New Vistas.' Although he covered the 'fundamental elements of a simulation study; models and behaviour; elements of a model-based simulation software system; synergies of simulation, software engineering, artificial intelligence, and general system theories; knowledge-based modeling and simulation systems; and highlights of desirable research in simulation,' Tuncer's main thrust, it seemed to me, was 'to place simulation in a central position for several scientific disciplines by a shift of paradigm' (McLeod 1982).

John McLeod's last observation about my motivation is still valid. In early 1980s, it was time to promote model-based approach; currently simulation-based approach appears to be a viable way to enhance many disciplines.

Not having a publication in a refereed journal in early 1970s did not help me to get non-trivial research grants from the Canadian research granting agency in natural sciences. Luckily, I realized that I didn't need funding to generate ideas.

However, after being accepted by other colleagues, the agency asked me later to review some of the submitted research proposals. I gladly did the reviews.

Another lack of understanding occurred in Switzerland. My colleague Prof. François Cellier organized an International Symposium on Simulation—SIMULATION '80, during June 25–27, 1980 in Interlaken, Switzerland and honored me by asking me to be the keynote speaker. I chose the following theme for my talk: "Computer-Aided Modelling Systems." After my presentation, a Canadian scientist criticized me heavily, refusing strongly the concept of computer-aided modeling. Since, this was an open criticism, I had the opportunity to reply with even stronger zeal. The book (Cellier 1982) and my contribution were published (Ören 1982a). Today, nobody needs to promote the concept of computer-aided modeling systems, since it is the only obvious way to do it. However, at that time, somebody had to promote the concept to programmer-simulationists. I am glad I did it.

## 4.4 The Trio

Bernie Zeigler, Maurice Elzas, and I often referred to ourselves as the trio. In one of our activities we also had George Klir. Let me outline how we got together and what we did:

During 1976–83, I was a member of the Board of Directors and Delegate to Canada of the International Association for Cybernetics (Namur, Belgium). During September 6–11, 1976, the association had its 8th International Congress on Cybernetics. Within the congress, I organized a symposium on the application of cybernetics and general system theoretic concepts to the simulation of large scale systems. As it can be seen in the proceedings of the symposium (Ören 1978b), I had several eminent scientists as speakers. One of them was Prof. Bernie Zeigler who had published his first seminal book (Zeigler 1976).

During my first sabbatical leave (1977–78), I was a senior research fellow at the Computer Science Department of the Agricultural University, Wageningen, The Netherlands. The chair of the department was Prof. Maurice Elzas whom I met in 1976 at the 8th AICA (International Association for Analog Computing) Congress which was held in Delft, the Netherlands. I had the pleasure of introducing Bernie to Maurice. The Trio organized four main conferences. The titles of the books reflect our themes:

1. First one was in 1978 in Rehovot, Israel while Bernie was at the Weizmann Institute of Science. He was the editor-in-chief of the volume titled: "Methodology in Systems Modelling and Simulation." In this conference, Prof. George J. Klir joined us (Zeigler et al. 1979).
2. Second one was a NATO Advanced Study Institute (ASI) held in 1982 in Ottawa. The book was titled: "Simulation and Model-based methodologies: An Integrative View" (Ören et al. 1984).
3. The third one was held in 1985 in Papendal, Arnhem, The Netherlands, very close where Maurice was (Wageningen). At the NATO Advanced Study Institute, we had funding from NATO. Furthermore, NATO sent an observer to our ASI. In this third conference, we didn't require funding and we invited the same observer all expenses paid. The title of the book was: Modelling and Simulation Methodology in the Artificial Intelligence Era" (Elzas et al. 1986).
4. The fourth conference was held in Tucson, Arizona, in 1987 while Bernie was established at the University of Arizona (my Alma Mater for my Ph.D. studies). The book was titled: "Modelling and Simulation Methodology: Knowledge Systems' Paradigms" (Elzas et al. 1989)

In addition to these four conferences and the books, we had several other joint activities. Some of them are:

In 1977, while Bernie was at the Weizmann Institute of Science in Rehovot, Israel, extended me an invitation to write a joint article. We decided the theme of the article in Rehovot and finished it in a week. However, the then editor-in-chief of the transactions of SCS could publish it only in 1979 (Ören and Zeigler 1979).

The article was an iconoclast; it started as follows:

> Conventional simulation techniques have three shortcomings when applied to large-scale modelling: They provide an inadequate man-machine interface, they provide a poor conceptual framework, and they lack needed tools for managing data and model. These shortcomings may be ameliorated by developing new simulation languages that differentiate the functional elements of simulation programs and by recognizing the goals of these functional elements. This paper provides concepts for the design and implementation of such advanced simulation methodologies.

In 1979, Bernie, Maurice, and I organized the Sorrento Workshop for the International Standardization of Simulation languages. The Workshop papers were published later (ACM SIGSIM Digest 1984). One of the articles, by Bongulielmi and Cellier: "On the usefulness of deterministic grammars for simulation languages" was categorically different than the then usual ad hoc definition of simulation languages (Bongulielmi and Cellier 1984).

One of the highlights of the Sorrento Workshop was that one of the invited colleagues was Dr. Harry M. Markowitz. His article was titled: "Proposal for the standardization of status description" (Markowitz 1984). Dr. Markowitz is known by the simulationists as the designer and developer of SIMSCRIPT, a discrete simulation language, widely used before Bernie developed long needed theoretical basis to specify discrete systems (DEVS—Discrete Event System Specification) (Zeigler 1984). Dr. Markowitz is the only simulationist who is a Nobel laureate – albeit in a different domain, namely in portfolio theory. His collected, selected works on portfolio theory, sparse matrices, as well as on the SIMSCIPT are available in a comprehensive volume (Markowitz 2009). Dr. Markowitz was the keynote speaker at the 1981 Winter Simulation Conference (Markowitz 1981).

## 4.5 Basics: Inputs and Data

In 2001, I was invited to deliver a plenary talk in a conference held in St. Petersburg, Russia. My talk was titled: "Software Agents for Experimental Design in Advanced Simulation Environments." In this talk (Ören 2001a), I also provided a classification of inputs which are normally considered to be exogenous, namely, generated outside of the system of interest. In this classification, I also introduced endogenous inputs, namely, internally generated inputs as outlined in Table 4.1.

Endogenous inputs are important especially in the simulation of cognitive systems. Any computationally intelligent system should be able to perform introspection; and generate questions, hypotheses, and goals.

Nowadays, the importance of big data analytics is properly acknowledged. In 2001, I was also invited to deliver a plenary talk at the Eurosim conference held in Delft, The Netherlands. My talk was titled: "Impact of Data on Simulation: From Early Practices to Federated and Agent-Directed Simulations." I started with a well-known milestone example and pointed out the importance of Tycho Brahe's meticulous data collection activities in the second half of the 16th century (Ören 2001b):

> Data is essential and provides a conceptually rich paradigm for many types of discourse including scientific inquiry. As a milestone example, one can cite the fact that relevant data had an impact in the history of ideas in Western civilization, as reflected in the works of Ptolemy and his predecessors to the works of Galilei; with Copernicus, Brahe, and Kepler, in between. As Riley summarizes: "Kepler's work is an example of the deduction of general laws from a mass of observations—the essence of science. But it was primarily his attempt to apply physical principles to astronomical data that marks his break with ancient astronomy" (Riley 1992). Claudius Ptolemy (100–175) who dominated the Western world for 15 centuries advocated the previously known earth-centric (i.e., Ptolemaic) world view. Nicolaus Copernicus (1473–1543), leading to the Copernican revolution, (i.e., sun-centric world view), argued just the contrary. However, his "methods of arguments were still distinctly medieval" (Hall 1992, p. 178). Thus, what Kepler (1571–1630) achieved was based on his master, Tycho Brahe's (1548–1601) relentless observations of the planetary system. (Brahe also advanced astronomical apparatus that was needed for the observations.) Furthermore, Kepler's abstraction of relevant data and Galileo Galilei's (1564–1642) own observations led Galileo to promote the Copernican world view with well known consequences.

My interest in the importance of data still continuous, especially with one of my diligent young colleagues; i.e., Dr. Mayank Singh (Singh et al. 2017; Gupta et al. 2019).

## 4.6 Quality and Failure Avoidance in Simulation

Considered from the knowledge processing perspective, simulation is a model-based knowledge processing activity; hence similitude of the simuland and a model to represent it is important. Validation in modeling and simulation deals with the assurance of goal-directed similitude of models. Another issue is verification which deals with the proper computerization of simulation models and associated

**Table 4.1** Internally generated inputs

(1) Active perception of endogenous inputs:
    Introspection □ perceived:
    - Internal facts, events; or realization of lack of them
(2) Generation of endogenous input:
    - Anticipated facts and/or events (anticipatory systems)
    - Internally generated questions
    - Internally generated hypotheses by:
        -- Expectation-driven reasoning (Forward reasoning, bottom-up reasoning, data-driven reasoning)
        -- Model-driven reasoning
    - Internal goals: internally generated goals

experimentation as well as model behavior generation conditions. Sargent and Balci contributed extensively to the validation and verification in simulation (Sargent 1991, 2011 (with extensive references to previous publications); Balci 1987, 1998; Balci and Sargent 1984). For a historic review of verification and validation theories, see: Sargent and Balci (2017) and Durst et al. (2017).

My activities on reliability and quality assurance in modeling and simulation (over 60 publications and over 20 activities in meetings) started in mid 1970s with a study of the syntactic errors of the original formal definition of CSSL 1967 (Ören 1975). In 1981, I published: "Concepts and criteria to assess acceptability of simulation studies: A frame of reference" (Ören 1981). During 1980–1996, I served as a member of the Technical Advisory Committee (TAC) of the Atomic Energy of Canada Ltd. (AECL) on the Nuclear Fuel Waste Management Program as the representative of the Canadian Information Processing Society (CIPS) and as Chairman of the System Analysis Subcommittee of TAC.

As a novel paradigm in reliability of modeling and simulation, my first article on failure avoidance was published in 2009: "Failure Avoidance in Agent-Directed Simulation: Beyond Conventional V&V and QA" (Ören and Yilmaz 2009a). Along this line, the following sources of errors in decision making should also be taken into consideration:

(1) dysrationalia –inability to think and behave rationally despite adequate intelligence (Stanovich 1993);
(2) cultural biases, i.e., the tendency for people to judge the outside world through a narrow view based on their own culture (Worldatlas; Hofstede 2001),
(3) as well as drawbacks of rule-based intelligent systems, and misunderstanding (Ören et al. 2013).

## 4.7 Ethics in Simulation

Ethics can be considered as the top link of reliability in simulation-based studies (as in many human activities), Several colleagues were influential in pointing out the need for development, and adoption of a code of ethics for simulationists. Ethics for simulationists was recommended for the first time by John McLeod (1983). (My special thanks and appreciation to Prof. Helena Szczerbiska for pointing it out.) A special symposium was organized in Wageningen, The Netherlands on the 65th birthday of Prof. Maurice S. Elzas (July 2, 1999). Maurice selected the theme as: Simulation and Ethics. As one of his close colleagues, I was invited to talk. This was my first talk on the topic. In 2001, Dr. Bruce Fairchild, the then President of the SCS asked me whether I would like to develop a code of ethics for simulationists. In 2002, two articles were presented at the Summer Computer Simulation Conference; one, on the rationale for a code of professional ethics for simulationists (Ören 2002b); the other, a code of professional ethics for simulationists (Ören et al. 2002). Our late and distinguished colleague, W. Waite was behind the resolution of SimSummit "that a Code of Professional Ethics should be one of the four pillars –along with Science, Technology and Applications– for Modeling and Simulation to be considered as a profession" (Ören 2014). The code is adopted by the SCS (SCS SimEthics) and for the member organizations of two umbrella organizations of SCS, namely, M&SNet and MISS, as well as by influential simulation organizations such as CMSP (Certified Modeling & Simulation Professional) certification program of M&SPCC (Modeling and Simulation Professional Certification program) (CMSP-Ethics).

## 4.8 On Machine Understanding

In late 1980s and early 1990s, a large Canadian company offered a substantial research funding for understanding simulation software. This very favorable funding was the initial impetus first to develop reverse engineering tools to understanding simulation programs (Ören et al. 1990). Table 4.2 depicts an outline of how my research on understanding evolved over the years. With my colleagues, I have published over 20 articles on machine understanding and had about the same number of presentations in meetings.

I consider machine understanding as one of the pillars of computational intelligence including computational emotional intelligence and wonder how a knowledge processing system without understanding ability could be considered intelligent.

## 4.9 Synergies

*The relationships of several entities can be one of the following types:*

1. They may co-exist without any reaction. This may be, for example, the essence of *co-existence pacific.*

**Table 4.2** An outline of how my research on understanding evolved over the years

| | |
|---|---|
| 1990-1995 | Understanding simulation software; Program understanding |
| 1997-2006 | Understanding systems in general |
| 2006-2008 | Systems with understanding ability and Understanding agents |
| 2009- | Agents with ability to understand emotions; Switchable understanding |
| 2011- | Avoidance of misunderstanding |
| 2015- | Enriching machine understanding paradigm |
| 2016- | Exploring the synergy of machine understanding and all three aspects of agent-directed simulation (ADS) |
| 2017- | Computational awareness; Computational consciousness |

2. They may be living together. This is the essence of *symbiotic relationship* (or *symbiosis*) which can be mutualistic, commensalistic, or parasitic relationship.
   2.1 In a mutualistic relationship "two organisms of different species exist in a relationship in which each individual fitness benefits from the activity of the other. Similar interactions *within* a species are known as co-operation. Mutualism can be contrasted with interspecific competition, in which each species experiences *reduced* fitness (Wikipedia—mutualism)
   2.2 In a commensalistic relationship "in which members of one species gain benefits while those of the other species neither benefit nor are harmed" (Wikipedia—commensalism)
   2.3 In a symbiotic relationship, some may exploit some others. This is the case of parasitic relationship.
3. They may unite to form a new entity, such as hydrogen and oxygen forming water. This is *systemic relationship* (Checkland 1993).
4. They may have two types of *synergistic relationship*.
   4.1 Synergy generally means working together to have an extended effect larger than the total of the individual effects.
   4.2 Another way of expressing "working together" can be one enhancing others; or entities may enhance each other. In this type of *synergistic relationship*, each entity conserves its identity, is enhanced by the influence of other entities, and may enhance other entities.

The difference between synergy and symbiosis lies in the fact that symbiosis implies living together. The difference between synergy and systemic relationship is important: In systemic relationship the component entities form a new entity with characteristics different than the characteristics of the component entities; while in a synergy, entities enhance each other while retaining their identities.

Over the years, I had (or initiated) occasions to elaborate on synergies of simulation with several disciplines (Ören 1978b, 1982b, 1984a, 1996; Ören and Yilmaz 2006, 2009b, 2012; Ghasem-Aghaee et al. 2017; Yilmaz and Ören 2009).

In early 1980s, while I was the editor of SIMULETTER, the publication of the ACM Special Interest Group on Simulation, I even elaborated on a possible forum for the synergy of ACM Special Interest Groups (Ören 1984b). I wrote: Simulation as an important model-based activity:
can have two-way interaction with almost all the fields of knowledge represented by ACM Special Interest Groups: … I would like to publish in SIMULETTER two articles for each SIG group under the following categories:

1. Contributions of SIGSIM to SIGxxx: One article would survey the current and possible contributions of simulation in the field represented by this particular SIG. An additional bibliography would be also very useful.
2. Contributions of SIGxxx to SIGSIM: This article would cover just the opposite, i.e., it would be a survey of the current and possible contributions of the specific

field of knowledge to any aspect of simulation. In this case also an additional bibliography would be very useful.

There was no interest to this suggestion. However, once done, it could document the central role simulation can play to empower many disciplines. After two publications, namely, "Guide to Simulation-Based Disciplines: Advancing our Computational Future" (Mittal et al. 2017) and "The Evolution of Simulation and its Contributions to Many Disciplines" (Ören et al. 2017), the time is ripe to finally elaborate on this topic.

## 4.10 Intelligence and Simulation

The synergy of simulation and computational (artificial, machine) intelligence played an important role even at the beginning of computational intelligence which started by simulation of human cognitive abilities in 1950s. My first publication on the synergy of simulation and computational intelligence was in 1982 (Ören 1982b). In the editorial of the January 1985 issue of SIMULETTER, I wrote:

> … As expected from the inertia of human intelligence, however, some of the shifts of paradigms were not easily achieved. … For a long time, simulation contributed to the field of artificial intelligence by making possible the cognitive simulation studies. Now, simulation can benefit from the advances in artificial intelligence (as well as from advances in general system theories, software and computer engineering, and mathematical modelling and experimentation techniques).

The question is not whether or not to have artificial intelligence in simulation, but rather how to have it? at which level? how reliably? how soon? and above all how intelligently? …

The quotation on the cover of this issue would well summarize an attitude: " … *unintelligent computerization is not enough* (Ören 1985)".

## 4.11 Agent-Directed Simulation

Agent-directed simulation (ADS) refers to the full possibilities of the synergy of simulation and agents as outlined in Fig. 4.1. The first publication about ADS was in 2000 (Ören et al. 2000). Afterwards Prof. Levent Yilmaz has been contributing tremendously to ADS (Yilmaz and Ören 2005, 2009). Another colleague who has been contributing to ADS is Prof. Yu Zhang (Zhang et al. 2010, 2012).

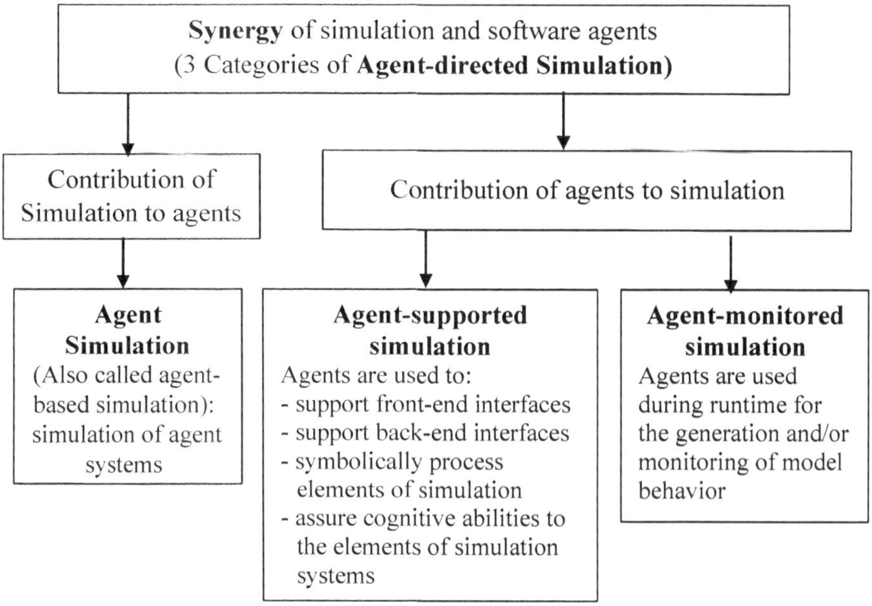


**Fig. 4.1** Synergy of simulation and software agents

## 4.12 Simulation Terminology

As it is also the case for simulation, the conceptual richness of a field can be seen in its terminology. My first elaboration about simulation terms was in late 1970s (Ören 1977b). The list included about 800 terms and at that time, it looked rather long.

In 2006, colleagues at the Université Paul Cézanne—Aix-Marseille III, and I had the pleasure of seeing in print, our first tri-lingual (English, French, and Turkish) modeling and simulation dictionary (Ören and the French team 2006). The French team comprised of 14 colleagues and Mme. Lucile Torres was the co-ordinator. The dictionary had about 4500 English terms that I gleaned over the years. The support of late Professor Norbert Giambiasi and Professor Claudia Frydman for the preparation of the French terms and the publication in Marseille as well as the valuable contributions of dedicated colleagues in the French team are gratefully acknowledged.

In 2012, with the contribution of 30 Chinese colleagues, the first version of the Chinese-English and English-Chinese modeling and simulation dictionary with about 9000 English terms were published (BoHu Li et al. 2012). Currently, a Chinese team under the leaderships of Prof. Bo Hu Li and Prof. Guanghong Gong are finalizing the second edition which comprises about 13 000 English terms that I compiled.

Time permitting, I will revise the French, Italian, and Spanish terms and continue co-operating with colleagues for the respective editions of the modeling and simulation dictionary.

Ontology-based dictionaries combine taxonomy of the relevant terms and their definitions. An example is ontology-based dictionary of about 80 machine understanding terms (Ören et al. 2007).

## 4.13 M&S Body of Knowledge (Bok)

My first publication about the Body of knowledge of modeling and simulation was published in 2005 (Ören 2005a). Over the years I had over 30 publications and/or presentations on modeling and simulation BOK as well as on integrative view of modeling and simulation.

Currently, an SCS Technical Committee is working to finalize the M&S Body of Knowledge (SCS-M&SBOK).

## 4.14 Big Picture of Simulation

Simulation provides a very rich paradigm (Ören 2010). And as clarified by Ören (2011a), M&S can be perceived from the following perspectives:

(1) Purpose of use
(2) Problem to be solved
(3) Connectivity of operations
(4) Types of knowledge processing and
(5) Philosophy of science

However, our perspective acts as a filter in our perceptions. Accordingly, nine focus areas of about 100 definitions of simulation are given in the sequel (Ören 2011a). Number of definitions are given within parentheses.

Group 1 experiment (24), training (experience) (9), game (experience) (1)
Group 2 modeling (12), model implementation/execution (14), technique (8)
Group 3 similarity/imitation (19), pretense/imitation (14), other (7)

A critical review of these definitions is given in (Ören 2011b). Seeing the big picture of simulation is essential to appreciate the many possibilities that it offers. A recent publication lists about 750 types of simulation (Ören et al. 2018).

## 4.15 Bigger Picture of Relationships: Similarity and Veracity

Simulation is based on the similarity relationship of a model (or a representation) with an—existing or non-existent– system; hence it is related to many other disciplines which are also germane to similarity. Relationship of a model (or a representation) with reality is the essence of veracity which is also relevant for simulation; hence validation and verification are essential aspects of simulation. Figure 4.2 outlines the relationships aspects of similitude and veracity.

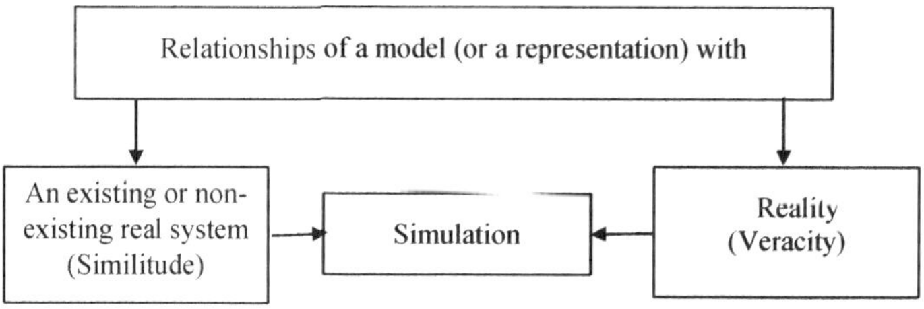


**Fig. 4.2** Relationship of simulation with similitude and veracity

**Table 4.3** Examples to similarity and veracity in different areas

| | | | |
|---|---|---|---|
| **--Similarity--** | imitate (v) | congruity | mimicry |
| affinity | imitation | congruous | verisimilar |
| alike | inverisimilitude | conjugate | verisimilitude |
| analog | likeness | endomorph | |
| analogical | pastiche | endomorphic | **--Indistinguish-** |
| analogous | replica | endomorphism | **ableness--** |
| analogy | verisimilitude | endomorphous | indistinguishable |
| consimilitude | | equivalence | indistinguishableness |
| correspondence | **--Similarity in** | equivalent | indistinguishably |
| homogeneity | **linguistics--** | homolog | undistinguishing |
| like | alternative | homologic | to be mistaken for |
| likeness | assimilation | homology | |
| match | equivalence | homomorph | **--Imitation--** |
| pose (v) | equivalent | homomorphic | copy |
| resemblance | homograph | homomorphism | imitate |
| resemble (v) | homographic | homomorphous | imitate (v) |
| resembled | homography | homomorphy | imitated |
| resembling | homonym | homothecy | imitation |
| sameness | homonymous | homothetic | imitative |
| self-similar | homonymy | homothetic | imitator |
| similar | homophon | transformation | |
| similarity | homophonous | homothetism | |
| similitude | homophony | homothety | **--Hiding similarity--** |
| simulacra | isomorph | isomorph | dissimular |
| simulacrum | isomorphism | isomorphic | dissimularity |
| verisimilitude | synonymous | isomorphism | dissimulate |
| | synonymy | isomorphous | dissimulate (v) |
| **--Behavioral** | tautology | map (v) | dissimulation |
| **similarity--** | | noncongruent | dissimulative |
| mimesis | **--Similarity in** | noncongruently | dissimulator |
| mimetic | **literature--** | | |
| mimicry | metaphor | **--Similarity in** | **--Dissimilarity--** |
| pantomime | metaphoric | **medicine--** | dissimilar |
| pretend (v) | pastiche | biosimilar | dissimilarity |
| role playing | pataphor | | dissimilarly |
| | pataphoric | **--To be alike--** | dissimilate (v) |
| **--Functional** | | assimilate (v) | dissimilation |
| **similarity--** | **--Similarity in** | assimilated | dissimilitude |
| emulate (v) | **mathematics--** | assimilatingly | non-similar |
| emulated | automorph | assimilation | semblance |
| emulating | automorphic | assimilationism | unalike |
| emulation | automorphism | homochromy | unique |
| emulative | automorphous | homotypy | uniqueness |
| emulator | congruence | mimesis | |
| | congruent | mimetic | |
| **--Similarity in art--** | congruently | mimetism | |

Table 4.3 displays similarity terms in many groups such as: similarity, behavioral similarity, functional similarity, similarity in art, similarity in linguistics, similarity in literature, similarity in mathematics, similarity in medicine, to be alike, indistinguishableness, imitation, hiding similarity, and dissimilarity.

The similarity concept in simulation and painting or sculpture are amazingly opposite. For a painter or a sculptor, the product of their activity is called a painting or a sculpture; the real system that they use as inspiration is often called "a model." For simulationists, the source of inspiration is often called the real system, while the product of their activity is simply a model. Other types of art often provide an esthetic experience (Devey 1934). As posited by Devey (1934, p. 272), "esthetic experience is imaginative." Thought experiments as non-computational simulations are also imaginative as expressed by Brown (2014): "Thought experiments are devices of the imagination used to investigate the nature of thing." In fiction (including science fiction), an existing or fictional reality is the source of inspiration for the author and the product provides an esthetic experience to the reader or the viewer if the topic of the book is visualized for theater, opera, movie, or TV.

Simulation used to provide experience can be used for training to enhance three types of skills: motor skills (virtual simulation, simulators), decision making skills (constructive simulation, gaming), or operational skills (operational simulation); as well as for entertainment as in simulation gaming. One can find many other similarities between simulation and other disciplines based on similarity and veracity of a model (or a representation).

## 4.16 My Aspiration for The Future of Simulation

Many advancements of simulation have been achieved by talented colleagues active in simulation in many continents. Simulation has been maturing for a long time at several fronts (Ören 2005b) including theoretical advancements and maturity (Zeigler 1976; Zeigler et al. 2018). Several groups expressed their views to explore enhancing use, usefulness and advancement possibilities of modeling and simulation. A recent book, based on a workshop (Jan. 13–14, 2016), where participation was by invitation only, documents recent challenges in modeling and simulation for engineering complex systems (Fujimoto et al. 2018). One of the important issues is still model reuse. In a very old article the concept of model bases was recommended for this purpose (Ören and Zeigler 1979).

After over 50 years of involvement, and closely following advancements such as quantum simulation (Bramüller 2018), in addition to some wishes expressed in previous sections, I am still hoping to see realization of the following possibilities:

1. To witness simulation-based decision making by advanced computationally intelligent systems, such as software agents, robots, advanced cyber-physical systems, and intelligent cities.

2. Use of personality, emotional, and cultural filters in human behavior simulation for decision making.
3. Simulation-based approach to be the base for large number of application areas in physical and social sciences, engineering, and technologies (Ören et al. 2017).
4. Widespread simulation education at every level for the preparation of future generations for whom simulation-based decision making may be common practice. A recent publication on this topic is by Niazi and Temkin (2017).
5. Extensive simulation-based education, including simulation-gaming for education at several levels to enhance education in several disciplines (Ören et al. 2017).
6. As the theory-based simulation for discrete systems DEVS (Discrete Event System Specification) formalism provides a solid background, its use with GEST (General System Theory implementor) (Ören 1971a, b; 1984c) to represent combined continuous and discrete-change systems.
7. Similar to the maturity levels of software engineering companies, establishment of maturity levels of simulation companies.
8. Establishment of SII (Simulation Industry Initiative) on Business Ethics and Conduct like DII (Defense Industry Initiative) on Business Ethics and Conduct (DII-ethics).

## 4.17 Epilogue

A professional autobiography may also necessitate a clarification of "why one did what one did?" Maslow's hierarch of needs (with several updated versions) has at the top level "Self-actualization" (Maslow 1998). I acted as if this level consists of three sub-levels:

(1) at the lowest sub-level, achieve one's full potential (this is why, in 1966, I quit a very successful career as an IBM systems engineer to start my Ph.D. studies and afterwards switch to an academic career);
(2) to assist others to achieve their full potential (I have been available to anyone who would ask my advice); and
(3) to contribute to my professional discipline (which happens to be simulation) to achieve its full potential.

Prof. Levent Yilmaz, with the volume he edited (Yilmaz 2015) and all the colleagues who contributed not only honored me but also proved that shared worthwhile ideas might be amplified and made available to others who might benefit from them.

I would like to finish this professional autobiography with what I wrote in 2005 on the publication of IBM Canada on their thoughtful consideration of honoring some of us as "IBM pioneers of computing in Canada":

The early days of computers remind me of the rich girl who was asked to write an essay on a poor family: She started her essay by stating, 'this family was so poor, even the driver, the cook, and the gardener were very poor.' … When I started my career, the world was so poor that nobody had personal computers, nobody had laser printers, and nobody had Internet, because even the concepts did not exist.

What a wonderful experience it was to witness all the developments and to have had the chance to contribute to some aspects of computerized modeling and simulation (IBM 2005).

## References


ACM SIGSIM Digest. (1984). *ACM SIGSIM Simulation Digest, 15*(1).

Al-Begain, K., & Bargiela, A. (Eds.) (2016). *Seminal Contributions to Modelling and Simulation—30 Years of the European Council of Modelling and Simulation*. Springer.

Balci, O. (Ed.) (1987). *Proceedings of the Conference on Methodology and Validation (1987 Eastern Simulation Conferences, Orlando, FL, April 6–9)*. Published as Simulation Series 19, 1 (Jan. 1988), The Society for Computer Simulation, San Diego, CA, 89 pp.

Balci, O. (1998). Verification, validation, and testing. In J. Banks (Ed.), *The handbook of simulation*. (August, Chapter 10, pp. 335–393). New York, NY: Wiley.

Balci, O., & Sargent, R. G. (1984). A bibliography on the credibility, assessment and validation of simulation and mathematical models. *Simuletter, 15*(3), 15–27.

BoHu, L., Ören, T. I., Zhao, Q., Xiao, T., Chen, Z., & Gong, G., et al. (2012). *Modeling and Simulation Dictionary: Chinese-English, English-Chinese—(over 9000 terms)*, Chinese Science Press, Beijing, P.R. of China. (With the contribution of 30 Chinese colleagues).

Bongulielmi, A. P., & Cellier, F. (1984). On the usefulness of deterministic grammars for simulation languages. *ACM SIGSIM Simulation Digest, 15*(1), 14–36.

Bramüller, J. (2018). Quantum simulation experiments with superconducting circuits. (Dissertation). Karsruher Insitute für Teknologie (KIT) Scientific Publishing, Karsruhe, Germany.

Brown, J. R. (2014). *Thought experiments, stanford encyclopedia of philosophy*. (First published Sat Dec 28, 1996; substantive revision Tue Aug 12, 2014).

Cellier, F. E. (Ed.). (1982). *Progress in modelling and simulation*. London, England: Academic Press.

Checkland, P. (1993). *Systems thinking, systems practice*. John Wiley & Sons.

CMSP-Ethics—Certified Modeling & Simulation Professional) certification program. Retrieved Oct 27, 2018, from http://www.simprofessional.org/.

Devey, J. (1934). Art as Experience. First Perigee printing, 1980. The Penguin Putnam Inc.

DII-ethics—Defense Industry Initiative on Business Ethics and Conduct. Retrieved Oct 27, 2018, from https://connect.dii.org/home.

Durst, P. J., Anderson, D. T., & Bethel, C. L. (2017). A historical review of the development of verification and validation theories for simulation models. *International Journal of Modeling, Simulation, and Scientific Computing, 6*(2).

Elzas, M. S., Ören, T. I., & Zeigler, B. P. (Eds.). (1986). *Modelling and simulation methodology in the artificial intelligence era* (p. 423). Amsterdam: North-Holland.

Elzas, M. S., Ören, T. I., & Zeigler, B. P. (Eds.). (1989). *Modelling and simulation methodology—knowledge systems' paradigms* (p. 487). Amsterdam: North-Holland.

Fujimoto, R., Bock, C., Chen, W., Page, E., Panchal, J. H. (Eds.) (2018). *Research challenges in modeling and simulation for engineering complex systems*. Springer.

Ghasem-Aghaee, N., Ören, T., & Yilmaz, L. (2017). Simulation, intelligence and agents: exploring the synergy. In F. N. Mofakham (Ed.), (Invited ebook chapter), *Frontiers in artificial intelligence—intelligent computational systems*. Bentham Science Publishers, pp. 1–58.

Gupta, P. K., Ören, T., & Singh, M. (Eds.) (2019–In Press). *Predictive intelligence using big data and the internet of things*. IGI Global.

Hall, D. E. (1992). Galileo galilei. In I. P. McGreal (Ed.), *Great thinkers of the western world*. (pp. 178–181). New York: Harper-Collins.

Hofstede, G. (2001). Cultural Consequences- Second Edition. Sage Publications,

IBM. (2005). Pioneers of computing 2005. Honouring those who influenced the history of computing in Canada. IBM (Canada) Centers for Advanced Studies, p. 49.

Markowitz, H. M. (1981). Barriers to the practical use of simulation analysis. In *Proceedings of the 1981 Winter Simulation Conference, held in Atlanta, Georgia*, Dec. 9–11, 1981. pp. 3–9.

Markowitz, H. M. (1984). Proposal for the standardization of status description. *ACM SIGSIM Simulation Digest, 15*(1), 37–55.

Markowitz, H. M. (2009). Harry Markowitz: selected works. world scientific publishing company. *World Scientific-Nobel Laureate Series, 1*.

Maslow, A. H. (1998). *Toward a psychology of being*. (3rd edn), Wiley.

McLeod, J. (1982). IF I KNEW THEN ..., Simulation in the Service of Society ($S^3$), October 1982, pp. IX–X.

McLeod, J. (1983). Professional ethics and simulation. In *Proceedings of the 15th Winter Simulation Conference*, pp. 371–374.

Mittal, S., Durak, U., & Ören, T. (Eds.). (2017). *Guide to simulation-based disciplines: advancing our computational future*. Springer.

Niazi, M. A., & Temkin, A. (2017). Why teach modeling & simulation in schools? Complex Adaptive Systems Modeling. *Springer, 5*, 7.

Ören, T. (1971a). *GEST: General system theory implementor, a combined digital simulation language*. Ph.D. Dissertation, 265 p. University of Arizona, Tucson, AZ.

Ören, T. I. (1971b). GEST: a combined digital simulation language for large-scale systems. In *Proceedings of the Tokyo 1971 AICA (Association Internationale pour le Calcul Analogique) Symposium on Simulation of Complex Systems*, Tokyo, Japan, September 3–7, pp. B-1/1–B-1/4.

Ören, T. I. (1975). Syntactic errors of the original formal definition of CSSL 1967. Technical Report 75-01, Computer Science Department, University of Ottawa, Ottawa, Ontario, 57 p. (Also, in the IEEE Computer Society Repository, No. R75-78).

Ören, T. I. (1977a–Invited Paper). Simulation—As it has been, is and should be. *Simulation, 29*(5), 182–183.

Ören, T. I. (1977b). A list of simulation terms. *ACM Simuletter, 8(Summer)*(4), 63–72.

Ören, T. I. (1978a). A Personal view on the future of simulation languages (Keynote Paper). In *Proceedings of the 1978 UKSC Conference on Computer Simulation*, April 4–6, 1978 (pp. 294–306). Chester, England: IPC Science and Technology Press.

Ören, T. I. (Ed.) (1978b). *Cybernetics & modelling & simulation of large-scale systems*. *International Association for Cybernetics*, Namur, Belgium, 191 p.

Ören, T. I. (1981). Concepts and criteria to assess acceptability of simulation studies: A frame of reference. *CACM, 24*(4), 180–189.

Ören, T. I. (1982a–Keynote Paper). Computer-aided modelling systems. In *progress in modelling and simulation*. In F. E. Cellier (Ed.), Academic Press, London, England, pp. 189–203. Ören, T. I. (1982b). Promising synergies of simulation model management, software engineering, artificial intelligence, and general system theories. In R. K. Ragade (Ed.), *General systems*, Vol. 26, pp. 203–209. Also published In *Proceedings of the 26th Annual Conference of the Society for General Systems Research*, Vol. 1, pp. 197–204.

Ören, T. I. (1982b). Promising synergies of simulation model management, software engineering, artificial intelligence, and general system theories. In R. K. Ragade (Ed.), *General systems*, Vol. 26, pp. 203–209. Also published In *Proceedings of the 26th Annual Conference of the Society for General Systems Research*, Vol. 1, pp. 197–204.

Ören, T. I. (1984a). Model-based information technology: computer and system theoretic foundations. *Behavioral Science, 29*(3), 179–185.
Ören, T. I. (1984b). A forum for the synergy of ACM special interest groups. *ACM Simuletter, 15* (2), 5–6.
Ören, T. I. (1984c). GEST—A modelling and simulation language based on system theoretic concepts. In T. I. Ören, B. P. Zeigler, & M. S. Elzas (Eds.), *Simulation and model-based methodologies: an integrative view* (pp. 281–335). Heidelberg, Germany: Springer.
Ören, T. I. (Ed.) (1985). *ACM Simuletter, 16*(1). editorial: p. 3.
Ören, T. (1995). Enhancing innovation and competitiveness through simulation. In *Preface of the Proceedings of 1995 Summer Computer Simulation Conf.*, Ottawa, Ont., July 24–26 (pp. vi–vii). San Diego, CA: SCS.
Ören, T. I. (1996). Towards simulative and intelligent problem-solving environments (position statement at a panel on: synergy of artificial intelligence and simulation). In *Proceedings of 1996 Summer Computer Simulation Conf.*, Portland, Oregon, July 21–25 (p. 596). San Diego, CA: SCS.
Ören, T. I. (2001a—Invited Paper). Software agents for experimental design in advanced simulation environments. In S. M. Ermakov, Yu. N. Kashtanov, & V.Melas (Eds.), *Proceedings of the 4th St. Petersburg Workshop on Simulation*, June 18–23, 2001, pp. 89–95.
Ören, T. I. (2001b—Invited Paper). Impact of data on simulation: from early practices to federated and agent-directed simulations. In A. Heemink et al. (Eds.), *Proceedings of EUROSIM 2001*, June 26–29, 2001, Delft, the Netherlands.
Ören, T. I. (2002a–Invited Paper). *SCS and Simulation: Fifty Years of Progress, 50th Anniversary Issue, Modeling and Simulation, 1*(3) (July-September), 32–33.
Ören, T. I. (2002b). Rationale for a code of professional ethics for simulationists. In *Proceedings of the 2002 Summer Computer Simulation Conference*, pp. 428–433.
Ören, T. I. (2005a—Invited Tutorial). Toward the body of knowledge of modeling and simulation (M&SBOK). In *Proceedings of I/ITSEC (Interservice/Industry Training, Simulation Conference)*. Nov. 28–Dec. 1, Orlando, Florida; paper 2025, pp. 1–19.
Ören, T. I. (2005b—Invited Keynote Article). Maturing phase of the modeling and simulation discipline. In *Proceedings of: ASC—Asian Simulation Conference 2005 (The Sixth International Conference on System Simulation and Scientific Computing (ICSC'2005)*, 2005 October 24–27, Beijing, P.R. China, International Academic Publishers—World Publishing Corporation, Beijing, P.R. China, pp. 72–85.
Ören, T. I. (2010). Simulation and reality: The big picture. (Invited paper for the inaugural issue) *International Journal of Modeling, Simulation, and Scientific Computing (IJMSSC) (of the Chinese Association for System Simulation - CASS)* by the World Scientific Publishing Co. China, Vol. 1, No. 1, 1–25.
Ören, T. I. (2011a). The many facets of simulation through a collection of about 100 definitions. *SCS M&S Magazine*, 2(2), 82–92.
Ören, T. I. (2011b). A critical review of definitions and about 400 types of modeling and simulation. *SCS M&S Magazine*, 2(3), 142–151.
Ören, T. I. (2012). Evolution of the discontinuity concept in modeling and simulation: From original idea to model switching, switchable understanding, and beyond. In *60th Anniversary Issue, Modeling and Simulation –The Transactions of SCS*. Vol. 88, Issue 9, Sept. 2012, pp. 1072–1079.
Ören, T. (2014). An invitation to modeling and simulation organizations to declare that they adhere to ethical principles. *SCS Newsletter*, June 2014 issue, pp. 1–3.
Ören, T. I., Abou-Rabia, O., King, D. G., Birta, L. G., & Wendt, R. N. (1990) (Plenary Paper). Reverse engineering in simulation program understanding. In A. Jávor (Ed.), *Problem solving by simulation, Proceedings of IMACS European Simulation Meeting, Esztergom*, Hungary, August 28–30, 1990, pp. 35–41.
Ören, T. I., Elzas, M. S., Smit, I., & Birta, L. G. (2002). A code of professional ethics for simulationists. In *Proceedings of the 2002 Summer Computer Simulation Conference*, pp. 434–435.

Ören, T. I. & the French Team (2006). Lucile Torres (coordinator), Frédéric Amblard, Jean-Pierre Belaud, Jean Caussanel, Olivier Dalle, Raphaël Duboz, Alain Ferrarini, Claudia Frydman, El-Amine Maâmar Hamri, David Hill, Aziz Naamane, Pierre Siron, Erwan Tranvouez, Gregory Zacharewicz. Modeling and Simulation Dictionary: English-FrenchTurkish. Marseille, France. With the support of Centre Nationale de La Recherche Scientifique, I3, Université Paul Cézanne—Aix-Marseille III, LSIS (300 pp). ISBN: 2- 9524747-0-2.

Ören, T. I., Ghasem-Aghaee, N., & Yilmaz, L. (2007). An ontology-based dictionary of understanding as a basis for software agents with understanding abilities. In *Proceedings of the Spring Simulation Multiconference (SpringSim'07)*. Norfolk, VA, March 25–29, 2007, pp. 19–27.

Ören, T. I., Kazemifard, M., & Yılmaz, L. (2013). Machine understanding and avoidance of misunderstanding in agent-directed simulation and in emotional intelligence. In *Proceedings of SIMULTECH 2013, 3rd International Conf. on Simulation and Modeling Methodologies, Technologies and Applications*. July 29–31, Reykjavik, Iceland. pp. 318–327.

Ören, T., Mittal, S., & Durak, U. (2017). The evolution of simulation and its contributions to many disciplines. In S. Mittal, U. Durak, & T. Ören (Eds.), *Chapter 1 of: Guide to simulation-based disciplines: advancing our computational future*, Springer, pp. 3–24.

Ören, T., Mittal, S., & Durak, U. (2018—Invited Chapter—In Press). Modeling and simulation: The essence and increasing importance. In the book: Modeling and Simulation of Complex Networks (M. A. Niazi, ed.), the IET Book Series on Big Data. (Appendix A: A list of over 750 types of simulation, Appendix B: A list of 900 types of models, and Appendix C: A list of 120 types of input).

Ören, T., Mittal, S., Turnitsa, C., & Diallo, S. Y. (2017). Simulation-based learning and education. In S. Mittal, U. Durak, & T. Ören (Eds.), *Chapter 13 of: Guide to Simulation-Based Disciplines: Advancing our Computational Future*, Springer, pp. 293–314.

Ören, T. I., Numrich, S. K., Uhrmacher, A. M., Wilson, L. F., & Gelenbe, E. (2000—Invited Paper). Agent-directed simulation: Challenges to meet defense and civilian requirements. In J. A. Joines et al. (Ed.), *Proceedings of the 2000 Winter Simulation Conference*, Dec. 10–13, 2000, Orlando, Florida, pp. 1757–1762.

Ören, T. I., & Yilmaz, L. (2006). Synergy of systems engineering and modeling and simulation. In *Proceedings of the 2006 International Conference on Modeling and Simulation—Methodology, Tools, Software Applications (M&S MTSA)*, pp. 10–17. July 31–August 2, 2006, Calgary, AL, Canada.

Ören, T. I., & Yilmaz, L. (2009a). Failure avoidance in agent-directed simulation: Beyond conventional V&V and QA. In L. Yilmaz & T. I. Ören (Eds.), *Agent-Directed Simulation and Systems Engineering. Systems Engineering Series*, Wiley-Berlin, Germany, pp. 189–217.

Ören, T. I., & Yilmaz, L. (2009b). On the synergy of simulation and agents: an innovation paradigm perspective. *Special Issue on Agent-Directed Simulation. The International Journal of Intelligent Control and Systems (IJICS)*, *14*(1), 4–19.

Ören, T., & Yilmaz, L. (2012). Synergies of simulation, agents, and systems engineering. *Expert Systems with Applications*, *39*(1), 81–88.

Ören, T. I., & Zeigler, B. P. (1979). Concepts for advanced simulation methodologies. *Simulation*, *32*(3), 69–82.

Ören, T. I., & Zeigler, B. P. (2012). system theoretic foundations of modeling and simulation: A historic perspective and the legacy of a. Wayne Wymore. In *60th Anniversary Issue, Modeling and Simulation –The Transactions of SCS*. Vol. 88, issue 9, Sept. 2012, pp. 1033–1046.

Ören, T. I., Zeigler, B. P., & Elzas, M. S. (Eds.) (1984). Simulation and model-based methodologies: An integrative view. NATO ASI Series. ( 651 p.). Berlin, Heidelberg, New York: Springer-Verlag.

Riley, M. T. (1992). Johannes Kepler. In I. P. McGreal (Ed.), *Great thinkers of the western world* (pp. 182–186). Harper-Collins, New York.

Sargent, R. G. (1991). Simulation model verification and validation. In *Proceedings of the 1991 Winter Simulation Conference*.

Sargent, R. G. (2011). Verification and validation of simulation models. In *Proceedings of the 2011 Winter Simulation Conference*, pp. 183–198.

Sargent, R. G., & Balci, O. (2017). History of verification and validation of simulation models. In *Proceedings of the 2017 Winter Simulation Conference* (Las Vegas, NV, Dec. 3–6) (pp. 292–307). Piscataway, NJ: IEEE.

SCi Software Committee (1967). *The continuous system simulation language (CSSL), simulation*, pp. 281–303, Dec. 1967.

SCS-AR – Awards and Recognition. Retrieved Oct 27, 2018, from http://scs.org/scs-awards-and-recognition/.

SCS-M&SBOK. Retrieved Oct 27, 2018, from http://scs.org/body-of-knowledge/.

SCS SimEthics. Retrieved Oct 27, 2018, from http://scs.org/ethics/.

Singh, M., Gupta, P. K., Tyagi, V., Sharma, A., Ören, T., Grosky, W. (Eds.) (2017). Advances in computing and data sciences. In: *First International Conference, ICACDS 2016, Ghaziabad, India*, November 11–12, 2016, Revised Selected Papers. Springer.

Stanovich, K. E. (1993). Dysrationalia: A new specific learning disability. *Journal of Learning Disabilities, 26*(8), 501–515.

Tolk, A., Fowler, J., Shao, G., & Yücesan, E. (Eds.) (2017). Advances in *Modeling and Simulation—Seminal Research from 50 Years of Winter Simulation Conferences*. Springer.

Wikipedia—commensalism). Retrieved Oct 27, 2018, from https://en.wikipedia.org/wiki/Commensalism.

Wikipedia—mutualism. Retrieved Oct 27, 2018, from https://en.wikipedia.org/wiki/Mutualism_(biology).

Worldatlas. Retrieved Oct 27, 2018, from https://www.worldatlas.com/articles/what-does-cultural-bias-mean.html.

Wymore, A. W. (1967). *A mathematical theory of systems engineering: The elements*. Huntington, NY: Krieger.

Yilmaz, L. (2015). *Concepts and Methodologies for Modeling and Simulation: A Tribute to Tuncer Ören*. Springer.

Yilmaz, L., & Ören, T. I. (Eds.) (2005). Editorial: Introduction to agent-directed simulation, special issue on agent-directed simulation. *Transactions of the Society for Modeling and Simulation International, 81*(7), 463–464.

Yilmaz, L., & Ören, T. I. (Eds.) (2009-All Chapters by Invited Contributors). *Agent-directed simulation and systems engineering. wiley series in systems engineering and management*. (520 p.) Berlin, Germany: Wiley.

Zeigler, B. P. (1976). *Theory of modeling and simulation*. New York: Wiley Interscience (2nd. ed. 2000).

Zeigler, B .P. (1984). *Multifacetted modeling and discrete event simulation*. Academic Press, London; Orlando.

Zeigler, B. P., Elzas, M. S., Klir, G. J., & Ören, T. I. (Eds.). (1979). *Methodology in system modelling & simulation* (p. 537). Amsterdam: North-Holland.

Zeigler, B. P., Muzy, A., & Kofman, E. (2018). *Theory of modeling and simulation – discrete event and iterative system computational foundations* (3rd edn). Academic Press.

Zhang, L., Zeigler, B. P., & Yuanjun, L. (Eds.) (In Press—2019). *Model engineering for simulation*. Academic Press.

Zhang, Y., Yilmaz, L., Ören, T., Madey, G., & Sierhuis, M. (Eds.) (2010). Introduction to the special issue on agent-directed simulation. *The International Journal of Agent Technologies and Systems (IJATS) by IGI Publishing*, *2*(3), pp. iiii. July-September 2010.

Zhang, Y., Yilmaz, L., Ören, T., Madey, G., & Sierhuis, M. (Eds.) (2012). Special issue on agent-directed simulation—selected best papers from the 2011 ADS conference. *Computational and Mathematical Organization Theory, 18*(4), December 2012.

# Chapter 5
# Limitations and Usefulness of Computer Simulations for Complex Adaptive Systems Research

**Andreas Tolk**

**Abstract** Complex adaptive systems in nature may produce something new, like structures, patterns, or properties, that arise from the rules of self-organization. These novelties are emergent if they cannot be understood as any property of the components but are a new property of the system. Emergence is a key property of complex systems. A popular method to better understand complex adaptive systems is the use of their computational representation, predominantly using the agent metaphor to produce emergence. The philosophy of science differentiates ontological and epistemological emergence. Ontological emergence produces something systemically new, while epistemological emergence produces new rules and laws, and as such can be reduced by gaining a better understanding of the system. The work presented here makes the case that emergence in computational complex adaptive systems cannot be ontological, as the constraints of computable functions do not allow for ontological emergence. As such, computer representations of complex adaptive systems are limited, as claims to produce systemically real emergence with computational systems contradicts some fundamental insights from computer science and philosophy of science. Nonetheless, they are useful to understand better the relationship of emergence and complex adaptive systems and conduct adductive research, which may be the best support of complex systems evaluation we can provide to complexity scientists to move the borderline between what is theoretically feasible to what is practically possible.

## 5.1 Introduction

The message of this contribution is strong, as the thesis is that *computer simulations of complex adaptive systems are not and can never be complex systems producing strong emergence in the sense of complexity science*. They can be highly complicated and have legions of nonlinear relations, but at the end of the day, they are a

A. Tolk (✉)
The MITRE Corporation, 903 Enterprise Parkway #200, Hampton, VA 23666, USA
e-mail: andreas.tolk@gmail.com

 
J. Sokolowski et al. (eds.), *Summer of Simulation*, Simulation Foundations, Methods and Applications, https://doi.org/10.1007/978-3-030-17164-3_5

composition of computable functions that transform data. They can produce emergent properties, but only in the context of simple or weak emergence, which will be defined in a later section of this contribution in more detail. Simulations are nonetheless utmost important tools to increase our understanding of complexity. They can be used to teach decision makers how highly interconnected, nonlinear system produce unexpected and sometimes counterintuitive system behavior. They can be used by scholars to gain new insight into such systems and provide a safe experimentation environment, but they cannot produce emergence in the sense that "something new" is created.

System science has evolved significantly over recent years. It started with linear approximations of relations between system components that were predictable and easy to engineer. Even nonlinear relations can be approximated using linear relations if the predictions are relatively short-term and short-distance related. Furthermore, the system could be understood by applying reductionism: by understanding the components, the system could be understood as well. The system behavior was congruent with the developers' design, engineering and governing the system were simple. With time, the systems became more complicated, relations had to be understood to be nonlinear, as time and space prediction extended. Part of the important system information was carried by the relations. These complicated systems could still be understood using the principles of reductionism, but nonetheless, it was necessary to look at the system holistically as well, as information was not only carried by the components but often by its nonlinear relations between components. The behavior was still predictable, but the experts needed increasingly sophisticated tools to be able to understand, engineer, and govern the systems. Over time, a new class of systems that was beyond being complicated was discovered: complex systems. They could no longer be understood by reductionism. Complexity science showed that new properties can emerge in such systems. The resulting behavior is generally not predictable, but in many cases, can be explained after it occurred using holistic and systemic complex system science methods. System thinking is now differentiating between simple systems with congruent behavior, complicated systems with predictable behavior, and complex systems with no predictable behavior. Some observations are even outside of this domain of system thinking, namely when emergent behavior is not even explainable within the context of system science after its occurrence. Malik called this progression cybernetic–evolutionary systems (Malik 2016). The borderlines between these system categories are not always clear, as the terms complexity and emergence are terribly overloaded, and the definitions can be circular, contradictory, and sometimes domain specific.

Nonetheless, the use of system science to understand natural systems better has a longstanding tradition. Buckley was among the first to use the term complex adaptive system (Buckley 1968). He applied systems research methods to understand behavior in social systems better, as he observed that the often used linear, categorical descriptions of processes and interactions did not sufficiently explain the complex nature of the subject of discourse. Of particular interest to Buckley was the observation of emergence in such systems. His examples were followed by many researchers of complex adaptive systems in the future.

The recent developments of computational methods supporting scientific research have led to the rise of a variety of computational science disciplines, and to the increased use of computational methods to evaluate complex adaptive systems, as discussed in significant detail in (Holland 1992; Miller and Page 2009). In particular, agent-based modeling is predominantly used to implement, computationally, complex adaptive systems. The claim is that such computational representations of natural complex adaptive systems allow observation of emergent macrolevel system behavior that is not formulated explicitly but rather results from the many microlevel interactions of the agents. Consequently, agent-based models became the tool of choice for many computational social scientists. As Bankes captures it in his introduction to the proceedings of the National Academy of Sciences on this topic:

> In social science, topics such as the emergence of cultural norms or institutions from the interaction of individual activity are indeed very important and not well addressed by competing modeling formalisms. So, the demonstrated ability of Agent Based Modeling to discover examples of such emergent dynamics from knowledge about the behavior of members of a society is potentially quite useful (Bankes 2002).

Unfortunately, however, many researchers do not have a formal computer science education and are not fully aware of many of the underlying principles from the philosophy of science. Therefore, they do not only use simulation as a reference to study complex adaptive systems but rather interpret observations of computationally instantiated complex adaptive systems to represent fully and equally their natural counterparts. Knowledge gained from the computational experiments is mapped directly to insights and applications of the real systems. Studying the computational representation becomes equivalent to studying the underlying system in the real world rather than as the generation of a sufficiency theorem related to the real-world system (Axtell 2000). Even many simulation experts share this view.

Based on the research conducted in support of (Tolk et al. 2018a), this contribution gives examples of natural and computational complex adaptive systems and observable emergence, introduces a critical review of the categories of emergence from the philosophy of science perspective, and concludes that there are significant epistemological constraints when computational methods are used for evaluating emergence. It uses these insights as extended in (Tolk et al. 2018b), and furthermore shaped by ideas exchanged with Dr. Bernie Zeigler, based on his observations on closure under coupling presented in (Zeigler 2018) as well as discussions with Saurabh Mittal and his ideas about emergence as discussed in (Mittal 2013a; Mittal and Rainey 2015).

## 5.2 Complex Adaptive Systems

As discussed in the introduction, the literature agrees that simple systems behave in a straightforward, mechanical, usually linear, and, most of all, easy to predict, congruent way. A complicated system is composed of many often nonlinearly interacting parts that can be studied using reductionist and probabilistic models and statistical methods. It is still predictable, but it usually requires experts who are highly educated and experienced and have a tailored tool set available. For the definition of complex systems, in (Sheard and Mostashari 2009) the authors propose the following definition after a review of systems engineering relevant literature from complexity theory:

> Complex systems are systems that do not have a centralizing authority and are not designed from a known specification, but instead involve disparate stakeholders creating systems that are functional for other purposes and are only brought together in the complex system because the individual "agents" of the system see such cooperation as being beneficial for them (Sheard and Mostashari 2009).

Complex systems are not predictable, and the principles of reductionism do not bear fruit when laboring to understand them, as system behavior emerges on all levels of the system. Although they are not fully knowable, within reason there may be some prediction possible, in particular when the system is in a stable state.

Complex adaptive systems add the element of adaptivity of some or all of its components and of the system itself. For its definition, the focus very often is the agent metaphor for the system components, as compiled in (Dooley 1996) and revisited in (Brownlee 2007). One of the insights drawn from these overview articles is that the diversity of definitions suggests one should focus on properties of such systems, as elaborated in detail within Holland's seminal contributions to the unified theory of complex adaptive systems (Holland 1995), which contained a particular focus on aggregation—complexity emerges from the interaction of smaller components, which themselves may be the products of systems—and nonlinearity—agents interact in dynamical and nonlinear ways.

The reason we use computational complex adaptive systems is that they help us to understand natural systems. There are legions of examples of natural complex adaptive systems. Without claiming completeness, some examples include society (Buckley 1968), the ecosystem and biosphere (Levin 1998), supply networks (Choi et al. 2001), human language (Beckner et al. 2009; Zeigler and Mittal 2018), product development environments (McCarthy et al. 2006), health care (Rouse 2008), climate change (Ingwersen et al. 2014), and urban hazard mitigation (Godschalk 2003). All these systems are not fully knowable and often quite hard to predict. Moreover, it is often difficult to even collect useful data about these systems. Unfortunately, all of these systems (and many others) are very important, so we cannot ignore them. We must try to get a better understanding of their dynamics and causal structures, and we want to provide decision-makers a better foundation with which to make informed decisions. Currently, the most powerful tool to do this is the use of computational representations of these systems and to simulate their dynamics (Buss et al. 1990).

As already discussed in the introduction, computational science disciplines, such as computational physics, computational biology, computational chemistry, computational social science, and many more, explore the use of computer models and simulation in direct support of their research (Tolk 2018). The discipline of complex systems research benefits significantly from these developments, as computers amplify our abilities to model, simulate, and evaluate the computational representations of the systems of interest. The principal steps of such a computational study of complex adaptive systems were captured in detail by (Holland 1992) and professionalized by many authors since then. The Santa Fe Institute and other similar organizations dedicate their work to the multidisciplinary study of the fundamental principles of complex adaptive systems, including physical, computational, biological, and social systems. These multidisciplinary scholars and students are experts in their fields and come together to use computational complex adaptive systems in support of their research, and the resulting studies are impressive, such as studies conducted to prevent collapse of tropical forests (Hébert-Dufresne et al. 2018), or new insights into how evolution works (Payne et al. 2018), just to name a few.

These studies are generally understood to prove the enormous value of computational complex adaptive systems, in particular when it comes to emergent behavior, which is a characteristic property of complex systems: system behavior that does not depend on its individual parts but on the multiple relations and interactions on all system levels. The next section will provide a short overview of the different forms of emergence from a philosophical, as well as, from a systems engineering perspective.

## 5.3 Categories of Emergence

As discussed before, there are legions of definitions for complexity and emergence. It is therefore important to understand the definition used. In his introduction chapter "What is emergence?" to the book on emergence in complex, cognitive, social, and biological systems, Pessa observes the following:

> A general theory of emergence should first start taking into account that this concept cannot be defined in an objective way. The word 'emergence' refers to a relationship between an observer, the models which he is equipped with, and certain results of observations, or of measure operations, in turn dependent on his mental schemata and on his technological apparatuses. Therefore, the emergence can be defined only relatively to a given observer, whose features should be specified in a suitable way (Pessa 2002).

This idea that emergence requires an observer who—as a rule—is not part of the system, but a user of the system conducting analyses, has already been formulated by Crutchfield, one of the pioneers in the complexity science domain.

> The emergence … is the product of both the complicated behavior of nonlinear dynamical systems and the limitations of the observer. … The newness … is in the eye of an observer: the observer whose predictions fail or the analyst who notes that the feature of statistical self-similarity captures a commonality across length scales (Crutchfield 1994).

However, Crutchfield (1994) also defines intrinsic emergence, where the system itself takes advantage of new emerging patterns by capitalizing on it without the explicit need of an external observer. Within this chapter, we understand emergence as a system-level behavior that dynamically arises from the spatiotemporal and multilevel interactions between the parts at the different system levels. As discussed in detail in (Mittal 2013a), these interactions may also be a constraint on various levels within the system, creating feedback loops of causality. Natural systems are open systems exposing emergent behavior on the macrolevel. This novel irreducible macrobehavior is systemic and adds further complexity to the system when it causes itself changes at lower levels, which then again can result in new behavior emerging on the higher levels. Overall, the system may adapt to a new environment by developing new multilevel interactions and feedback loops (Kauffman 1992).

Philosophy has dealt with the challenge of emergence for more than a century, starting with George Henry Lewes foundational work (Lewes 1877), long before there was any computational system to support them. Philosophers distinguish between epistemological and ontological emergence. Of the two, ontological emergence became a far more active research thread than did epistemological emergence (Silberstein and McGeever 1999). We will start with a closer look at the definitions and ideas.

## *5.3.1 Epistemological and Ontological Emergence*

Epistemology is the theory of gaining knowledge, its methods, validity, and its scope. Knowledge is understood as the scientifically justified belief in something. In the epistemological view, emergent properties and laws are systemic features of complex systems. This system is governed by true, law-like generalizations within a special science that is irreducible to fundamental physical theory for conceptual reasons. As such, this view characterizes the concept of emergence in terms of limits on human knowledge of complex systems and its foremost laws. The unpredictability is a matter of human knowledge about these laws. The novel emergent properties are not part of any component alone but come into being from the interplay of components following the laws governing the systemic features. For philosophers of science, epistemological emergence is also understood as a matter of practice that complex systems cannot be described in terms of their component units because of our limitations, that is, our inability to do the computations. If we understand the laws and compute them, there are no more surprises to the systems engineer.

Ontology is the theory of the nature of being, of what is, and what we know about it. Coming from the field of metaphysics, ontology also deals with how entities that exist can be grouped, compared, etc. The ontological view is quite different from the epistemological view. Essentially, the ontological view of emergent properties is premised upon the idea that they are independent of the human knowledge about them. Instead, they are novel, fundamental types of properties in and of themselves. Something new emerges that was not there before, and that cannot be explained by the components and their interactions and relations alone. Furthermore, their unexplainability is not a matter of practice, but a matter of principle. The occurrence of emergent properties is not in any sense constituted by the occurrence of more fundamental properties and relations of the object's parts.

A detailed discussion of the philosophical views of emergent properties and their interpretation under epistemological and ontological viewpoints is presented in (O'Connor and Wong 2015). For the purposes of this paper, we will use a rather simple view, specifically: that epistemological emergence can be reduced, while ontological emergence cannot. Epistemological emergence results mainly from an insufficient understanding, a lack of knowledge. Once we understand from what the observation of emergent behavior originates, it becomes predictable. Ontological emergence may in some cases be explained after the fact, but remains unpredictable. It simply emerges from within the complex system.

### 5.3.2 Maier's Emergence Categories

As simulationists, this way of philosophical thinking is not necessarily part of our daily tasks. Therefore, a more systems engineering approach that may help to communicate the same ideas will be helpful. Mark Maier is known for his contributions to the discussion of systems of systems. In (Maier 2015), he defines four categories of emergence from a systems science perspective, depending upon how well the emergence observed in the natural system can be reproduced and explained through a simulation system.

- *Simple* emergence: The emergent property or behavior is predictable by simplified models of the system's components.
- *Weak* emergence: The emergent property is readily and consistently reproduced in the simulation of the system but not in reduced complexity non-simulation models of the system, that is, simulation is necessary to reproduce it.
- *Strong* emergence: The emergent property is consistent with the known properties but, even in simulation, is inconsistently reproduced without any justification of its manifestation.

- *Spooky* emergence: The emergent property is inconsistent with the known properties of the system and is impossible to reproduce in a simulation of a model of equal complexity as the real system.

As implied by the terms used above, systems engineers normally do not like emergent phenomena as they result in behavior that is unforeseen and unpredictable. Consequently, Maier's viewpoint is written from the position that emergence is be avoided or, at least, controlled. In contrast, as engineers of complex adaptive systems, we seek to enable and leverage emergence (Mittal and Rainey 2015; Norman et al. 2018).

In (Tolk et al. 2018a) we observe that simple emergence corresponds with simple computable systems, and weak emergence with complicated computable systems, where both system categories are predictable, at least in hindsight. Strong emergence falls into the category of unpredictable systems that are not fully reproducible by simulation systems. Finally, Maier's "spooky" emergence lies even outside of our system thinking boundaries, referencing real ontological emergence phenomena. Figure 5.1 shows this mapping, using qualitative parameters on its axis. On the vertical axis, the behavior of the system is described, as we discussed it earlier in the contribution: congruent, predictable, not predictable, and not explainable. These behavior categories are accompanied by the system types on the horizontal axis: simple, complicated, and complex system, plus observations that are not explainable in our system thinking framework. Maier's "spooky" emergence is not explainable because it lies outside the systems thinking realm. As Maier states it, a spooky property is inconsistent with the known properties of the system and is impossible to reproduce with computational means.

There is no clear mapping established in the literature so far that connect epistemological and ontological emergence to the four emergence categories defined by Maier, but the point can be made that simple and weak emergence or

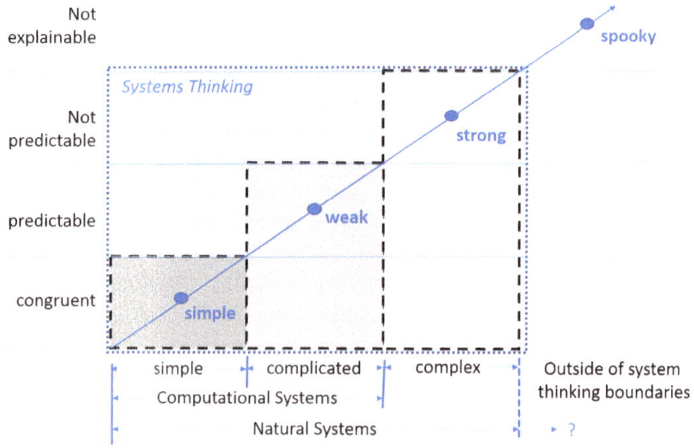


**Fig. 5.1** Mapping emergence types to system types

epistemological, while strong and spooky are ontological in nature. However, as we will show in the next sections, strong emergence may play a double role.

This figure also directs us to the next point of the discussion, namely that the sphere of systems thinking and emergence is bigger than computational support. The following section will make the argument why this is the case, using existing knowledge from the computer science body of knowledge.

## 5.4 Epistemological Constraints of Computational Systems

In the light of this discussion, it seems to be fair to look at how we can gain insight and knowledge from the application of computational systems, such as modeling and simulation, and, in particular, agent-based approaches. Are they truly equal to the natural complex adaptive system they represent, as often at least implicitly assumed by domain experts? Can we produce the kind of emergence we are interested in as a scientist? Can we at least get sufficient support in discovering, understanding, and managing emergence in complex systems as envisioned in (Tolk et al. 2018a)?

The famous researcher Huber L. Dreyfus is well known for his two books on the limits and constraints of computers: "What Computers Can't Do: The Limits of Artificial Intelligence" (Dreyfus 1972) and "What Computers Still Can't Do: A Critique of Artificial Reason" (Dreyfus 1992). In his essays on the constraints of cognitive simulation and artificial intelligence, among other arguments, Dreyfus points to known limits that are rooted in the nature of computers as captured in the works of Turing, Church, Gödel, and other pioneers of computer science that were often overlooked by his colleagues. These constraints are, of course, still valid for computational platforms in general despite the advances in hardware and software that have been made over the years. Furthermore, these constraints extend to systems represented within a computational platform, including representations of complex adaptive systems. What computers generally can't do can also not be done by computational complex adaptive systems. What is computationally not decidable can also not be decided by simulation systems (Tolk 2015). These computational constraints are valid for all paradigms, including agent-based approaches, as agent-based systems with no runtime interaction with external systems are Turing computable, as shown among others in (Epstein 2006).

The essential constraint remains: computers transform input parameters into output parameters using computable functions to do so, no matter if this is done using traditional programming paradigms or interaction-based agent models. As a result, computers cannot create something new out of nothing, as everything that is produced by a computer must be in the input data or the transforming algorithm. As Boden points out in her paper, we can model combinational, exploratory, and transformational forms of creativity (Boden 2009). However, all these forms of creativity are discovering through rearrangement and transformation, not creating something new that was not there before. Even surprising behaviors, like Conway's cellular automata based "Game of Life" being able to produce prime numbers (Guy 1983), have been proven to be completely defined by input data and computable transfer functions.

These foundational insights allow us now to answer the question can computational systems indeed be complex in the sense of producing emergence? They can be complicated, even extremely complicated, but in principle, it boils down to the transformation of input parameters into output parameters using computable functions. That is true for all computations, including agent-based models. Which means that they cannot produce something new! For quantum computing, Kundu and colleagues observe the following:

> A functional quantum computer will provide much faster computation in a number of key areas, including: searching large databases, solving complicated sets of equations, and modeling atomic systems such as biological molecules and drugs. This means they'll be enormously useful for finance and healthcare industries, and for government, security and defense organizations (Kundu et al. 2016).

However, they also observe that "the fundamental concept of quantum computing and information—the qubit—does not carry more information than a classical bit." It just allows evaluation of many possible permutations of a problem all in parallel, calculating and storing all possible outcomes and storing the results in one pass. While the foundations of the computational complexity of operations and what can be solved based on resource and time constraints will be shaken by quantum computing, the findings presented in this section will still be valid for functional quantum computers as well.

## 5.5 Implications

As a result of these observations we do know that computational complex adaptive systems can only address simple and complicated systems, and therefore also can only produce emergent behavior following the same constraints, which is epistemological—following rules and laws that may be hidden, are in principle knowable—and as such lead to simple or weak emergence in Maier's spectrum. What we are more interested in as a scientist, however, is ontological emergence, but that lies outside the computational realm, and maybe even outside system thinking, as implied in Fig. 5.1. So, what are computational complex adaptive systems good for in this context?

### 5.5.1 Research and Education

The Complexity Primer for systems engineers by the International Council on Systems Engineering (INCOSE) identifies the need for an extended set of orchestrated tools to support coping with complex system challenges (Sheard et al. 2015). These tools comprise recommended methodologies that combine methods and approaches from a variety of disciplines that have to cope with complexity, in particular, artificial intelligence, data science, and modeling and simulation, such as agent-based computational complex adaptive system representations of the system of interest.

One of the main objectives of building a simulation of the reference system is to gain a deeper knowledge of the system. As already observed by Rouse in (Rouse 2003), the borders between the system components and categories—simple, complicated, and complex—are often somewhat fuzzy and may depend on the education and experience of the team. What looks complex and unpredictable to a novice may turn out to be complicated at best for an expert team. As stated in (Crutchfield 1994), the emergence lies in the perception of the observer, and the perception is constrained by education and cognitive abilities of the observer. This view reflects our objective to decrease epistemological emergence: if we increase the system knowledge, we reduce the epistemological emergence; we reduce strong emergence to weak emergence, that is explained and—as the underlying law and rules are understood—can be replicated and predicted. In the best case, we close all knowledge gaps. If we are able to design a computational complex adaptive system that can mimic and predict all observed emerging in the natural complex adaptive system, then we actually show that the system of interest is not really complex, but rather just pretty complicated.

Following the arguments given in (Mittal 2013b), in the general case that the natural systems cannot be reduced, a novel emergent property is a multivariable and multidimensional phenomenon within the space–time continuum. Multivariable implies an often-large number of variables with sometimes incomplete knowledge about their interdependencies. Multidimensional implies possibly multiple vantage points that are dependent on the frame of reference when observing the phenomenon. These vantage points are not mutually exclusive, but rather focus on different facets. Furthermore, the phenomenon may manifest over time as well as over space, requiring methods allowing for spatiotemporal analysis instead of exclusively looking at local snapshots. Szabo and Birdsey give several examples and methods to discover and identify emerging behavior using computational methods (Szabo and Birdsey 2017), showing the power of simulation to contribute to the better understanding, managing, and eventually even engineering of complex adaptive systems.

Humphreys is one of the leading philosophers of science. In his work on computational science and the scientific method, he shows how we use computational means to extend our abilities to gain knowledge and produce new scientific insights (Humphreys 2004). This is the case for the use of computational complex adaptive systems as well. This insight does not come by exactly reproducing natural complex adaptive systems, but it helps to reduce epistemological emergence by increasing our knowledge. This can be done using deductive, inductive, and abductive approaches.

Deductive reasoning moves from the general rule to the specific applications. If we know the general rules well, such as is the case in classical mechanics or physics, we can develop a simulation system that implements these rules for the simulated entities. As discussed before, the results can be very complicated computational systems. Such systems allow the evaluation of new rules, constraints, or regulations for agents to find out what ultimately will occur with the system. Berger's use of agent-based models to evaluate technology diffusion, resource use changes, and policy in agriculture is an example of this category: the agents follow established rules and react accordingly to new rules for the system of interest (Berger 2001).

Inductive reasoning moves from specific observations to general rules, which can be understood as a bottom-up approach. This is the classical empirical approach going back to Sir Francis Bacon. When we are using data to instantiate our agents, which means defining their properties and relations from data, and also initiate our agents, which means providing the data for the properties as well as the overall agent population, we support inductive reasoning, which today can be supported by Big Data and Deep Learning, such as described among others in (Lee et al. 2013). These ideas are also giving a new push to the next generation of dynamic data-driven applications systems, as introduced to the simulation community a decade ago (Darema 2004).

Abductive reasoning is a relatively new form of reasoning that is master tailored to the application of simulation. An intuitive explanation of abductive reasoning is developing a model that explains the observations better than any other model! It is a little bit of a mix of deductive regulations and inductive rules. It allows experimenting with solutions driven from both sides of traditional reasoning and using a model to judge the results. Yilmaz uses a generative parallax simulation to support the abductive discovery of new insight (Yilmaz 2018).

In summary, when using computational science, we are creating "sufficiency theorems" (Axtell 2000). We are deducing an outcome from an input and set of transformation rules, these rules may or may not be how nature actually works. Essentially, we can use computational methods to create deductions that can then be collected for inductive conclusions to solve abductive problems. But, just like all analytic methods, we are working with models of the real system in question. In the context of the definition of soft and hard Operations Research methods provided in (Pidd 1997), computational tools should be understood as soft methods to structure, understand, and commit to the problem, but they may be insufficient to reproduce the system of interest sufficiently enough to provide the capabilities needed for hard methods actually solving the problem.

### 5.5.2 *Emergence in Simulations of Complex Adaptive Systems*

Emergence can for sure be observed in simulations, but they are epistemological in nature. As shown in this contribution, due to the nature of simulations being computer programs, there will always be not only an explanation for an emergent property but—as the explanation will uncover the underlying law and rules that did lead to the explanation—these emergent properties are inherently predictable. But many researchers observe and publish about observations of emergence in simulation systems. They follow the definitions given in (Das et al. 1994) who understand computational emergence as "the appearance in a system's temporal behavior of information-processing capabilities that are neither explicitly represented in the system's elementary components or their couplings nor in the system's initial and boundary conditions." However, we already have seen that computational functions are purely transformational, so the source for the emergence must be in the algorithm or the input data, which include initial and boundary conditions, geometry of the initial system, etc.

As we already observed, simulations of complex adaptive systems can become very complicated and can provide everything but intuitive results. Such counter-intuitive results that were not planned by the designer of the simulation regarding its functional characterization. When using simulations to provide numerical insight into the dynamics of the complex adaptive system, the discovery of unintended behavior is often the objective of the experiment, but the simulation engineer needs to be aware if the observation is a correctly reflected behavior of the original system reproduced by the simulation, or if the observations results from unintended behavior within the computational system. In other words, whether the observed emergence is a characteristic of the simulated system correctly reproduced, or a computational artificiality originating in the simulation. Bedau (2008) observes in his work the following.

> Note that these patterns and regularities produced by computational systems are not mere simulations of emergent phenomena. Rather, they are computational embodiments of real emergent phenomena. That is, the computer produces something that is weakly emergent in its own right. If the computer happens to be simulating some natural system, that natural system might also exhibit its own emergent phenomena. Further, if the simulation is accurate in the relevant respects, it might explain why and how the natural system's phenomena is weakly emergent. But the computer simulation itself, considered as an object it its own right, is also exhibiting emergent behavior (Bedau 2008, p. 454).

If the simulation correctly reproduces an emergent characteristic of a natural complex adaptive system, we increase our knowledge about the system and can in the future not only discover, but actually manage this kind of emergent property. If we can be sure that an observed emergent behavior of the natural system cannot be reproduced by the simulation, we also learned something about the system, namely that this emergent property must by strong and ontological. In both cases, simulation systems are helpful tools.

However, if the simulation produces emergent behavior because it is too big and complicated to be fully understood by the researcher who applies it, the use of such a system will lead to questionable results, as the observed emergence may be a computational artificiality, or the result of misinterpretations of an insufficiently educated researcher. As I captured it for the general case in my work on ethics "promoting the reliable and credible use of modeling and simulation is a tenet in the code of ethics for simulationists." (Tolk 2017) Simulationists must therefore be aware of the different application constraints of simulation systems when it comes to the evaluation and management of emergence in natural complex adaptive systems.

### 5.5.3 Decision Support and Management

General Stanley A. McChrystal became action officer for the Army Special Operations in 1990, working in the Joint Special Operations Command. In 1991, he saw action in the Desert Shield and Desert Storm tours. He was commander of the Joint Special Operations Command from 2003 to 2008, and he became top commander in Afghanistan in 2009. After he resigned in 2010, he summarized his lessons learned in the book "Team of Teams: New Rules of Engagement for a Complex World." The guiding idea of this work is that of an ecosystem, emphasizing the change of the role of leadership:

> The temptation to lead as a chess master, controlling each move of the organization, must give way to an approach as a gardener, enabling rather than directing. A gardening approach to leadership is anything but passive. The leader acts as an "Eyes-On, Hands-Off" enabler who creates and maintains an ecosystem in which the organization operates (McChrystal et al. 2015).

These insights are pretty close to the metaphor of the watchmaker and gardener often used by INCOSE in their introductions to the new role of engineers in the era of complex systems. The same insight is also published by Keating, who focuses on the need for governance of complex systems, not controlling them (Keating 2014). It also aligns well with the recommendations given by the NATO Net-enabled Capability (NEC) Command and Control (C2) Maturity Model (N2C2M2) (Alberts et al. 2010): the complexity of the decision sphere requires a paradigm shift in command and control enabling rapid and agile decision-making processes based on seamless and transparent information sharing. Smart and social components of each system will have access to all information they need to make the decision, regardless of where they are, which components they use to gain access to the information, or where the information came from.

More requirements may exist for the data describing the observations, such as that the components must be aware that the information will be—or already was—collected, temporal constraints of the observation is, and they must be aware of the uncertainties associated with the collection. If the pedigree and path are important,

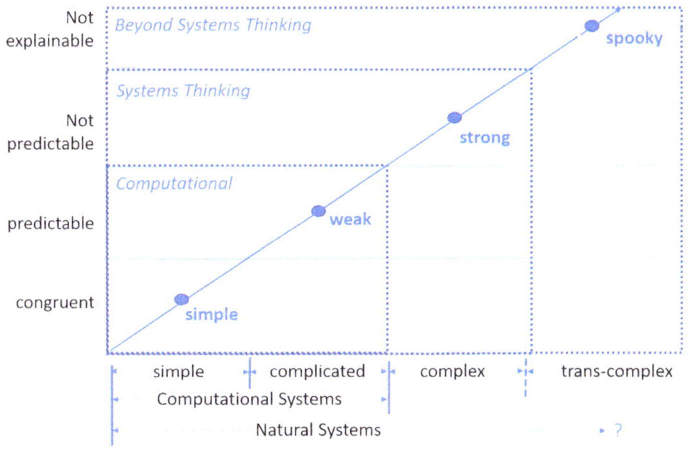


**Fig. 5.2** Mapping decision and management types to system types

the meta-qualities of the information itself must be preserved as well. These observations and recommendations regarding decisions support and management are tightly connected with the insights covered in this contribution. Let's revisit Fig. 5.2 and add the areas of computational, system thinking, and general human decision-making to it in Fig. 5.2.

Computer simulations of complex adaptive systems can cover simple and complicated systems. They can discover, understand, manage, and even engineer emergent properties in these categories, as these systems are fully predictable. The reason is that computational systems, no matter how complicated they get, in principle are predictable. As just explained in more detail in the previous section: if we are surprised by an emergent property of our computer system, we do not yet understand it completely.

In the case of complex systems exposing strong emergence, computational systems can still help with discovering the emergence, as the emergent property is consistent with the known properties of the system. We may actually be able to understand it, as it is possible in some cases to explain the emergence retrospectively, but we cannot predict it. As such, it lies outside the realm of computation or simulation-based decision and management. And spooky emergence lies completely off the realm of computational support, only accessible to humans.

However, if the retrospective explanation of strong emergence leads to a mathematical model that now can be used to make future predictions, we created new knowledge that reduces epistemological emergence: the former strong emergence is reduced to be weak emergence and will be covered by future applications of the simulation support.

This observation has significant implications for the applicability of computational support in complex and trans-complex system environments, as computational support must fall short, as already observed in (Tolk et al., 2018b) for the methods of traditional systems engineering when coping with complex systems engineering challenges. Failure to understand the nature of the decision environment will lead to management mistakes, as observed in (McChrystal et al. 2015). At the same time, underestimating the power of computer simulation, such as recently covered in (Mittal et al. 2017), will lead to similarly suboptimal decisions.

### *5.5.4 Moving the Boundaries of Possibilities*

Philosophy of science, as we applied in in this contribution in the earlier section, is focused on understanding what is principally impossible and what is feasible. As pointed out by Humphreys (2004), this is a very valuable contribution, as something that is theoretically impossible can never be implemented in any practical implementation. The viewpoint changes dramatically, however, when we are focusing on the difference of what is theoretically feasible and what is practically achievable. This is where computational sciences contribute a lot.

When Humphreys talks about the use of computers to extend ourselves, he compares computational means to comparable to microscopes and telescopes: they do not change what we can observe in principle, but practically they allow us to observe things not accessible by humans only using the naked eye. In the same way computers allow us solving mathematical challenges describing complex adaptive systems and simulate them within the constraints of the possible. They do not overcome the limits of theoretically solvable problems, the principles discovered by Church, Turing, Gödel, and others do still apply, but the amount of challenges we can address using computer simulations increases significantly.

As such, the use of computer simulation directly contributes to the progress of science by pushing the boundary of from what is only theoretically possible towards what is now practically applicable for scientists in their tool boxes.

## 5.6 Conclusion

Within this contribution, we started with the increasing interest in complex adaptive systems in the various application domains. The driving common components across all these domains is the possibility of emergent properties that can no longer be explained by the traditional tools of systems engineering as well as systems management. From a philosophy of science perspective, such emergent behavior can be epistemological or ontological. Both classes describe emergent as being irreducible to component behavior, novel observations on the systems level, and having a possible feedback loop on the component behavior. However, ontological emergent properties are something really new that were not there before while epistemological emergence attributes the emergence to not yet fully understood laws governing the system. Computer simulations are increasingly used to represent and study complex adaptive systems. However, due to their epistemological constraints, they can produce, discover, explain, and manage epistemological emergence, but they cannot produce ontological emergence.

So, what are computer simulations of complex adaptive systems good for? Exactly for what has been highlighted in the last paragraph: producing, discovering, explaining, and managing epistemological emergence to allow the engineering of such properties in systems. They can be used as a soft method to understand, structure, and commit to a challenge, but they may be insufficient to provide an actual solution. As shown in (Sheard et al. 2015), the tools used to develop computational complex adaptive systems are very much the same as those that are used to support complex systems engineering. If we can explain all observed emergence using simulation systems, then we have shown that the system of interest is not complex, but rather very complicated, but nonetheless predictable by experts equipped with the right knowledge and supporting tools.

But what does this mean for a complex, or even trans-complex, environment that exposes strong or spooky emergence as defined in (Maier 2015)? First, it implies that computational tools have limited applicability in this domain providing solutions. As discussed in this contribution, a simulation system cannot fully reproduce the behavior of such systems. It can still provide valuable insight and help to discover strong emergent properties in the spatiotemporal dimension of the system, and the metrics known since (Crutchfield 1994) are an example of how to do this, but it cannot produce them. It also means that it is pivotal for managers and decision-makers to realize that their environment changed and methods and tools that were successfully applied before as hard methods may no longer work, as described in (McChrystal et al. 2015). Failure to recognize this will lead to wrong decisions resulting in suboptimal plans that may in the worst-case result in the failure of the mission.

This is no call to throw the baby out with the bath water. Like the fundamental insights of Turing and Gödel on the limits of computability did not imply we should not use computers, so the insights provided in this contribution should not discourage from the use of simulations of complex adaptive systems to support research on emergence. As described in (Mittal et al. 2017; Tolk 2018), simulation supports more and more disciplines, and in the form of computational science is gaining even more influence. Simulation pushes back the boundaries of what can be applied and solved in practice, addressing challenges which before could only be solved theoretically.

Understanding the potential as well as the limits of our discipline is, therefore, more important than ever before. Discussions like in this contribution must make it into the curriculum to ensure that we are neither overpromising nor undervaluing the capabilities of simulation.

**Acknowledgements** Many discussions with colleagues at MITRE as well as at conferences and other meetings helped to shape this contribution, and unfortunately not all of them can be mentioned here. Part of the research and the related discussions were supported by the MITRE Innovation Program. The following individuals deserve recognition for their contributions, peer reviews, and constructive comments and recommendations for improvements: Paul Davis (RAND Corporation), Saikou Diallo (Old Dominion University), Frank DiBonaventuro (MITRE), Matt Koehler (MITRE), Saurabh Mittal (MITRE), Mike Norman (MITRE), Ernie Page (MITRE), Anne Rainville (Vtrek), John Selby (MITRE), Paul Silvey (MITRE), E. Dante Suarez (Trinity University), Levent Yilmaz (Auburn University), and Bernie Zeigler (RTSync Corp).

**Disclaimer**

The author's affiliation with The MITRE Corporation is provided for identification purposes only and is not intended to convey or imply MITRE's concurrence with, or support for, the positions, opinions, or viewpoints expressed by the author. This paper has been approved for Public Release; Distribution Unlimited; Case Number 18-2167-3.

## References

Alberts, D. S., Huber, R. K., & Moffat, J. (2010). The NATO net enabled capability command and control maturity model. *Technical report*, Office of the Assistant Secretary of Defense, Command and Control Research Program, Washington, DC.

Axtell, R. (2000). *Why agents? On the varied motivations for agent computing in the social sciences.* Center on Social and Economic Dynamics Brookings Institution.

Bankes, S. C. (2002). Agent-based modeling: A revolution? *Proceedings of the National Academy of Sciences, 99*(suppl 3), 7199–7200.

Bedau, M. A. (2008). Is weak emergence just in the mind? *Minds and Machines, 18,* 443–459.

Beckner, C., Blythe, R., Bybee, J., Christiansen, M. H., Croft, W., Ellis, N. C., et al. (2009). Language is a complex adaptive system: Position paper. *Language Learning, 59*(s1), 1–26.

Berger, T. (2001). Agent-based spatial models applied to agriculture: a simulation tool for technology diffusion, resource use changes and policy analysis. *Agricultural Economics, 25*(2–3), 245–260.

Boden, M. A. (2009). Computer models of creativity. *AI Magazine, 30*(3), 23.

Brownlee, J. (2007). *Complex adaptive systems.* Complex Intelligent Systems Laboratory, Centre for Information Technology Research, Faculty of Information Communication Technology, Swinburne University of Technology: Melbourne, Australia.

Buckley, W. (1968). Society as a complex adaptive system. In *Modern Systems Research for the Behavioral Scientist.* Aldine Publishing Company.

Buss, S., Papadimitriou, C. H., & Tsitsiklis, N. (1990). On the predictability of coupled automata: An allegory about chaos. In *Foundations of Computer Science*, 1990. Proceedings, 31st Annual Symposium on, pages 788–793. IEEE.

Choi, T. Y., Dooley, K. J., & Rungtusanatham, M. (2001). Supply networks and complex adaptive systems: control versus emergence. *Journal of operations management, 19*(3), 351–366.

Crutchfield, J. P. (1994). The calculi of emergence. *Physica D: Nonlinear Phenomena, 75*(1–3), 11–54.

Das, R., Mitchell, M., & Crutchfield, J. P. (1994). A genetic algorithm discovers particle-based computation in cellular automata. In *International Conference on Parallel Problem Solving from Nature*, pp. 344–353. Springer.

Darema, F. (2004). Dynamic data driven applications systems: A new paradigm for application simulations and measurements. In *International Conference on Computational Science*, pp. 662–669. Springer.

Dooley, K. (1996). Complex adaptive systems: A nominal definition. *The Chaos Network, 8*(1), 2–3.

Dreyfus, H. L. (1972). *What computers can't do: the limits of artificial intelligence*. New York, NY: Harper & Row.

Dreyfus, H. L. (1992). *What computers still can't do: A critique of artificial reason*. Boston, MA: MIT Press.

Epstein, J. M. (2006). *Generative social science: Studies in agent-based computational modeling*. Princeton University Press.
Godschalk, D. R. (2003). Urban hazard mitigation: Creating resilient cities. *Natural hazards review, 4*(3), 136–143.
Guy, R. K. (1983). Conway's prime producing machine. *Mathematics Magazine, 56*(1), 26–33.
Hébert-Dufresne, L., Pellegrini, A. F., Bhat, U., Redner, S., Pacala, S. W., & Berdahl, A. M. (2018). Edge fires drive the shape and stability of tropical forests. *Ecology letters*.
Holland, J. H. (1992). Complex adaptive systems. *Daedalus*, pp. 17–30.
Holland, J. H. (1995). *Hidden order: How adaptation builds complexity*. Addison Wesley Publishing Company.
Humphreys, P. (2004). *Extending ourselves: Computational science, empiricism, and scientific method*. Oxford University Press.
Ingwersen, W. W., Garmestani, A. S., Gonzalez, M. A., & Templeton, J. J. (2014). A systems perspective on responses to climate change. *Clean Technologies and Environmental Policy, 16*(4), 719–730.
Kauffman, S. A. (1992). The origins of order: Self-organization and selection in evolution. In *Spin glasses and biology*, pp. 61–100. World Scientific.
Keating, C. B. (2014). Governance implications for meeting challenges in the system of systems engineering field. In *9th International Conference on System of Systems Engineering (SOSE)*, pp. 154–159. IEEE.
Kundu, S., Kundu, R., Kundu, S., Bhattachaijee, A., Gupta, S., Ghosh, S., & Basu, I. (2016). Quantum computation: From Church-Turing thesis to qubits. In *2016 IEEE 7th Annual Ubiquitous Computing, Electronics Mobile Communication Conference (UEMCON)*, pp. 1–5.
Lee, J., Lapira, E., Bagheri, B., & Kao, H.-A. (2013). Recent advances and trends in predictive manufacturing systems in big data environment. *Manufacturing Letters, 1*(1), 38–41.
Levin, S. A. (1998). Ecosystems and the biosphere as complex adaptive systems. *Ecosystems, 1*(5), 431–436.
Lewes, G. H. (1877). Problems of life and mind. *Trübner & Company*.
Maier, M. W. (2015). The role of modeling and simulation in system of systems development. In *Modeling and simulation support for system of systems engineering applications*. Wiley.
Malik, F. (2016). *Strategy: Navigating the complexity of the new world*. Campus Verlag.
McCarthy, I. P., Tsinopoulos, C., Allen, P., & Rose-Anderssen, C. (2006). New product development as a complex adaptive system of decisions. *Journal of Product Innovation Management, 23*(5), 437–456.
McChrystal, G. S., Collins, T., Silverman, D., & Fussell, C. (2015). *Team of teams: New rules of engagement for a complex world*. Penguin.
Miller, J. H., & Page, S. E. (2009). *Complex adaptive systems: An introduction to computational models of social life: An introduction to computational models of social life*. Princeton University Press.
Mittal, S. (2013a). Emergence in stigmergic and complex adaptive systems: A formal discrete event systems perspective. *Cognitive Systems Research, 21*, 22–39.
Mittal, S. (2013b*). Netcentric complex adaptive systems. Netcentric system of systems engineering in DEVS unified process*. (pp. 639–661). CRC Press.
Mittal, S., Durak, U., & Ören, T. (2017). *Guide to simulation-based disciplines: Advancing our computational future*. Springer.
Mittal, S., & Rainey, L. (2015). Harnessing emergence: The control and design of emergent behavior in system of systems engineering. In *Proceedings of the conference on summer computer simulation* (pp. 1–10). Society for Computer Simulation International.
Norman, M. D., Koehler, M. T., & Pitsko, R. (2018). Applied complexity science: Enabling emergence through heuristics and simulations. *Emergent Behavior in Complex Systems Engineering: A Modeling and Simulation Approach*, 201–226.
O'Connor, T., & Wong, H. Y. (2015). Emergent properties. In E. N. Zalta (Ed.), *The Stanford encyclopedia of philosophy*. Metaphysics Research Lab, Stanford University, summer 2015 edition.

Payne, J. L., Khalid, F., & Wagner, A. (2018). RNA-mediated gene regulation is less evolvable than transcriptional regulation. *Proceedings of the National Academy of Sciences*, 201719138.
Pessa, E. (2002). What is emergence? In *Emergence in Complex, Cognitive, Social, and Biological Systems* (pp. 379–382). Kluwer Academic Publishers.
Pidd, M. (1997). Tools for thinking—Modelling in management science. *Journal of the Operational Research Society, 48*(11), 1150.
Rouse, W. B. (2003). Engineering complex systems: Implications for research in systems engineering. *IEEE Transactions on Systems, Man, and Cybernetics, Part C (Applications and Reviews), 33*(2), 154–156.
Rouse, W. B. (2008). Health care as a complex adaptive system: Implications for design and management. *Bridge-Washington-National Academy of Engineering, 38*(1), 17.
Sheard, S., Cook, S., Honour, E., Hybertson, D., Krupa, J., McEver, J., McKinney, D., Ondrus, P., Ryan, A., & Scheurer, R. et al. (2015). *A complexity primer for systems engineers.* INCOSE Complex Systems Working Group White Paper.
Sheard, S. A., & Mostashari, A. (2009). Principles of complex systems for systems engineering. *Systems Engineering, 12*(4), 295–311.
Silberstein, M., & McGeever, J. (1999). The search for ontological emergence. *The Philosophical Quarterly, 49*(195), 201–214.
Szabo, C., & Birdsey, L. (2017). Validating emergent behavior in complex systems. In *Advances in Modeling and Simulation*, pp. 47–62. Springer.
Tolk, A. (2015). Learning something right from models that are wrong: Epistemology of simulation. In *Concepts and methodologies for modeling and simulation*, pp. 87–106. Springer.
Tolk, A. (2017). Code of Ethics. In *The profession of modelling and simulation: Discipline, ethics, education, vocation, societies, and economics*, pp. 35–51. Springer.
Tolk, A. (2018). Simulation and modeling as the essence of computational science. In *Proceedings of the 50th Summer Computer Simulation Conference*, Bordeaux, France. Society for Computer Simulation International.
Tolk, A., Diallo, S., & Mittal, S. (2018a). Complex systems engineering and the challenge of emergence. *Emergent Behaviour in Complex Systems Engineering: A Modeling and Simulation Approach*, pp. 79–97. Society for Computer Simulation International.
Tolk, A., Koehler, M. T. K., & Norman, M. D. (2018b). Epistemological constraints when evaluating ontological emergence with computational complex adaptive systems. In *Unifying Themes in Complex Systems IX: Proceedings of the 9th International Conference on Complex Systems* (pp. 1–10), Springer Proceedings in Complexity, Springer Nature Switzerland.
Yilmaz, L. (2018). Generative parallax simulation: Creative cognition models of emergence for simulation-drive model discovery. *Emergent Behaviour in Complex Systems Engineering: A Modelling and Simulation Approach*, pp. 59–75. Wiley.
Zeigler, B. P. (2018). Closure under coupling: Concept, proofs, DEVS recent examples (WIP). In *Proceedings of the Theory of Modeling and Simulation Symposium*, TMS '18 (pp. 7:1–7:6). San Diego, CA, USA. Society for Computer Simulation International.
Zeigler, B. P., & Mittal, S. (2018). System theoretic foundations for emerging behavior modeling: The case of emergence of human language in a resource-constrained complex intelligent dynamical system. In *Emergent Behavior in Complex Systems Engineering: A Modeling and Simulation Approach* (pp. 35–57). Wiley.

# Chapter 6
# Cloud-Based Simulation

**Lin Zhang, Fei Wang and Feng Li**

**Abstract** In order to accommodate the development and application of simulation systems in network environments, modeling and simulation technology embraced increasingly web-based to cloud-based solutions. This chapter describes the development from the early application of web services, the use of simulation grids, towards modeling and simulation as a service. A current architecture for cloud simulation platforms is presented and key technologies for its implementation are identified. The chapter deals with big data challenges as well as digital twins and provides some applications of cloud-based simulation. It closes with the conclusion that the trend of simulation technology will be cloud-based and intelligent and motivates an intelligent cloud in support of simulation.

## 6.1 The Development of Modeling and Simulation: From Web to Cloud

Simulation research has been going through several stages in recent decades. While the application of Modeling and Simulation (M&S) technology have become more and more complex, varied resources are involved in the M&S systems (Zeigler et al. 2000). In order to accommodate the development and application of simulation systems in network environment, Web Services (WS) technology was brought in M&S field and Web-based Simulation (WBS) emerged. The basic idea of WBS is to encapsulate M&S resources in the form of web services and expose M&S services on the Web. As the IEEE M&S standard, the High-Level Architecture

---

L. Zhang (✉) · F. Wang · F. Li
School of Automation Science and Electrical Engineering, Beihang University, Beijing 100191, China
e-mail: zhanglin@buaa.edu.cn

F. Wang
e-mail: wangfeitime@163.com

F. Li
e-mail: li_feng_china@163.com

 
J. Sokolowski et al. (eds.), *Summer of Simulation*, Simulation Foundations, Methods and Applications, https://doi.org/10.1007/978-3-030-17164-3_6

(HLA) has also been updated to provide Web Services support on the basis of original HLA functionality.

Compared to classical simulation systems, WBS facilitates the sharing of M&S resources and improves data accessibility, interoperability and user experience (Fortmann-Roe 2014). However, web service has some drawbacks such as the service statelessness, which makes it difficult to maintain the state of models exposed as services. Grid technology was used in M&S to be complementary to WBS and support the management of M&S services because of its advantages in distribute resource management and collaboration. Simulation grid overcomes some short-comings existing in traditional distributed modeling and simulation systems.

Cloud Computing has been an extremely popular paradigm in IT industry in the past ten years. By right of the advantages of cloud technologies, Cloud-Based Simulation (CBS) shows superiority over WBS with lower costs and easier ways to develop M&S systems (Wang and Wainer 2015, 2016). Its concept of M&S as Service (MSaaS) attracts an increasing number of M&S practitioners to conduct simulations in the cloud. Later, with some new technologies appear, such as Cyber Physical Systems (CPS), Internet of Thing (IoT), cloud simulation which connect physical objects with cloud environment was proposed and gotten a great development (Zhang et al. 2010).

## 6.2 Web-Based Simulation

### *6.2.1 Definition of Web-Based Simulation*

Web-based Simulation (WBS) is the integration of Web Services (WS) technologies with modeling and simulation (Wang and Wainer 2016; Yingping and Gregory 2005; Byrne et al. 2010). In WBS, existing models, simulation functions, their simulation environment and other M&S resources are exposed as web services (Byrne et al. 2010). Thus, the M&S resources that are originally accessible on a single computer could be shared through the Internet and experiments could be done using M&S services (Byrne et al. 2010). The user requests with simulation parameters could be sent to the simulator through web servers. The results return to the user after simulator finishes experiments remotely.

The early research on WBS began in 1995. Web-front ends were provided to run simulations as Common Gateway Interface (CGI) scripts/programs. They were developed based on Java-based simulation packages, systems and environments and could run anywhere on the Web (Miller et al. 2000; Miller et al. 2001). One of the early researches about WBS (Paull 1996) presented some issues and concepts on Web-based simulation. Since then, WBS attracted much attention of scientists and industries (Bencomo 2004) and a large amount of projects about M&S WSs were developed. These projects can be divided to two categories, that is, SOAP-based WS (Saurabh 2013) and RESTful WS (Al-Zoubi and Wainer 2013). DEVS (Discrete

Event System Specification) M&S formalism (Bernard et al. 2000) was used to develop many simulators (i.e., DEVS/SOA (Saurabh 2013; Al-Zoubi and Wainer 2013) and RISE (Al-Zoubi and Wainer 2013)) in the form of web services.

### 6.2.2 *The Advantages and Disadvantage of WBS Compared to Classical Systems*

The main advantages of Web-based simulation compared to classical systems have been researched and classified by many researchers, which are concluded as follows (Guan et al. 2016; Dhananjai et al. 2000; Tsai et al. 2006; Thomas 2001; Whitman et al. 1998; Veith et al. 1999; Chen and Heath 2005; Ashu 2000):

- Ease of use. The M&S services provides standard Web interfaces for users to invoke them.
- Model reuse. The fine granularity and common data access protocols facilitate the reuse of existing simulation models.
- Cross-platform capability. The M&S services could be opened by Web browsers without the limitation of operating system.
- Wide availability. The M&S services are accessible with an Internet connection from anywhere.
- Integration and interoperability. The M&S services could integrate and interoperate with others over the web easily.

On the other hand, the disadvantages of Web-based simulation over classical systems are listed as follows:

- Loss in speed. The network communication delay results to loss in speed when users interact with M&S services over the web.
- Web-based simulation application stability. The network environment seriously affects the running and interaction of Web-based simulation applications.
- Graphical user interface limitation. The graphical interface of Web-based simulation applications is constrained by the technology limitation of web service.
- Security vulnerability. M&S applications bear the risk of malicious attacks from other web users.

## 6.3 Extensible Modeling and Simulation Framework (XMSF)

Extensible Modeling and Simulation Framework (XMSF) is defined as a composable collection of standards, profiles and application instruction for WBS (Brutzman et al. 2002). This work was led by researchers from the Naval

Postgraduate School, Old Dominion University, George Mason University, and SAIC in early 2000 to meet the demand of The Department of Defense (DoD) for web-based simulation. XMSF aims to provide a web-based framework for distributed simulation with great scalability and support the reuse and composition of models. Web-based technologies used in XSMF facilitate the emergence, development and interoperation of M&S systems. Extensible Markup Language (XML)-based languages enables the compatibility of the future M&S requirements and the existing M&S technologies (Blais et al. 2005).

XMSF intended to identify different description methods of M&S resources to guarantee the interoperability of heterogeneous resources. Besides, XMSF must support the reuse and composition of heterogeneous M&S resources, which requires the further researches on ontologies and semantic network (Brutzman et al. 2002).

Many applications of XMSF has been carried out. An XMSF example is the Flexible Asymmetric Simulation Technologies (FAST) program, which integrates the combat simulations, databases, and computational tools for military training and analysis (Blais 2004). XMSF Profiles is studied to define the specification of XMSF. It includes the details of the structure and application of XMSF, protocol standards, composability guidelines and so on.

## 6.4 HLA Updating Towards Web

High level architecture (HLA), which is a systematic architecture for distributed interactive simulations (DIS), is one of important area of research on simulation technology today (Chi and YU 2015). It was developed in the early '90 s within the U.S. Department of Defense (DoD) to increase interoperability and reusability for simulations. HLA was accepted as an open international standard by IEEE in 2000, called IEEE 1516 (Zhang et al. 2010). The HLA standard has achieved widespread application. More than 200 parts were revised and supplemented to form new HLA Evolved standard in 2010 (Chi and YU 2015).

HLA Evolved is the new version of the HLA (Björn et al. 2008). HLA Evolved provides a lot of improvements for simulation developers and users on the basis of original HLA. The main updates focus on the development, deployment and net centric capacities of HLA federation. Simulation developers and modelers could obtain HLA functionality with Web Services communication frameworks over networks. Distributed federates connect with each other using Web Services to compose HLA federation. A wide range of programming languages are supported to develop HLA federations, such as C, C ++, Java, Fortran, ADA, Perl. Besides, the Web Services support enables HLA to meet the demand of distributed simulation over the web through the new Web Services Description Language application program interface (WSDL API). WSDL is used to describe M&S services as collections of network endpoints, or ports based on XML (Extensible Markup Language).

Run-Time Infrastructure (RTI) is the software developed according to HLA standard. To adapt to the new version of HLA, new RTI software adds Web Services Provider RTI Component (WSPRC) to support the Web Services API. WSDL federates can create and join federation using their URL (Björn et al. 2008).

## 6.5 Simulation Grid

The Internet and Web Technology has realized the connection of computers and Web pages. The Grid technology, however, attempts to realize the comprehensive connection of various distributed resources on the Internet, including computing resources, storage resources, software resources, information resources, knowledge resources and so on. Its goal is to integrate the entire Internet into one huge supercomputer. Combining modern network technology with networked M&S technology (e.g. HLA), simulation grid was proposed as a new infrastructure for networked M&S in early 2000.

Research projects combining simulation and Grid technology has been carried out. For instance, SF-Express solved such issues as resource allocation and dynamic fault-tolerance based on grid technology. CrossGrid studied the realization of RTI in accordance with HLA standard from RTI Layer, Federation Layer and Federate Layer. DS-Grid, NessGrid and Federation X Grid systematically studied the simulation-oriented application grid.

Simulation grid overcomes some shortcomings existing in traditional distributed modeling and simulation systems in dynamic sharing of simulation resources, autonomy and security mechanism. It not only supports the implementation of simulation system engineering, but also extends simulation application pattern and provides a new simulation method in the following two aspects.

(1) Simulation models in the form of Grid services

Grid service is a kind of Web service essentially. The simulation model of grid in the form of Grid service has all the advantages of the simulation models in the form of Web service, such as self-inclusion, loose coupling. Furthermore, Grid service addresses some inherent problems of Web service, such as statelessness and non-temporary of instances. A web service can't maintain data and state between invocations, which is adverse to simulation application. Web Service Resource Framework (WSRF) is a Grid resource infrastructure. It provides some functions which support M&S services to realize stateful interaction.

(2) Service-oriented simulation resources integration

Simulation resources include all software and hardware related to modeling and simulation activities, such as computing resources, model resources, storage resources, simulator resources, tool resources. Service-oriented resources integration requires several operations like resources servitization, resources deployment and resources virtualization. The main issues of resource servitization are the content and the form of servitization. The content of servitization means "what is to be a service". Concerning the current technology, three main forms of servitization are Web Service, GT3 Grid Service and WSRF. Resource deployment means to deploy a developed service into a service container. Resource virtualization is the management of resources after servitization. Services are registered in the management organization, such as information server, resource routing and UDD.

However, from the view of application, the simulation grid also requires improvement in the following directions:

(1) Capability to reusing fine-grained M&S resources.
(2) Capability to fully supporting multi-users.
(3) Collaboration capability of all kinds of simulation resources.
(4) Fault tolerance capability of simulation system.
(5) Security mechanism of application.
(6) Capability to get modeling and simulation services on demand wherever and whenever users are through network.

While grid computing was facing more and more challenges, a new computing technology, cloud computing drawn wide attention because of its outstanding capabilities of resource on-demand sharing and collaboration. Together with the development of cloud computing, by combing the philosophy and methodologies of web-based simulation and grid simulation, a new simulation paradigm, cloud-based simulation, is emerging.

## 6.6 Cloud-Based Simulation

Cloud-based Simulation (CBS) has attracted a lot of attention in recent years. It integrates WBS and cloud computing technology to manage various simulation resources and build different simulation environments (Erdal 2013).

Cloud computing has shown its increasing importance in many fields. It exposes virtualizing hardware and software as services over a network, which generates the new concepts of infrastructure as a service (IaaS), platform as a service (PaaS), software as a service (SaaS). Inspired by these concepts, a cloud-based simulation technology called "cloud simulation" was proposed (Bo-Hu et al. 2009). Besides, the utilize of web services in CBS has received the name of Modeling and Simulation as a Service (MSaaS). MSaaS is a special form of SaaS, as it hides the details of M&S resources, which are exposed as services to build simulation systems.

Cloud-based simulation generally has such features as: M&S resources are virtualized and stored in a cloud resource pool and users can easily obtain them over the Internet (Tolk and Mittal 2014). The cloud-based scenarios are as follows (Onggo 2014).

(1) Simulation users run simulation applications on cloud infrastructure simulation as a Service).
(2) Simulation modelers build and modify models using simulation development tools on cloud infrastructure (Modelling as a Service).
(3) Simulation modelers configure simulation development tool itself by mixing and matching the components of the tool (similar to PaaS).
(4) Simulation users may operate over storage, execution platform and middleware (similar to the IaaS).

Cloud computing is believed to bring M&S into the Cloud-based Simulation era primarily due to: (1) The performance degradation of simulation applications running in clouds could be ignored with the development of high-performance techniques. (2) CBS owns many advantages such as the centralized software deployment and the sharing of M&S resources on demand. It could solve some difficult problems encountered by M&S community, including expensive cost in hardware but low utilization and the repetitive development of similar simulation application.

But research on CBS is still in the beginning phase. Many researchers consider CBS as a big challenge in the following perspectives (Taylor et al. 2015, 2013; Siegfried et al. 2014; Taylor et al. 2012):

(1) Technical perspective. The requirements of infrastructures, protocols, information exchange formats need to be explored in a cloud.
(2) Governance perspective. The consistent management, cohesive policies, guidance, processes and decision processes should be defined.
(3) Security perspective. The person, organization or service has the authentication required on the cloud platform.
(4) Business perspective. The fair share of the financial burden of setting up and conducting a distributed simulation event, clear rules and value assessment remains to be addressed.
(5) Conceptual perspective. The conceptual alignment of the models to support composability should be uniform in addition to the means required for interoperability of the simulation components to compose and execute services on the cloud.

### 6.6.1 *Modeling and Simulation as a Service (MSaaS)*

"Modeling & Simulation as a Service" is the combination of service-based approaches and ideas taken from cloud computing (Siegfried et al. 2014). Over the past years, the concept of MSaaS has been investigated by NATO Modeling and Simulation Group MSG-131. They also collected national perspectives and experiences regarding MSaaS. MSG131 defines M&S as a Service as follows:

"M&S as a Service (MSaaS) is a means of delivering value to customers to enable or support modelling and simulation (M&S) user applications and capabilities as well as to provide associated data on demand without the ownership of specific costs and risks." A service in the MSaaS concept can be a model service or a simulation tool service, such as a aerodynamics model service or a tool for building simulation environment. MSaaS contains the following connotations:

(1) MSaaS as a cloud service model;
(2) MSaaS using cloud service models;
(3) MSaaS as a Service Oriented Architecture;
(4) MSaaS as a business model.

Perspective 1 describes the methods of accessing M&S applications. Perspectives 2 and 3 focus on how to compose M&S applications by the cloud service models. Perspective 4 describes the application mode of MSaaS as an organizational or professional service (Fig. 6.1).

As such, MSaaS features the loosely coupled services and the ability to reuse M&S services. The objectives of MSaaS can be concluded into two kinds, i.e. effectively and efficiently supporting operational requirements (like executing an exercise) and improving development, operation and maintenance of M&S applications. MSaaS is a special form of SaaS but different from SaaS. MSaaS provides developers M&S services on demand. The services are selected and composed according to their functionality and quality of service (QoS) maintained by service providers.

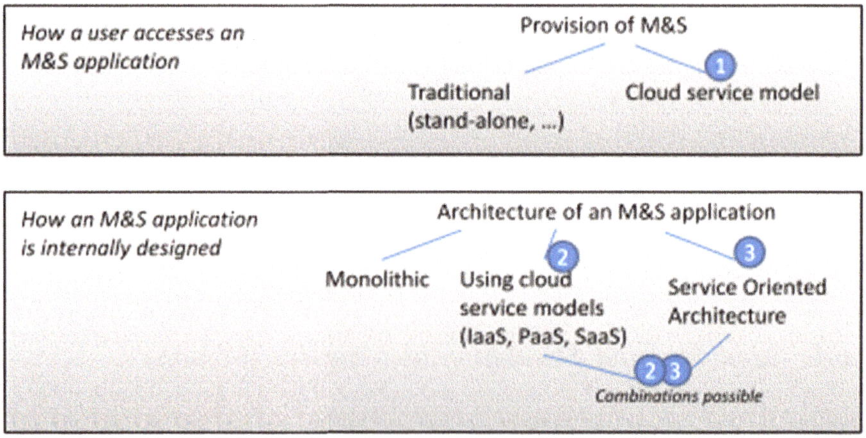


**Fig. 6.1** MSaaS perspectives (Siegfried et al. 2014)

## 6.7 Cloud Simulation

### 6.7.1 Concept of Cloud Simulation

Based on the research of simulation grid and further integration of virtualization technology, the concept of "cloud simulation" was proposed in 2009 (Bo-Hu et al. 2009). Cloud simulation is a simulation mode to provide simulation services on user's demand from simulation resource cloud pool with network and cloud simulation platform. Its main idea is to encapsulate M&S resources into services and form a cloud service pool, so that they can be easily shared and reused on-demand in a simulation system and support all activities of systems simulation.

A cloud simulation platform can automatically discover the required resources, and dynamically build a simulation system on-demand based on service-oriented composition and scheduling.

Furthermore, a new networked M&S platform "cloud simulation platform" is built to improve the capabilities of existing networked M&S platform. It utilizes network and cloud technologies to compose the M&S services in network on demand, and to provide various simulation services to users.

### 6.7.2 Architecture of Cloud Simulation Platform

The architecture of Cloud Simulation Platform is illustrated in Fig. 6.2. It consists of 4 layers: resource layer, cloud simulation service layer, application portal and support tools layer, and application Layer.

Resource layer provides network and various M&S resources encapsulated by virtualization technology, including model resources, tools and software resources, computing resources, storage resources, model/data resources, knowledge resources and various types of simulators, scientific instruments and so on.

Cloud simulation service layer provides cloud simulation oriented core services, including multi-user-oriented resource scheduling and management services, pervasive co-simulation services, virtual simulation resources information management services, intelligent resource discovery services, co-simulation scheduling and composition services, simulation resources adaptive and fault-tolerant migration services, collaborative visualization engine services, Simulation resource management services based on web service/grid technology, web-based HLA/RTI distributed interactive simulation services and simulation resource dynamic management and optimization deployment services based virtualization technology and so on.

Application portal and support tools layer provides browser and desktop portals/tools for users to logon in "CSP" and carry out simulation activities. These portals/tools include project management tool, simulation database/model base/knowledge base management tool, multidisciplinary virtual prototype problem solving

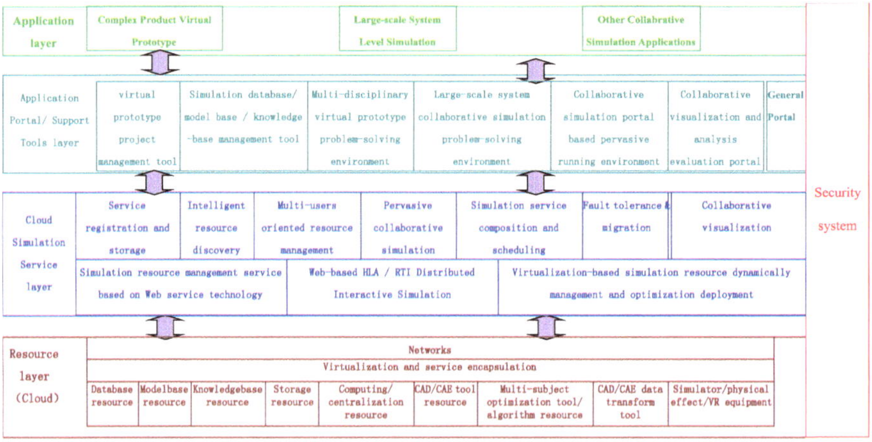


**Fig. 6.2** Architecture of cloud simulation platform (Bo-Hu et al. 2009)

environment, large-scale system level collaborative simulation problem solving environment, pervasive collaborative simulation portal, collaborative visualization and analysis evaluation portal and general portal.

Application Layer includes multidisciplinary virtual prototype collaborative simulation applications, large-scale system level collaborative simulation applications and other Collaborative Simulation Applications.

## 6.8 Key Technologies for Cloud-Based Simulation

There are many enabling technologies for cloud-based simulation. Some of them are introduced as follows.

### *6.8.1 Virtualization and Service Encapsulation of Model and Simulation Resources*

Virtualization layer, based on the simulation resources layer, maps physical simulation resources into a set of virtual machine (VM) templates. According to domain-specific simulation requirements, various simulation resources can be re-organized and encapsulated into VM templates with particular functions. During the run-time simulation, distributed and heterogeneous resources are transparent to users (Ren et al. 2012).

A virtualization management middleware is needed to expose functional interfaces to manage the simulation VMs. The interfaces include the encapsulation interface to package simulation resources into VMs, the register interface to register VMs in the simulation VM templates pool, the discovery interface to match suitable simulation VM templates according to simulation requirements, the monitor interface to monitor running status of both virtual and physical simulation resources, the dispatch interface and instantiate interface to instantiate and deploy a simulation VM (Ren et al. 2012).

### 6.8.2 Simulation Environment, Management and Interactive Simulation Technology

First, problem-solving environment technology (PSE) is an important part of application portal/support tool layer, which supplies a top-level tool integration environment. Technically, PSE provides a Top Level Modeling Language which adopts meta-model technology, presents the structure and behavior of simulation system, and describes experiment instruction of virtual prototype design and system-level simulation. Secondly, complex product project management technology is also important for cloud simulation. Existing project management tools can hardly meet the project management requirements for complex products. Project management technology mainly includes the planning, organizing, controlling and communication techniques during the whole project life-cycle, which covers project decision, design, implementation and evaluation. Project management model supplies the same view for persons joining in project management and gives an effect of orientation and improvement to their management practice. Thirdly, simulation management technology based on Web Service/Grid Technology mainly provides simulation resources management services for resources deployment/register, batch schedule, monitor and directory. "Cloud Simulation Platform" can be compatible with kinds of resource management middleware to provide management services for M&S resources. Fourthly, the simulation on the "cloud simulation platform" differs from other networked computing tasks. It mainly reflects on multiple iterative calls for services in the simulation application and the strict requirements of consistency of time and space for a wide range of services participated in simulation running. The main idea of Web-based HLA/RTI Distributed Interactive Simulation Technology is to solve collaboration problems among services in simulation application by HLA/RTI technology, and to solve the dynamic scheduling and management of simulation services on LAN and WAN by grid computing and Web Service technology (Bo-Hu et al. 2009).

## 6.9 Resource Discovery, Composition and Scheduling Technology

Semantics-based simulation model resource discovery technology is a key technology of cloud-simulation oriented services layer. The traditional keyword-comparison discovery method only performs matching based on syntax. Implementation of semantic-based model resource discovery by introducing semantic ontology in cloud simulation platform improves the accuracy, efficiency and intelligence of the resource discovery mechanism.

In simulation cloud, simulation tasks are always complicated which are composed dynamically by distributed simulation models. Traditional service composition technology can hardly be used to realize the description and execution of complex interactive operation because they commonly describe, schedule and implement service composition based on workflow. In most cases, more than one simulation tasks exist at the same time. Different tasks could use the same resource at the same time. Hence scheduling process is an essential step in this situation. Networked Simulation Resource Scheduling Problem has following Features: (1) Resource selection takes the goal of completing simulation tasks with the most proper resources; (2) Most of the cloud simulation applications involve collaboration among multiple computers; (3) Node computing performance and inter-node network communication speed and delay must be taken into consideration (Laili 2012; Ren et al. 2012).

In the cloud simulation environment, models and the related tools are encapsulated into simulation services. User requirements submitted the simulator can be divided into single-service requirement and multi-service requirement. For the single-service requirement, the platform picks up the most appropriate service to perform the task from a large number of cloud services, which is called cloud service optimization. In terms of the multi-service requirement, the platform decomposes the task to many subtasks and searches the services to support the subtask. Then platform picks up the most appropriate group of services to perform the task, which is called cloud service composition. When more than one multi-service requirement exists, scheduling process of services to fulfill multiple requirements is essential. The number and types of resource services in practical cloud simulation are numerous. How to correctly incorporate multiple services together to accomplish a multi-disciplinary simulation task and how to tackle multiple tasks requirements are two important concerns not only for users, but also for the cloud center itself. Feng et al. (2018) investigated the problem of service composition and scheduling for cloud simulation and proposed a new service network-based method. In this method, the number of composition steps was uncertain before to be executed. The execution of service composition and scheduling was based on a service network-based model. In this model, nodes represented tasks or services while the weight of edges was the performance of services. Through the built model, different composition paths can be obtained for a certain requirement and a near-optimal solution can be established with optimization algorithms.

## 6.10 Big Data for M&S

Cloud-based simulation will produce large amount of data, such as the simulation service information, simulation results and so on. As the volume of data is increasing exponentially, big data technology is used to process data that are huge in terms of volume, velocity, variety and veracity. Therefore, these data are stored in the cloud and all the analysis is done in the cloud by big data technology (Dasoriya 2018).

Big data challenges to modeling and simulation in many aspects including basic theory of simulation, modeling method, simulation method and tools (Feng et al. 2018). At the meantime, big data also provides opportunities to M&S. For example, based on big data and machine learning, a new type of model of the system can be established, which is called data model. Although it is a "black box" model, it can be optimized by continuously learning from the real system. This kind of method might be available for the simulation of complex systems that are difficult to be modeled with traditional methods. Based on the analyses of big data, simulation process, such as simulation resource assignment, simulation service matching and scheduling, can be more intelligent. Big data can also be used to enable quantitative evaluation of model credibility and assessment of simulation system risks.

## 6.11 Digital Twin

Digital Twin is the virtual representation of a physical object and can interact with the physical object. Digital twin was descripted by NASA (Dasoriya 2018) as an integrated multi-physics, multi-scale simulation of physical objects to reflect the behavior of the corresponding physical objects. It can be used in design and optimization, remote monitoring and control, cloud planning and scheduling, fault prediction and tracking, etc. Digital twin is defined by what you want to analyze, simulate, predict related to the physical product. It involves engineering data, operation data and dynamic behaviors of the models (Rosen et al. 2015).

A Digital Twin is actually a Cyber-Physical Dynamic Model that is generally stored and managed in a cloud. It connects the model to the real physical system and evolves the real-time data along the full life cycle of the system. Based on digital twins we can realize online simulation to provide more precise analyses and prediction to get better decision support.

Digital Twin is considered as a great improvement in the field of M&S (Rosen et al. 2015; Weyer et al. 2016). Its realization is significant for the development of manufacturing industry. It will simplify the simulation of the manufacturing processes and provide operation support along the life cycle (Boschert and Rosen 2016).

## 6.12 Applications of Cloud-Based Simulation

The cloud-based simulation can be used in manufacturing as well as design of a product. Two case studies are introduced in the following sections.

### *6.12.1 Cloud-Based Simulation for Manufacturing*

Cloud-based simulation plays important role for manufacturing. Theoretical simulation analysis for manufacturing requirements is needed before the truly processes are executed. Some manufacturing problems, such as manufacturing recourses collaboration, matching and composition of cloud services, cloud scheduling, etc., can be simulated in a cloud platform. Paper (Zhao et al. 2017) proposed a conceptual model of the simulation platform based on service as shown in the Fig. 6.3. Many researches have adopts agent to support the simulation process. For example, an agent-supported meta-level interoperation architecture is proposed to address the interoperation and dynamic composability of disparate simulations (Yilmaz and Paspuleti 2005). Service agent combines the function of service and the intelligence of agent (Liu et al. 2013). In this model, each manufacturing enterprise in the cloud environment acts as an independent agent. The relationship between them can form a service agent network. The simulation platform includes two libraries named data interface and function interface which can realize some management function in the cloud environment. The architecture of the manufacturing simulator is divided into five layers: the data layer, the lower tool layer, the management layer, the upper tool layer, application layer. This five-layer correlation, from data generation to analysis of final results, constitutes a complete manufacturing service platform prototype system. Each simulation requirement for a manufacturing task can be completed in the service platform. Three typical application scenarios are introduced in the following:

(1) Simulation of Manufacturing Resources cooperation

There are many kinds of manufacturing resources in the manufacturing environment. One of the relation between the resource and the service: several manufacturing resources provide a service. When a service gets a task, several manufacturing resources of service can perform the task coordinately. The simulation platform built above can support this application scenario.

(2) Simulation of Service Composition

In the field of manufacturing, according to user requirements or task granularity, user requirements can be divided into single task requirement and multiple tasks

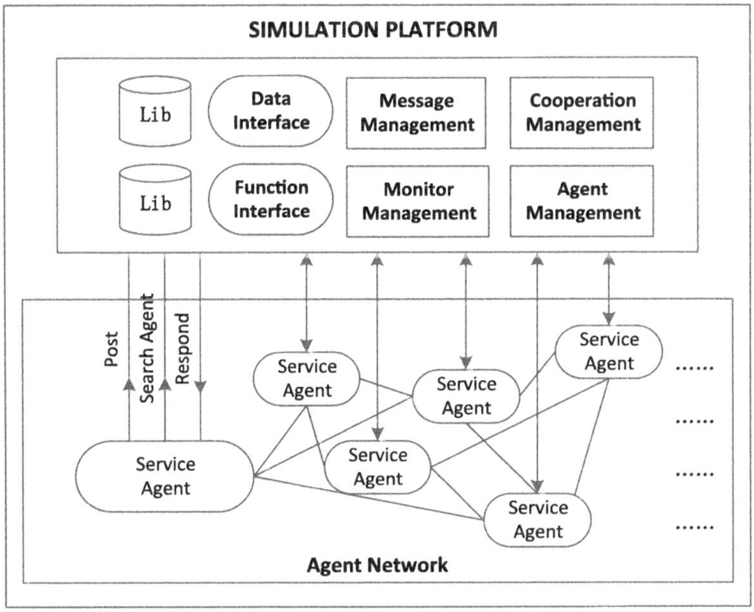


**Fig. 6.3** A conceptual model of the simulation platform

requirement. For the single task requirement, the platform picks up the most appropriate service to perform the task from a large number of cloud services. In terms of the multiple task requirement, the platform decomposes the task to many subtasks and searches services to support the subtask. Then platform picks up the most appropriate group of services to perform the task, which is called cloud service composition. The simulation platform can support this kind of scenario, and generates different services in a certain range and constraints. The user can verify their service composition algorithm in simulation platform.

(3) Simulation of Service Collaboration

In the platform, each enterprise user can finish a specific task or subtask independently, and adapt itself to the environment initiatively. In addition, they can communicate, and cooperate, and compete for each other. In the simulation platform, each enterprise is a service agent. Different enterprises in the platform can establish a multi-agent network. Based on the network, the researcher can describe, explain, predict, analyze the behavior, and collaboration model, and rule among the enterprises in the cloud manufacturing complex system.

## 6.13 Cloud-Based Simulation for Optimized Design

In the design process of a product, the design parameters should be optimized through multi-disciplinary collaborative simulation. Cloud simulation platform eases the building of simulation system to realize the above simulation optimization tasks.

The simulation system includes 3D view federate design, control federate design, hydraulic federate design and dynamic federate design is submitted into cloud-based simulation platform. There are different kinds of interdisciplinary simulators in the cloud platform, such as visualization services, dynamics system simulators, control system simulators and hydraulic system simulators. There are more than one candidates for each kind of resource.

The problem is to choose one simulation resource from the candidates that registered as services in the cloud. The platform uses service matching and composition algorithms to form a simulation process and assign suitable computing resources for each simulation service. The simulation resources and simulation system built by the cloud platform is shown in Fig. 6.4.

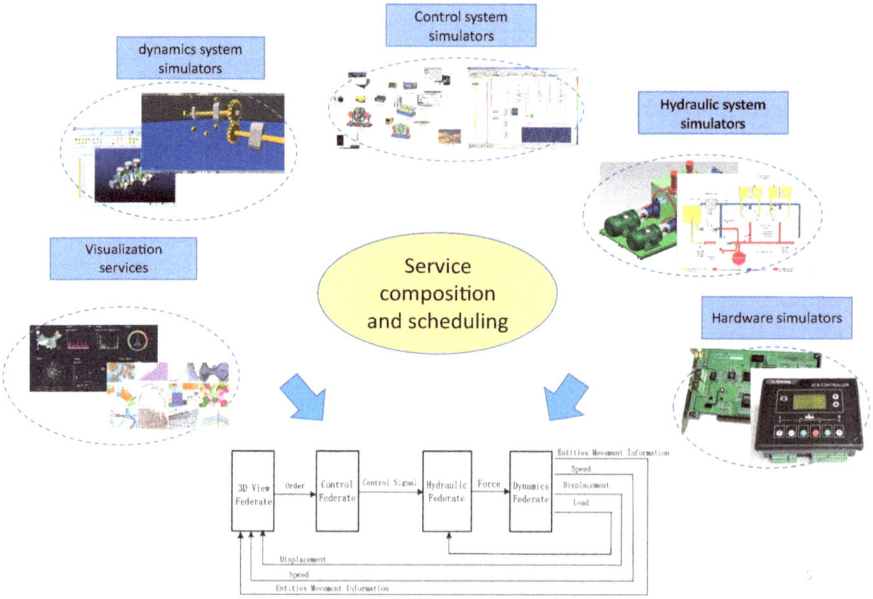


**Fig. 6.4** Collaborated simulation for design optimization

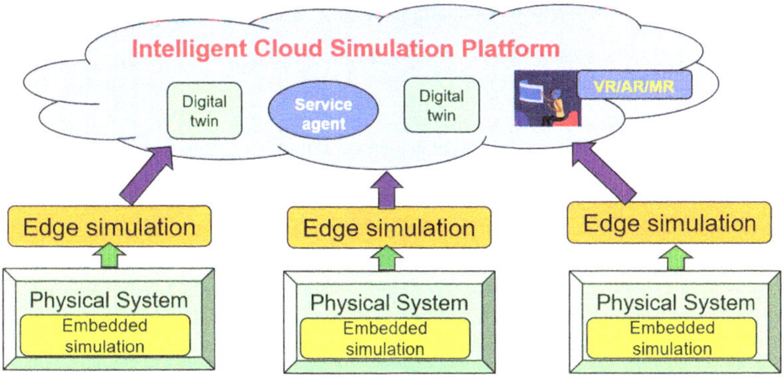


**Fig. 6.5** The cloud-based intelligent simulation system

## 6.14 Conclusion and Future Work

Simulation research has developed from Web-based Simulation to Cloud-based Simulation. MSaaS provides M&S resources and capabilities on demand in the form of Web service. Cloud Simulation has transplanted traditional simulation system to cloud platform so that M&S resources can be easily shared and reused on-demand in a simulation system and support the activities in the full life cycle of systems simulation. Other technologies such as Big Data, Digital Twin and Edge Computing are also important enabling technologies of promoting the development of cloud-based M&S.

The trend of simulation technology will be cloud-based and intelligent. Figure 6.5 give a picture of an envisioned intelligent simulation system, in which an intelligent cloud-based simulation platform will be the core part of the system while the edge simulation will be used to do quick analysis and optimization for high real-time processes. In some cases, the embedded simulation is needed to participate in the operation of the physical system to help the physical system to make a better decision quickly. The physical systems are modeled as digital twins and integrated into intelligent cloud simulation platform through IoT. The intelligent cloud simulation platform finally supports the collaborative simulation of distributed and heterogeneous physical systems.

## References

Al-Zoubi, K., & Wainer, G. (2013). RISE: A general simulation interoperability middleware container. *Journal of Parallel and Distributed Computing, 73,* 580–594.

Ashu, G. (2000). A web-based interface for storing and executing simulation models. In *Proceedings of the 32nd Conference on Winter Simulation*, Orlando, Florida.

Bencomo, D. S. (2004). Control learning: present and future. *Presented at the 15th Triennial World Congress, Barcelona, Spain.*

Björn, M., Katherine, L., Morse, Mike, L., Reed, L., & Robert, L. (2008). HLA evolved—A summary of major technical improvements. *Fall Simulation Interoperability Workshop.*

Blais, C. (2004). Extensible modeling and simulation framework (XMSF) exemplars in analytic combat modeling. *Paper 04S-SIW-142, Proceedings of the Spring 2004 Simulation Interoperability Workshop, Orlando, FL.*

Blais, C. L., Brutzman, D., Drake, D., Moen, D., Morse, K., Pullen, M., et al. (2005). *Extensible modeling and simulation framework (XMSF) 2004 project summary report.* Monterey, CA: Naval Postgraduate School Report.

Bo-Hu, L. I., Chai, X. D., Hou, B. C., Tan, L. I., Zhang, Y. B., & Hai-Yan, Y. U. et al. (2009). Networked modeling & simulation platform based on concept of cloud computing—cloud simulation platform. *Journal of System Simulation.*

Boschert, S., & Rosen, R. (2016). *Digital twin—The simulation aspect*: Springer International Publishing.

Brutzman, D., Zyda, M., Pullen, J., & Morse, K. L. (2002). Extensible modeling and simulation framework (XMSF): Challenges for Web-based modeling and simulation. Naval Postgraduate School Report, Monterey, CA.

Byrne, J., Heavey, C., & Byrne, P. J. (2010). A review of web-based simulation and supporting tools. *Simulation Modelling Practice and Theory, 18,* 253–276.

Chen, J. Q., & Heath, R. D. (2005). Web application development methodologies. In *Web Engineering: Principles and Techniques* (pp. 76–96). IGI Global.

Chi, X., & Yu, L. (2015). On the discussion of new problems in HLA based on high-efficiency simulation cloud. *Presented at the 2015 International Conference on Software, Multimedia and Communication Engineering (SMCE 2015).*

Dasoriya, R. (2018). A review of big data analytics over cloud. In *IEEE International Conference on Consumer Electronics-Asia.*

Erdal, C. (2013). Modeling and simulation as a cloud service: A survey. In *Proceedings of the 2013 Winter Simulation Conference*, Savannah, GA.

Feng, L., Yuanjun, L., Lin, Z., & Xiaolin, H. (2018). Service composition and scheduling in cloud-based simulation environment. *Presented at the SpringSim-Mod4Sim, Baltimore, Maryland, USA.*

Fortmann-Roe, S. (2014). Insight maker: A general-purpose tool for web-based modeling & simulation. *Simulation Modelling Practice and Theory, 47,* 28–45.

Guan, S., De Grande, R. E., & Boukerche, A. (2016). An HLA-based cloud simulator for mobile cloud environments. In *2016 IEEE/ACM 20th International Symposium on Distributed Simulation and Real Time Applications (DS-RT),* (pp. 128–135). IEEE.

Laili, Y. (2012). A study of optimal allocation of computing resources in cloud manufacturing systems. *International Journal of Advanced Manufacturing Technology, 63,* 671–690.

Liu, J., Zhang, L., & Liu, Y. (2013). Service-orientated model composition. *Sciencepaper Online* (in Chinese).

Miller, J. A., Seila, A. F., & Tao, J. (2000). Finding a substrate for federated components on the web. In *Proceedings of the 32nd conference on Winter simulation* (pp. 1849–1854). Society for Computer Simulation International.

Miller, J. A., Fishwick, P. A., Taylor, S. J. E., Perakath, B., & Boleslaw, S. (2001). Research and commercial opportunities in web-based simulation. *Simulation Practice & Theory, 9,* 18.

Onggo, B. S. (2014). The need for cloud-based simulation from the perspective of simulation practitioners. In *Operational Research Society Simulation Workshop.*

Paull, F. (1996). Web-based simulation some personal observations. In *Proceedings of the 1996 Winter Simulation Conference.*

Rao, M. D., & Philip, A. (2000). Wilsey "Dynamic component substitution in web-based simulation. In *Proceedings of the 2000 Winter Simulation Conference.*

Ren, L., Zhang, L., Tao, F., Zhang, X. L., Zhang, Y., & Zhang, Y. (2012). A methodology towards virtualization-based high performance simulation platform supporting multidisciplinary design of complex products. *Enterprise Information Systems, 6*, 267–290.

Rosen, R., Wichert, G. V., Lo, G., & Bettenhausen, K. D. (2015). About the importance of autonomy and digital twins for the future of manufacturing. *Ifac Papersonline, 48*, 567–572.

Saurabh, M. (2013). *Netcentric system of systems engineering with DEVS unified process*.

Siegfried, R., van den Berg, T., Cramp, A., & Huiskamp, W. (2014). *M&S as a service: Expectations and challenges*. Orlando, FL: Simulation Interoperability Workshop.

Taylor, S. J. E., Fishwick, P. A., Fujimoto, R., Uhrmacher, A. M., Page, E. H., & Wainer, G. (2012). Panel on grand challenges for modeling and simulation. In *Proceedings of the 2012 Winter Simulation Conference*, NJ, USA.

Taylor, S. J. E., Khan, A., Morse, K. L., Tolk, A., Yilmaz, L., & Zander, J. (2013). Grand challenges on the theory of modeling and simulation. In *Proceedings of the Symposium on Theory of Modeling and Simulation*, San Diego, CA.

Taylor, S. J. E., Khan, A., Morse, K. L., Tolk, A., Yilmaz, L., Zander, J., et al. (2015). Grand challenges for modeling and simulation: simulation everywhere—from cyber infrastructure to clouds to citizens. *Simulation, 91*, 648–665.

Thomas, W. (2001). Simulation application service providing (SIM-ASP). In *Proceedings of the 2001 Winter Simulation Conference* Arlington, Virginia, USA.

Tolk, A., & Mittal, S. (2014). A necessary paradigm change to enable compostable cloud-based M&S services. In *Simulation Conference*, 2014, pp. 356–366.

Tsai, W. T., Fan, C., Chen, Y., & Paul, R. (2006). DDOS: A dynamic distributed service-oriented simulation framework. In *Proceedings of the 39th Annual Symposium on Simulation (pp. 160–167). IEEE Computer Society* (pp. 160–167).

Veith, T. L., Kobza, J. E., & Koelling, C. P. (1999). Netsim: Java™-based simulation for the world wide web. *Computers & Operations Research, 26*(6), 607–621.

Wang, S., & Wainer, G. (2015). A Mashup architecture with modeling and simulation as a service. In *Proceedings of the International Conference on Web Information Systems Engineering* (pp. 247–261). Springer, Cham.

Wang, S., & Wainer, G. (2016). Modeling and simulation as a service architecture for deploying resources in the cloud. *International Journal of Modeling, Simulation, and Scientific Computing, 07*, 1641002.

Weyer, S., Meyer, T., Ohmer, M., Gorecky, D., & Zühlke, D. (2016). Future modeling and simulation of CPS-based factories: An example from the automotive industry. *Ifac Papersonline, 49*, 97–102.

Whitman, L., Huff, B., & Palaniswamy, S. (1998). Commercial simulation over the web, *1*, 335–339.

Yilmaz, L., & Paspuleti, S. (2005). Toward a meta-level framework for agent-supported interoperation of defense simulations. *Journal of Diagnostic Medical Sonography, 2*, 161–175.

Yingping, H., & Gregory, M. (2005). Autonomic web-based simulation. In *Proceedings of the 38th Annual Symposium on Simulation* (p. 8).

Zeigler, P. B., Herbert, P., & Kim, G. T. (2000). *Theory of modeling and simulation integrating discrete event and continuous complex dynamic systems*.

Zhang, Y., Li, B., Ren, L., Chai, X., & Hou, B., Research on virtualization-based simulation environment dynamically building technology for cloud simulation. In *2010 International Conference on Information Security and Artificial Intelligence*.

Zhao, C., Zhang, L., Liu, Y., Zhang, Z., Yang, G., & Li, B. H. (2017). Agent-based simulation platform for cloud manufacturing. *International Journal of Modeling Simulation & Scientific Computing*.

# Chapter 7
# Converting High Level Models into DEVS Modeling and Simulation Applications

**Gabriel A. Wainer**

**Abstract** We discuss a number of methods for converting high level modeling formalisms and languages into lower level discrete-event systems specifications using the DEVS formalism. We present the implementation of such methods in the CD++ open source toolkit, and discuss different case studies. We focus on a variety of methods, ranging from Petri Nets and Finite State Machines up to Modelica and advanced Traffic modeling languages, showing the generality of DEVS based solutions, and the definition of user libraries in different domains.

## 7.1 Introduction

As discussed in the Introduction chapter, I have attended SummerSim since 1999, presenting a variety of results of our research to the Modeling and Simulation community. Besides making lifetime friends and colleagues, I was able to witness (and collaborate with) research on different modelling and simulation methodologies. In the beginning, the conference focused in a number of methodologies and applications for continuous and discrete-event simulation, which were mature and well established. In parallel, there were a large number of research efforts focusing on modeling formalisms, and a number of those investigations focused on the transformation of models into DEVS (Zeigler et al. 2000). In this chapter, we introduce our efforts in this field, which are summarized in Fig. 7.1 (the reader can find information about these topics at http://cell-devs.sce.carleton.ca/ars).

G. A. Wainer (✉)
Department of Systems and Computer Engineering, Carleton University, 1125 Colonel By Dr, Ottawa, ON K12 5B6, Canada
e-mail: gwainer@sce.carleton.ca



J. Sokolowski et al. (eds.), *Summer of Simulation*, Simulation Foundations, Methods and Applications, https://doi.org/10.1007/978-3-030-17164-3_7

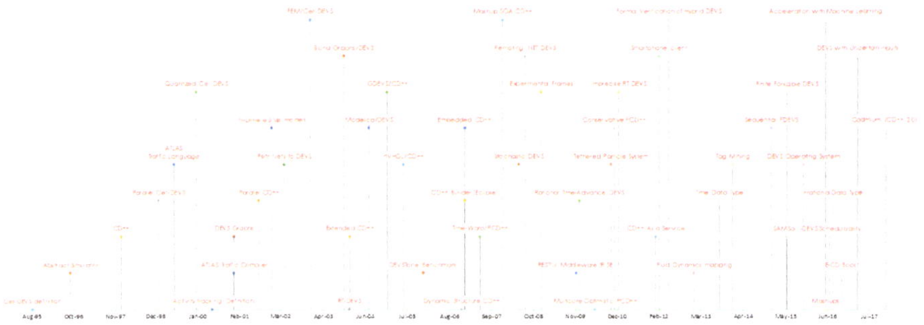

**Fig. 7.1** Modeling methodologies at the advanced real-time simulation lab

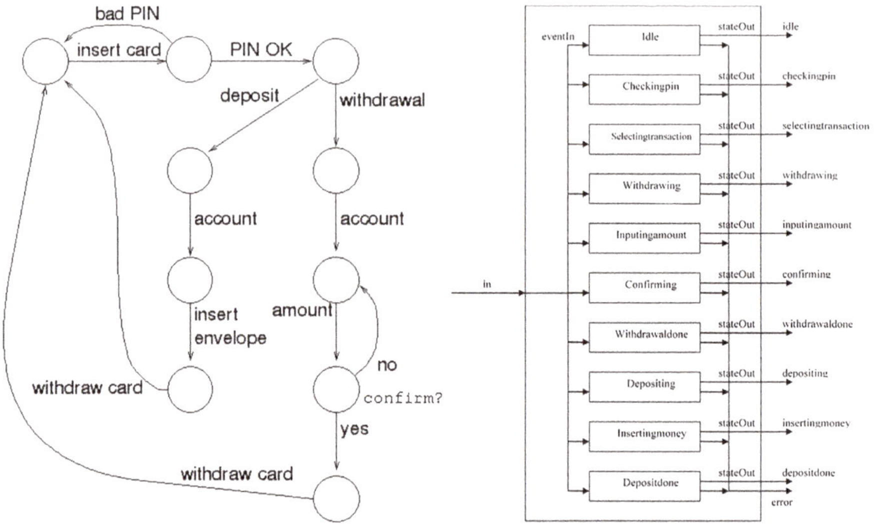


**Fig. 7.2** A model of an ATM machine, and the corresponding DEVS coupled model

Although the research in modeling methodologies in our laboratory includes numerous areas, ranging from Graphical User Interfaces (CD++Builder), Real-Time Simulation (RT-DEVS, Embedded CD++, E-CD Boost, I-DEVS, I-DEVS Schedulability, E-CDBoost), Parallel and Distributed Simulation (Parallel CD++, Conservative and Time-Warp/PCD++, Mashup SOA CD++, Remoting .NET DEVS, RESTful middleware, multi-core optimistic PCD++, DEVS as a service,

SamSaas Mashups), our main focus has been on theory of modeling and simulation. This included the definition of Cell-DEVS and its abstract simulator and its extension to Parallel Cell-DEVS and Quantized Cell-DEVS, as well as the first definition of Activity Tracking as a method to reduce cell activation in Cell-DEVS models, and the extension to N-Dimensional Cell-DEVS models. We defined the ATLAS traffic language and the ATLAS traffic simulation compiler. We also worked in the transformation of different methodologies into DEVS and Cell-DEVS. For instance, we showed how to transform Petri Nets (Jacques and Wainer 2002), Finite State Machines (Zheng and Wainer 2003). We also worked in various extensions to the DEVS formalism and proved important properties with the Stochastic DEVS formalism (Castro et al. 2010), how to check models based on the Rational Time-Advanced DEVS (Saadawi and Wainer 2010), and how to model real-time systems with constraints using Imprecise RT DEVS. We also introduced formal verification of Hybrid DEVS models, presented finite forkable DEVS and DEVS with uncertain inputs. We were also able to show the transformation of other methods, including Finite Element (FEM), Lattice Gas and Computational Fluid Dynamics. As research on modeling of continuous and hybrid systems advanced, the concept quantization of the state variables obtaining a discrete event approximation of the continuous system proved to be an important method to allow the integration of continuous and discrete models. This method that could be defined using DEVS (Kofman and Junco 2001), and was later extended to the family of QSS methods (Pietro et al 2018). We used these methods to define models of continuous systems, including VHDL-AMS, and later Bond Graphs (D'Abreu and Wainer 2006) and Modelica (Chechiu and Wainer 2005).

In this chapter, we summarize some of these efforts, presenting a tutorial on different modeling methods and their translation into DEVS. Most of the research presented has been discussed in numerous articles and books (Wainer 2009), and this serves as a generic introduction for the readers interested in advancing in this research area. The tools are open source, and all the models presented here are publicly available.

We first discuss the definition of simple Finite State Machines and its definition as DEVS models. Then, we introduce Timed Petri Nets, and show a library of DEVS models for modeling and simulating them. After, we show a method we defined, in which we have Timed Automata models converted into DEVS (and vice-versa), allowing advanced model checking using tools like UPPAL, in particular using hybrid models that can be defined through quantization of the input/output signals. We then elaborate on the definition of Bond Graph models and their implementation using Quantized DEVS models. Bond Graphs were used to build a

Modelica compiler, which is described next. We also built a library to build digital circuits using VHDL-AMS, which is described next. Finally, we present ATLAS, a specification language defined to depict city sections and to model and simulate traffic flow. Road segments are defined by their size, number of lanes, traffic direction, maximum speed, etc. Once the city section is outlined, the constructions are translated into Cell-DEVS models, and the traffic flow is automatically set up. The rule generation for describing the traffic behavior is based on macro templates, enabling changes in the model implementation in a flexible way.

## 7.2 Modeling Finite State Machines in DEVS

Finite State Machines have been used in systems engineering applications as they can represent complex artificial devices in an abstract fashion. They are characterized as abstract mathematical entities that can take a finite number of states, and they respond to external inputs to trigger transitions between states. A deterministic FSN can be formally defined, using a systems theoretical notation, as follows (Hopcroft and Ullman 1979):

$$\mathrm{FSM} = <\mathrm{S}, \mathrm{X}, \mathrm{Y}, \delta, \lambda>$$

Where

X: finite input set
Y: finite output set
S: finite state set
δ: $X\ x\ S \rightarrow S$ the next state function
λ: $\mathrm{S} \rightarrow \mathrm{Y}$ is the output function (Moore machine) or
$X\ x\ S \rightarrow Y$ (Mealy machine)

In (Zheng and Wainer 2003) we showed how to define FSM as DEVS. The basic idea is to represent the behavior of a generic state as an Atomic model, and then to build the FSM by combining a number of those atomic models into a coupled model representing the FSM. All the states in a FSM are encoded as integers, and a unique global value, *stateCode,* is assigned to each state. A *phase* is used to indicate if a given state is active or passive (and only one state is can be active at a time, according to FSM semantics). The events are encoded as integers as well.

The formal DEVS specification for the FSM for the atomic model *State* is:

$$\mathrm{FSM} = < X, Y, S, \delta ext, \delta int, \lambda, ta>$$

```
X = {eventIn, transitionIn}
Y = {stateOut, transitionOut }
S = { phase, events[],isEvent stateCode, stateValue, nextActiveState }
δext ((phase, events[],isEvent stateCode, stateValue, nextActiveState), e, x)
    case phase
        active:
           if x is from eventIn
                get nextActiveState from events[ ];
                  isEvent = true;
                  ta(active, 0); //trigger an immediate internal event for output
        passive:
                  if x is from transitionIn
      isEvent = false;
      ta(passive, 0);
}

δint (events[ ], stateCode, phase, stateValue, nextActiveState, isEvent)
{
      if (isEvent)  // always passivate after receiving an event
                passivate();
      else      // new state activated by the input from the transition
                ta(active, Infinity);
}

λ(events[ ], stateCode, phase, stateValue, nextActiveState) {
      if (isEvent)          // inform the next active state
         send nextActiveState to port transitionOutput
      else                              // indicate the current state
         send stateValue to port stateOutput
}
```

The state variable *events[ ]* is an array that records the legal events of the state and the associated *nextActiveState*; and *stateCode* includes the unique state code of the state (0 represents the initial state in a FSM). The *phase* can be active or passive; *stateValue* is the assigned output of the state; *nextActiveState* is the state code of the next active state; *isEvent* indicates if an event is received or if a transition signal is received. The *eventIn* receives encoded numbers representing external events. If a state receives a legal event listed in *events[ ]* when it is active, the *stateOut* port sends out the current *stateValue*, and the *transitionOut* port sends out the *nextActiveState* signal to all the *transitionIn* ports in the FSM reporting which state will be active in the next step. A state becomes active if the encoded number received from *transitionIn* port is same as its *stateCode*.

| | |
|---|---|
| 00:00:10 in insertCard | 00:00:010 checkingpin |
| 00:00:30 in PIN-OK | 00:00:030 selectingtransaction |
| 00:00:40 in withdrawal | 00:00:040 withdrawing |
| 00:00:50 in choose_account_type | 00:00:050 inputingamount |
| 00:00:60 in choose_amount | 00:00:060 confirming |
| 00:00:70 in confirm_no | 00:00:070 inputingamount |
| 00:00:80 in choose_amount | 00:00:080 confirming |
| 00:00:90 in confirm_yes | 00:00:090 withdrawaldone |
| 00:00:100 in withdraw_card ... | 00:00:100 idle ... |

**Fig. 7.3** Execution of the ATM machine model in CD++

In order to create an FSM we connect various *States*:

- All *transitionOut* and *transitionIn* ports should be connected together inside the FSM.
- All *eventIn* ports of *States* should be connected to an input port of a FSM.
- Each *stateOut* port of *State* could either be connected to an individual output port of a FSM or connected together as one output port of a FSM.
- All the states in a FSM are encoded as integer numbers (*stateCode*) during simulation. The state with the *stateCode* 0 is the initial state in the FSM.

After executing this model in CD++ with a set of external events, we obtain the results in Fig. 7.3. The left column shows the inputs, and the right column, the outputs obtained when we test it.

As we can see, initially we insert a card, which produces a *checkingpin* output. When we receive *PIN-OK,* we trigger an output in *selectingtransaction* (in this case, a *withdrawal*, which generates a *withdrawing* output). We then choose the account type and the amount. As we can see, the user does not confirm the amount (*confirm_no*), as they made a mistake. This is selected again at 00:80, after which we confirm, withdraw the card, and start the process all over again.

The library allows the users to define FSM and to use DEVS as the intermediate modeling and simulation language, obtaining an equivalent model that can be executed and combined with other models easily. For instance, it could be combined with a Petri Net model like the one described in the next subsection.

## 7.3 Petri Nets

Petri Nets (PN) originally defined by C.A Petri, is a modeling formalism developed to study concurrent systems (Peterson 1977). They became very popular as they can represent the models both using a formal mathematical notation which is adequate for formal proofs and software development, as well as a graphical representation, which is well suited for communication and analysis.

The static properties of PN are defined by bipartite graphs in which the nodes are called *Places* (represented graphically as bubbles) and *Transitions* (represented as bars), and the links in the graph connect places to transitions (and vice-versa). PN are executed by *firing* transitions that are *enabled*, one at a time, for as long as there is at least one enabled transition. This dynamic behavior is observed by using *tokens* that are added to places, and by the instantaneous firing of the transitions that are enabled to do so. A transition is enabled when they have at least one token on each input link to the transition. When a transition fires, a token is removed from each one of its input places and another token is deposited in each one of the output places. When more than one transition is enabled, the one that fires is selected in nondeterministic fashion.

Through the years, numerous extensions to PN have been introduced, allowing the modeler to define complex behavior (Liu and Zhang 2018). Our PN library in DEVS, allows the users to define PN and to run them using a DEVS tool like CD++. Our library supports some of these extensions.

- Inhibitor arcs: a special arc is placed between a place and a transition, and this new arc enables the transition only if the place is empty, as opposed to containing at least one token.
- Multiple Arcs: they indicate that the number of tokens being transferred when firing is more than one.
- Time: PN are *logical time* models, meaning that only the behavioral aspects are considered (deadlock, livelock, concurrent execution, etc.). Timed PN (TPN) introduced the concept of time, and therefore can be used for non-behavioral analysis (performance, throughput, timeliness, etc. (Bowman and Gomez 2006)). Timing can be represented as delays associated to the model execution.

We built a library of Petri Nets in CD++, whose main components are based on our FSM model in the previous section (Jacques and Wainer 2002). The library includes an atomic model to represent a place, and another to represent a transition. The Place atomic model can receive token inputs from transition models, and when a transition fires, we generate an input to tell the place to remove tokens (we support multiple arcs, and timed PN, as the DEVS message also includes a timestamp for the incoming tokens). We report the number of tokens the Place contains so transitions can determine if they are enabled (including TPN, in whose case we only consider enabled transitions if timestamps are equal to or lower than the current time). We consider two kinds of delays (Fig. 7.4):

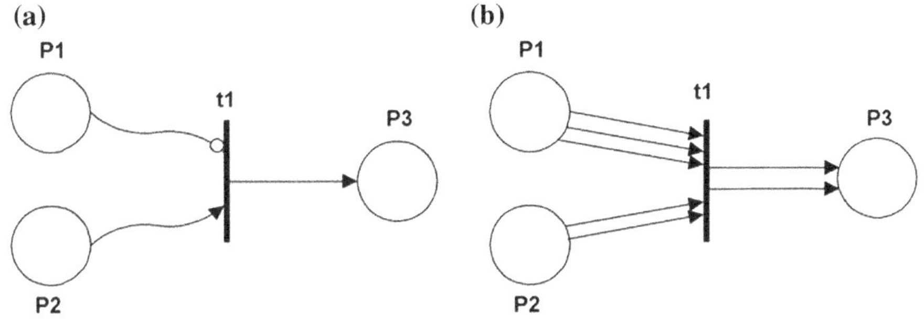


**Fig. 7.4** **a** Inhibitor Arc; **b** Multiple Arcs

- Delayed Transition Firing: these are special transitions where the firing condition must be maintained for a certain time before the actual firing occurs.
- Timestamped Tokens: we can use timestamps for the tokens in a place. The firing condition needs enabled tokens whose timestamp is equal to or lower than the current time. When a transition fires, the output tokens should ALSO include a timestamp.

The TPN Place atomic model can be defined as:

$$TPN\_Place = < X, S, Y, \delta_{int}, \delta_{ext}, ta, \lambda >$$

$X = \{IN \in \mathbf{N}^{+} x\ \mathbf{R}\}$
$Y = \{OUT \in \mathbf{N}^{+}\}$
$S = \{\{tokens \in \mathbf{bag(R)}\} \cup \{id \in \mathbf{N}^{+}\} \cup \{phase \in \{active, passive\} \cup \{id \in \mathbf{R_0}^{+}\}$
$\}$ where *tokens* is the list of tokens with their timestamps contained in the place, and *id* is the identifier of the place as assigned by the simulator.

$\delta_{ext}$ *( (number of tokens N, id, timestamp T)* $\in S$*, e, x* $\in X$*) {*
   ***case** id*
      ***0:***   *// generic message*
       ***add** 'N' tokens with timestamp 'T' to tokens;*
       ***k** = max(0 , min_timestamp(tokens) – e );*
       ***ta** (active, k);*

      ***!= 0:*** *// specific message*
         *id matches id of this place?*
           *no:*     *discard the message*
      *yes: **if** there are enough enabled tokens*
             *decrement tokens by the number specified*
        ***ta**(active, 0)*
*} **// end of*** $\delta_{ext}$

$\delta_{int}$ ***(****(number of tokens N, id, timestamp T)* $\in$ *S**)** {*
   ***if** there are tokens with timestamp larger than the current time*
      *no: passivate*
      *yes: ta (active, min_timestamp(tokens) – e)*
*} } // end of* $\delta_{int}$

$\lambda$ ***(****(number of tokens N, id, timestamp T)* $\in$ *S**)** {*
  ***send** ( (id, number of tokens), out);  }*

The external transition function receives new tokens with timestamps, and adds them to the list of tokens, or decrement the number of tokens (only the enabled ones based on their timestamp) to fire a transition. After this, an internal transition is scheduled; an output combining *id* and *tokens* state variables is generated and transmitted on the **OUT** port. The number of advertised tokens is that of enabled tokens, whose timestamp is equal to or lower than the current simulation time. The internal transition function keeps scheduling events until all tokens have been enabled.

The TPN transition atomic model is defined as:

$$TPN\_transition = < X, S, Y, \delta_{int}, \delta_{ext}, ta, \lambda >$$

$X = \{IN0, IN1, IN2, IN3, IN4 \in \mathbf{N}^{+}\}$

$Y = \{OUT1 \in \mathbf{1xR}, OUT2 \in \mathbf{2xR}, OUT3 \in \mathbf{3xR}, OUT4 \in \mathbf{4xR}, FIRED \in \mathbf{N}^{+}\}$

$S = \{\{inputs \in N\} \cup \{enabled \in bool\} \cup \{active \in bool\}\}.$

```
δext (s,e,x) {
   case port
        IN0: arc width = 0;        IN1: arc width = 1;
        IN2: arc width = 2;        IN3: arc width = 3;
        IN4: arc width = 4;
   extract the id of the place sending the message;
   if this is the first message we get from this id, inputs++;
   save (id, arc width);
  extract number of tokens in the place that sent the message and save it;
  if all input places have enough tokens to enable the transition
 if transition is enabled enabled = true
            if transition is active
               ta(active, nextChange)
           else
               active = true;
               ta(active, DELAY_FIRE);
else
      enabled = active = false
      passivate
} end of δext


δint (s) {
      if inputs = 0 // transition is a source,
         ta (active, random() ).
      else
         passivate
      active = false
}


λ (s) {
      firing_time = now + DELAY_TOKEN;
      send ((1, firing_time), OUT1);       send ((2, firing_time), OUT2);
      send ((3, firing_time), OUT3);       send ((4, firing_time), OUT4);
      send a message to every input place via the FIRED port;
} // end of λ (s)
```

The model uses a number of input and output ports:

- **IN1**: it is used to receive the number of enabled tokens in the places connected to this input. Places that connect to this port have a single connecting arc; if the transition fires, only one token will be removed from the input places.
- **IN2-4** are used for multiple arcs (2 to 4 arcs).
- **IN0**: inhibitor arc. The input place must contain zero tokens (including those enabled by time for TPN) for the transition to fire and when it does, no token is removed from the place.
- **OUT1-4** are used to transmit 1–4 tokens to the corresponding ports.
- **FIRED** is used to remove tokens from the input places which must have their **IN** port connected to this output port in addition to being connected to one of the input ports.

The *inputs* state variable contains the number of input places for the transition, and *enabled* indicates if the transition is enabled or not. The model uses the DELAY_TOKEN parameter to determine delayed firing (i.e., the tokens will have a timestamp which is the current time plus DELAY_TOKEN). After being enabled DELAY_FIRE time, the transition actually fires.

One factor that is complex when simulating TPN in a DEVS simulator like CD++ is that when two or more transitions are enabled, one must be chosen and fired in a non-deterministic fashion. This implies that a controlling agent, aware of the state of all transitions in the model, would be required to determine which one should fire. In order to do this, the transition model schedules its own firing including a random amount of time after the transition is enabled. Given that all transitions do the same, this result in a near non-deterministic decision process.

When simulating the TPN results, we need to see the evolution of the marking, as seen in the following figure. The first two lines list the name of the places and transitions that make up the model. Then, we show the initial marking of the TPN counting the number of enabled tokens. We see transition *t2* firing at time 00:00:05, which makes places *p2* and *p3* update their token count. Then 2 seconds later, *p1* and *p4* do the same (as this is a TPN) producing the marking (1, 4, 4, 1).

```
Petri Net places: p1 p2 p3 p4
Petri Net transitions: t1 t2
[00:00:000]        p1: 0 enabled tokens  p2: 5 enabled tokens
                   p3: 5 enabled tokens  p4: 0 enabled tokens
   (0,5,5,0)
       |
t2 [00:00:05:000]
       |
       V
[00:00:05:000] p2: 4 enabled tokens    p3: 4 enabled tokens
[00:00:07:000] p1: 1 enabled token     p4: 1 enabled token

   (1,4,4,1)...
```

Let us assume now that we want to study the behavior of an elevator. The elevator starts at the first floor, and it waits for people. Once people are inside, it closes the doors and it moves to the second floor, where opens its doors, waits for people leaving the elevator, closes the doors, and it returns to the first floor. A part of the TPN of this elevator model can be seen in Fig. 7.5. All places use time-stamped tokens. For instance, the *PeopleWaiting* place, each token represents one person, and their timestamp represent their time of arrival to the elevator. In the transition T1, the DELAY_FIRE parameter represents the time that the doors remain open if the button is not pressed and no one is entering the elevator. The DELAY_TOKEN parameter represents how long it takes to close the door. In T3, DELAY_TOKEN represents how long it takes to open the door. In T6, DELAY_FIRE models the time it takes to enter the elevator. In T7, DELAY_FIRE is how long it takes to start moving after the doors are closed. Finally, in T8 DELAY_FIRE represents how long it takes to reach the second floor.

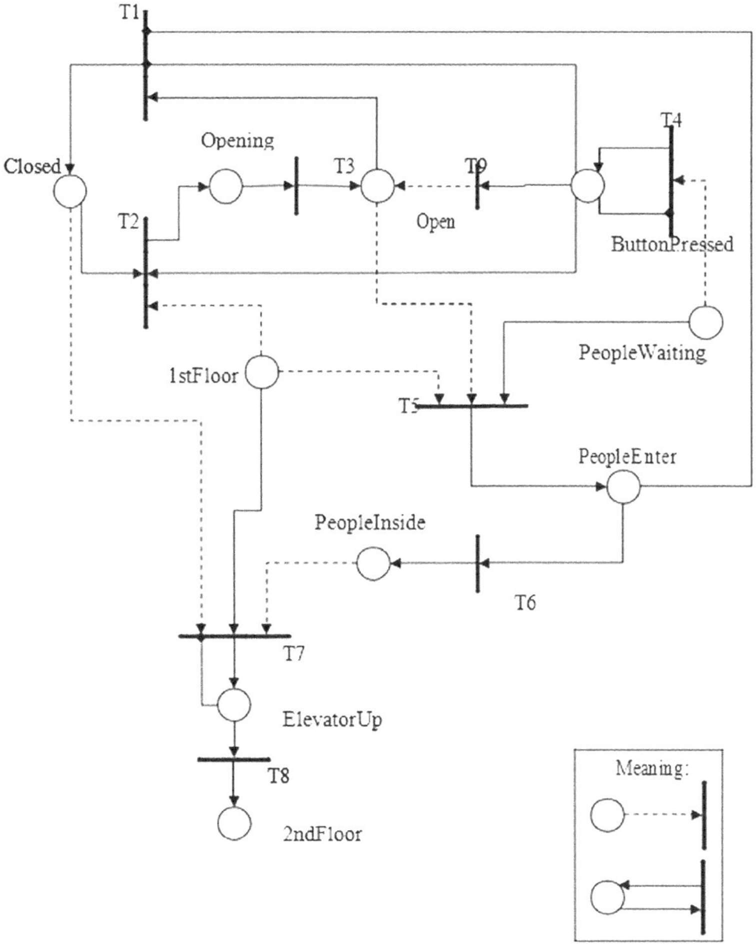


**Fig. 7.5** A TPN for modeling an elevator

The following test case shows a simulation run in which *PeopleWaiting* initially holds 3 tokens with timestamps 00:00:05, 00:00:10 and 00:00:15, representing the arrival of three persons at those times.

```
Petri Net places: Closed Opening Open ButtonPressed
                  PeopleWaiting PeopleEnter PeopleInside
                  PeopleArrived 1stFloor ElevatorUp
                  2ndFloor leavingElevator
Petri Net transitions: t1 t2 t3 t4 t5 t6 t7 t8 t9 t10
                       t11 t12 t13 t14

[00:00:000]     Closed: 1 enabled  Opening, Open,
                PeopleWaiting, PeopleEnter, ElevatorUp,
                PeopleInside, PeopleArrived: 0 enabled
                1stFloor: 1 enabled
                2ndFloor, leavingElevator: 0 enabled
[00:00:05:000]  PeopleWaiting: 1 enabled

(1,0,0,0,1,0,0,0,1,0,0,0)
            |
            t4 [00:00:05:017]
            |
            V
[00:00:05:017]  ButtonPressed, PeopleWaiting: 1 enabled

(1,0,0,1,1,0,0,0,1,0,0,0)
        |
        t2 [00:00:05:031]
        |
        V
[00:00:05:031]  Closed: 0 enabled ; Opening: 1 enabled
                ButtonPressed: 0 enabled ; 1stFloor: 1 enabled
(0,1,0,0,1,0,0,0,1,0,0,0)
        |
        t3 [00:00:05:047]
        |
        V
[00:00:05:047] Opening: 0 enabled


...

[00:00:49:044] ElevatorUp, Closed: 0 enabled; Opening: 1 enabled;
                 PeopleInside: 3 enabled; 2ndFloor: 1 enabled
(0,1,0,0,0,0,3,0,0,0,1,0)
       |
       t3 [00:00:49:060]
       |
       V
[00:00:49:060] Opening: 0 enabled ; Open: 1 enabled
(0,0,1,0,0,0,3,0,0,0,1,0)
       |
       t10 [00:00:54:083]
       |
       V
[00:00:54:083] Open: 0 enabled ; PeopleInside: 2 enabled ;
                     PeopleArrived: 1 enabled
...
[00:01:00:153] Open, PeopleInside: 0 enabled ; 2ndFloor: 1 enabled
```

As we can see, initially the elevator is in Floor 1, and the door is closed. No transition is enabled. However, at 05:000, the first person arrives, enabling one token in the *PeopleWaiting* place. Transition *t4* is enabled, and it fires, representing that the elevator button has been pressed (and it takes 17ms to react). Transaction *t2* is enabled 14ms after that, which fires, triggering the opening of the door, and disabling the button pressed. At 05:047, *t3* is enabled, and now the door is open. A while after the three individuals have arrived, the elevator has moved to the second floor. We can see that at 49:044 the elevator is in the second floor, it has stopped moving up and there are three individuals waiting to leave the elevator. After 14 ms, the door is enabled to open, and it opens, allowing the individuals to leave the elevator. They leave one by one, until 01:00:153, where the elevator is empty in the second floor.

As we can see, TPN can be simulated using DEVS atomic models for PN transitions and places with an unmodified DEVS simulator, including a method in the internal transition function enabling timed Petri Nets. The DEVS specification allows transforming the methods with ease, and building user libraries for developing multimodels. In the next section, we discuss an extension for more advanced models based on Timed Automata.

## 7.4 Timed Automata and DEVS

New theoretical advances in model checking have allowed guaranteeing properties about models of real world systems using a formal approach. Model checking techniques can be automated, and Timed Automata (TA) theory (Alur and Dill 1994), in particular, has provided many practical results in this area. However, these formal methods are difficult to apply, and in many cases, they do not scale up well. Instead, using Modeling and Simulation (M&S) to gain confidence about the model correctness can be used to improve the study of experimental conditions during model definition, experimenting with virtual sys-tems, explore options, including those cases where testing under actual operating conditions may be impractical. Nevertheless, no practical, automated approach exists to perform the transition that exists between the modeling and the development phases, and this often results in initial models being abandoned. Simultaneously, M&S frameworks are not as robust as their formal counterparts are. If the models used for M&S are formal, their correctness would also be verifiable, and a designer could see the system evolution and its inner workings even before starting a simulation (Saadawi and Wainer 2010). In order to deal with these issues, we showed a mechanism to represent TA with DEVS, and extended it to model hybrid systems using QSS methods, hence enabling formal verification of hybrid models within DEVS formalism.

We showed that, in order to be able to transform a DEVS model to a TA, we need that:

(a) the TA variables bounded integers, in order to guarantee the finiteness of state space and hence the termination of the reachability algorithm (nevertheless, for QSS, state variables are real numbers), and
(b) the time of the next event be approximated to an integer number (in doing so we need to preserve the original behavior of QSS).

The first issue was handled by converting rational real numbers to integers by multiplying all values by the least common multiple of all the denominators. For any irrational values, we introduced a new method in (Saadawi and Wainer 2010). For the second issue, we use abstraction by over-approximation. With this technique, we approximate the real value of the event time $t_i$ with a bounded time interval such that $t_c \in [T_L, T_H]$. This interval is bounded by floor($t_i$) and ceiling($t_i$) respectively. To obtain a TA that contains the behavior of a QSS model, we need a simulation relation with the QSS model (i.e., we need to show that the TA simulates QSS). To do so, each state in QSS would be simulated by a corresponding state in the TA, and each target state in QSS simulated by a corresponding target state in the TA. We will show these aspects using an example for an elevator controller originally introduced in (Saadawi and Wainer 2010). A summary of this case study is given below (Fig. 7.6).

In this model, whenever the elevator receives a command to stop, it synchronizes with a braking elevator motion model (*applyBrake!*). The elevator waits in state

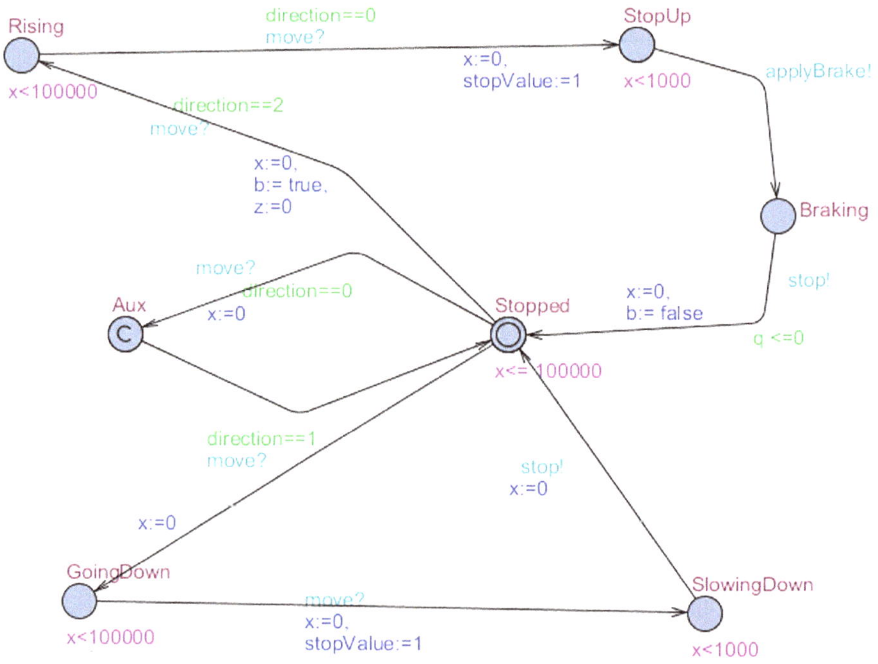


**Fig. 7.6** Elevator TA model

*Braking* for the quantized speed $q$ value to reach zero. Once the elevator speed reaches zero, the transition from *Braking* to *Stopped* would be enabled and executed, then the elevator sends *stop!* to the elevator-controller. This hybrid model allows the designer to verify the control system with different parameters of the elevator physical system such as different braking values of de-accelerations, different elevator initial speeds, or other parameters in a more detailed QSS model. This is an important addition to the elevator system verification as relevant physical factors to the controller performance can be identified and formally verified during design phase. The elevator de-acceleration motion and its speed are described by:

$$\frac{dv}{dt} = a \qquad v = at + v_i$$

where $v$ is the elevator speed, $a$ is the acceleration constant, and the speed is $v$ at any point in time $t$, with $v_i$ the initial elevator speed before applying the brakes.

To simulate and verify this hybrid model, we obtained a discrete representation of the elevator braking model, and we employed DEVS and QSS. We use a quantized variable $q$ related to $v(t)$ system variable by quantization. To enable the formal verification, we transformed the QSS model to an equivalent TA model, resulting in the TA shown in Fig. 7.7.

Here, *S1* represents the initial elevator speed (4 m/s), the quantum value is $dQ$ = 0.5 m/s, and *sigma* is the time interval between the outputs of two successive quantized values. When this model receives a synchronization input event (?) on the *applyBrake* channel, it changes to the state *S2* and starts a loop *S2*−*S3*−*S2*… in which we calculate the next quantized output $q$ and the next values of *sigmaL* and *sigmaH*. When the quantized speed $q$ reaches zero, the model moves back to *S1* and waits for another *applyBrake* event. As discussed earlier, sigma (the time advance) is over-approximated with an integer interval $\sigma \in [sigmaL, sigmaH]$.

Figure 7.8 shows the execution of the simulated hybrid DEVS model with braking de-acceleration equals −0.12 m/s$^2$. In this case, the time needed for the

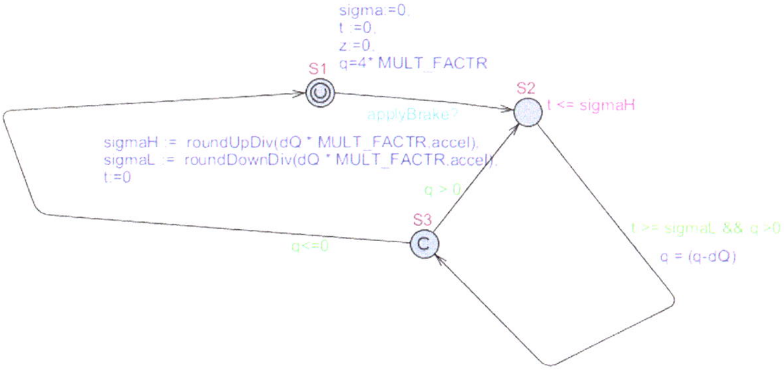


**Fig. 7.7** TA model of braking elevator motion

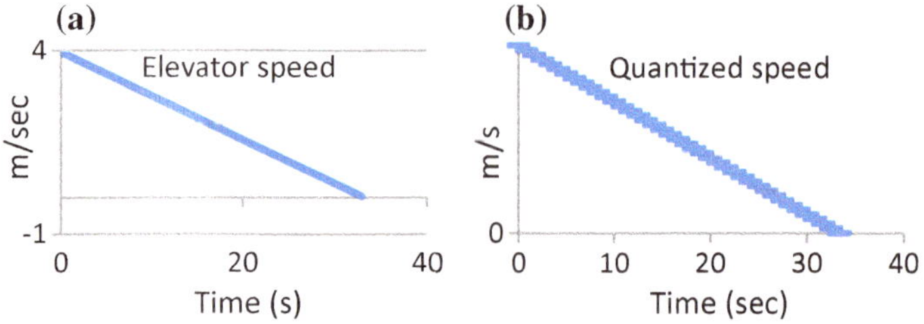


**Fig. 7.8** **a** Elevator speed; **b** Quantized speed, acceleration = −0.12 m/s2

elevator to stop is approximately 33 seconds. This would contradict the user requirements, as the user expects the elevator to reach third floor within 27 seconds at most, and after this time the requirement for the elevator controller to be ready to accept another as shown on the transition *S5* → *S6*. However, the slow-braking elevator would not be able to fulfill the second request in time, hence we have a time lock and the model cannot progress beyond *S5*.

It is important to count with advanced methods for modeling hybrid systems (where continuous and discrete phenomena interact), as they are found in many natural and artificial systems. Methods like those that we presented in this section can include verification of cyber-physical systems, which usually include discrete-event controllers interacting with a continuous plant. The combination of RTA-DEVS, hybrid Timed Automata and QSS allows verifying real-time hybrid systems modeled by DEVS. We showed a methodology to verify hybrid DEVS models. Some limitations, however, for this method of over approximation is that for systems described with nonlinear derivatives, it can lead to a wide flow pipe around the actual system trajectory. Other limitation is the inherit problem with model checking technique of state-space explosion that limits the ability to scale verification to larger models. In the next section, we show how to model these kind of hybrid systems using the Bond Graphs formalism and its transformation to DEVS.

## 7.5 Bond Graphs

The Bond Graphs formalism (BG) is a mathematical modelling method that focuses on the representation of continuous dynamic systems that can be described hierarchically (Karnopp et al. 1990; Vila and Rico 2018). BG represents the physical systems as a directed graph with hierarchical components, and its basic theory is based on the energy conservation law and the use of a lumped approach, separating dynamic system properties from each other, using submodels, and then linking those using ideal connectors. These connectors represent energy flow, and, it is assumed they have power continuity, and that no energy is generated or dissipated

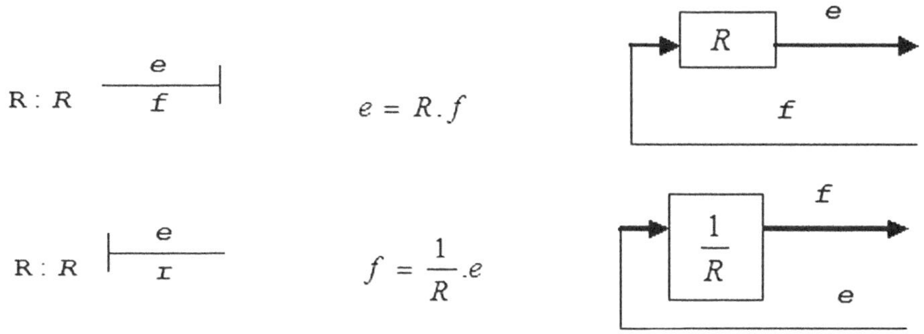


**Fig. 7.9** Top: R element with **flow-in causality**, equations and block diagram representation; bottom: R element with **effort-in causality**, equations and block diagram

in the ideal connections. The physical systems they represent are modeled using directed graphs whose nodes define the physical processes and whose links (bonds) represent the ideal exchange of energy between them. Energy (or its time derivative, power), is the fundamental exchange between elements of the system. Power is the product of *flow* and *effort* (and they have no specific semantics; it can be used in translation mechanics, as *force* and *velocity*; in electrical systems, as *voltage* and *current*, etc.). The energy flow is represented elements exchanging effort and flow through the bonds. In order to represent the exchange of power between elements, we need to show the flows between components and their causality, as no component can determine the two power variables (effort and flow) at the same time. Given a pair of elements connected through a bond, their causality determines which causes flow and which causes effort.

There are a number of BG elements, including: *Capacitor* (*C*), *Inductor* (*I*), *Resistor (R), Effort source* (Se), *Flow source* (*Sf*), *Transformer (TF), Gyrator (GY), 1-junction, 0-juction.* To illustrate how they are defined, let us consider the *R* elements, which can represent resistors in the electrical domain, dampers the in mechanical context, etc. Their constitutive equation is defined by an algebraic equation relating *flow* and *effort*: $e = r.\ (f)$. The electrical resistor is mostly linear, and the corresponding equation is $u=R.i$, where $R$ is the resistance's constant (Fig. 7.9).

Based on BG concepts, we defined a discrete event library for Bond Graphs using DEVS (D'Abreu and Wainer 2006), in which we define Quantized BG (QBG), that is, a BG where all the storages and sources are quantized elements. This method combines BG and DEVS models with QSS by adding quantizers equipped with hysteresis to the integrators output. The library consists of a number of atomic DEVS models developed on CD++, which implement QSS and QBG. The following code snippet shows a part of the quantizer model equipped with hysteresis implemented in CD++.

```
External transition
if ( msg.port() == in ) {
 if ( state() == passive ) {
   double inputValue = msg.value();     // gets message value
   currQValue = hquantize( inputValue); // applies QSS
   holdIn(active, Time::Zero); // schedule instantaneous transition }
}
Internal transition
  passivate();   //waits the reception of external messages

Output function
  if ( firstValue || ( currQValue != lastQValue ) ) {
      firstValue = false;
      lastQValue = currQValue;
      sendOutput( msg.time(), out, currQValue ) ;  }
// If first value received or boundary crossed, Sends quantized value

hquantize function
  if ( firstValue ) return QMethod->quantize( value );

  if ( ( value > lastQValue ) || ( value <= lastQValue - ∊ ) ) {
     return QMethod->quantize( value );  // hysteresis window size ∊
  }
  return lastQValue;
```

The model receives external inputs, and then computes the quantization function (with hysteresis) of the value received. Then, transmits the calculated value only if a boundary is crossed. The model parameters included the *Quantization method* (we support *uniform* and *intervals* quantization), their *parameters* (quantum size, lower and upper saturation values), the *intervals quantization method* (array of intervals and length of array of intervals, and the *hysteresis window size.*

A coupled DEVS representing a QBG model can be formally defined as:

$$CQBG = <X_{self}, Y_{self}, D, \{M_i\}, \{IC\}, select>$$

$X_{self} = \{\varnothing\}$ (no external inputs)
$Y_{self} = \{\varnothing\}$ (no external outputs)
$D$ is the set of elements representing BG components, and for each i in D,
$M_i$ is a DEVS atomic model representing a QBG component
$IC$ is the internal coupling set defined as: $IC = \{ice_{ui,\ vj}\} \cup \{icf_{vj,\ ui}\}$ where $ice_{ui,\ vj}$ and $icf_{vj,\ ui}$ represent the coupling between effort and flow ports on $u$ and $v$, being the effort calculated by element $u$.

$$ice_{ui,vj} = \begin{cases} ((u, out_{ei}), (v, in_{ej})) & \textit{if v is not a source (flow source)} \\ \varphi & \textit{otherwise} \end{cases}$$

*if u is a serial junction then i = 1 (only one effort-out port)*

$$icf_{vj,ui} = \begin{cases} ((v, out_{fj}), (u, \text{in}_{fi})) & \textit{if u is not a source (effort source)} \\ \varphi & \textit{otherwise} \end{cases}$$

*if v is a parallel junction then j = 1 (only one flow-out port)*

*select* gives priority to structural components (junctions, transformer, gyrator).

The library includes the following models: *QBGCapacitorFlowIn, QBGInductor EffortIn, QBGResistanceFlowIn, QBGResistanceEffortIn, QBGSourceEffort_Constant, QBGSourceEffort_Step, QBGSourceEffort_Sine, QBGSourceEffort_Pulse, QBGSourceFlow_Constant, QBGSourceFlow_Step, QBGSourceFlow_Sine, QBGSourceFlow_Pulse, QBGTransformer, QBGGyratorFlowIn, QBGGyrator EffortIn, QBGSerialJunction, QBGParallelJunction.*

In order to show how to use the library, we define a mechanical model of the response of mass against the application of effort, shown in Fig. 7.10.

Following, we show two cases of execution of the model, in which we use different frequencies. Figure 7.12 the evolution of mass speed over time. It can be seen how speed is modified every time that effort is imposed over M and how it tries to return to its original value. We use I = 40; C = 2; SE: signal = pulse, period = 200ms, pulse duration = 2ms, and different resistance. We can see that the variation of resistance produces changes in the number and speed of oscillations, as expected (Fig. 7.11).

Using our BG library, we then conducted an extended effort, building the first open source Modelica compiler, which was implemented on top of the QBG library, and will be described in detail in the following section.

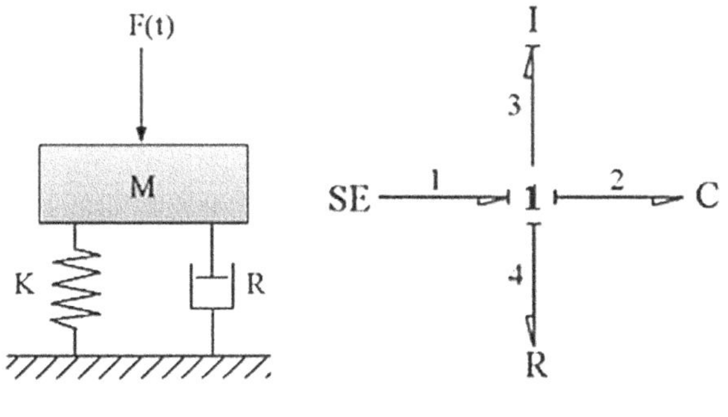


**Fig. 7.10** Mechanical circuit and its Bond-Graph representation

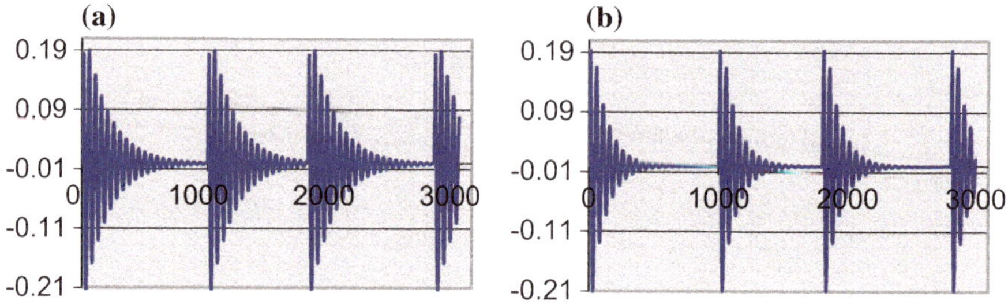


**Fig. 7.11** Mass speed; **a** R = 1.5; **b** R = 3

## 7.6 Modelica

Modelica is an object-oriented language, defined for modeling physical systems and built to support library development and model exchange. Modelica models are described using differential, algebraic and discrete equations, and there are various libraries of standard components using ODEs, block diagrams, electrical and mechanical formalisms (Fritzson 2004).

We defined a compiler that understands models from the *electrical library* in Modelica (D'Abreu and Wainer 2006). A source file in Modelica is compiled into an equivalent Bond Graph representation of the circuit. The generated BG constitutes the output, which is used for simulation. The BG generation algorithm is based on Karnopp's circuit construction method (Karnopp et al 1990). The approach is to build a BG that resembles the circuit structurally, and then to simplify it based on selected properties, as follows:

- `For each node with a distinct potential add a 0-junction:` we used transitive closure applied to every node on the graph (we check for paths between arbitrary nodes *x* and *y*, given only adjacency information). As there are different ways to specify the parallel coupling between elements in the circuit, we need to ensure that the *0-junction* elements are correctly inserted (independently from the definitions in the Modelica file).
- `Insert each 1-port circuit element by adjoining it to a 1-junction, inserting 1-junctions between the appropriate pair of 0-junctions (C, I, R, Se, Sf elements):` we add a *1-junction* to each *1-port* element, inserting it between the corresponding pair of *0-junctions*.
- `Assign power directions to all bonds:` a standard convention assumes positive direction of power when it flows out of sources (*Se* and *Sf*) and into *C*, *I* and *R* elements. For two-port elements, *TF* and *GY*, we consider that power flows into the elements. We use a power propagation algorithm that traverses the graph and assigns power using the standard conventions and the information in the Modelica file. At the end of this step we obtain a directed BG is obtained.

- `Erasing ground potential:` all the explicit ground potentials are deleted from the graph; if there is no explicit ground potential, we delete the 0-junction nearest to each source element. The 0-junctions selected are only those associated with the negative pin of every source's port.
- `Simplification:` a junction between two bonds with through power direction can be deleted; likewise, a bond connecting two junctions of the same type can be deleted and the junctions joined.

In the final BG, we check for algebraic loops and singularities (elements that have discontinuities e.g. diode), and we generate an optimized QBG corresponding to the electrical circuit, which is used to generate a coupled DEVS model in CD++.

```
model circuit
  Modelica.Electrical.Analog.Sources.PulseVoltage
                        V(V=10,width=50,period=2);
  Modelica.Electrical.Analog.Basic.Resistor R1(R=10);
  Modelica.Electrical.Analog.Basic.Capacitor C(C=50);
  Modelica.Electrical.Analog.Basic.Ground Gnd;
equation
  connect(V.p, R1.p);            connect(R1.n, C.p);
  connect(C.n, V.n);             connect(V.n, Gnd.p);
end circuit;
```

(a)

```
[top]
components : $SJ1@QBGSerialJunction C@QBGCapacitorFlowIn
               R1@QBGResistanceEffortIn V@QBGSourceEffort_Pulse

link : e1n@$SJ1 e1p@R1         link : f1p@R1  f1n@$SJ1
link : f2n@$SJ1 f1p@C          link : e1p@C   e2n@$SJ1
link : f3p@$SJ1 f1n@V          link : e1n@V   e3p@$SJ1

[C]
quantum : 0.1  hystWindow : 0.01     C :     50     initialLoad :  0

[R1]
R :      10

[V]
quantum : 0.1    hystWindow : 0.01    signal : Pulse    offset : 0
startTime : 0    amplitude : 10       period : 2        width : 50
```

(b)

Figure 7.12 shows a Modelica model of a circuit, first, and the translation to QBG after. As we can see, we generate a coupled model for each of the Modelica components. The corresponding QBG are built for each of the components, and, in this case, they are connected by the Serial Junction SJ1 (which is automatically generated by the compiler using the procedure explained above). The equations are

**(a)**

```
model circuit
  Modelica.Electrical.Analog.Sources.PulseVoltage
                     V(V=10,width=50,period=2);
  Modelica.Electrical.Analog.Basic.Resistor R1(R=10);
  Modelica.Electrical.Analog.Basic.Capacitor C(C=50);
  Modelica.Electrical.Analog.Basic.Ground Gnd;
equation
  connect(V.p, R1.p);          connect(R1.n, C.p);
  connect(C.n, V.n);           connect(V.n, Gnd.p);
end circuit;
```

**(b)**

```
[top]
components : $SJ1@QBGSerialJunction C@QBGCapacitorFlowIn
               R1@QBGResistanceEffortIn V@QBGSourceEffort_Pulse

link : e1n@$SJ1 e1p@R1        link : f1p@R1   f1n@$SJ1
link : f2n@$SJ1 f1p@C         link : e1p@C    e2n@$SJ1
link : f3p@$SJ1 f1n@V         link : e1n@V    e3p@$SJ1

[C]
quantum : 0.1  hystWindow : 0.01      C :     50     initialLoad :  0

[R1]
R :      10

[V]
quantum : 0.1    hystWindow : 0.01    signal : Pulse    offset : 0
startTime : 0    amplitude : 10       period : 2        width : 50
```

**(c)**

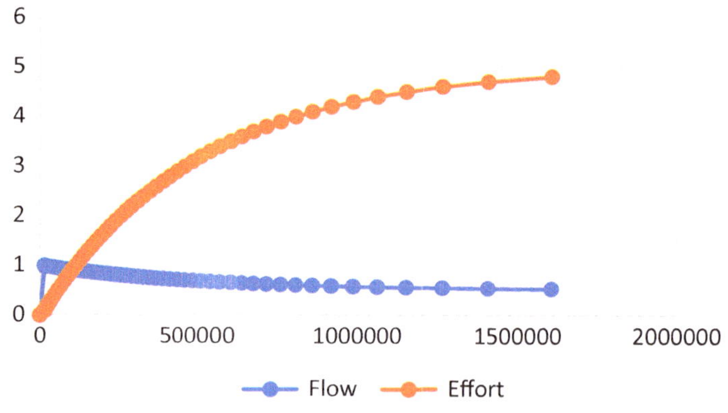


**Fig. 7.12** **a** Modelica model **b** QBG translation **c** Model execution

converted into couplings in the DEVS coupled model. After the coupled model definition, we can see the initialization values for each of the components, which also follow the Modelica specification. The Capacitor and the Pulse Voltage generator are approximated using QSS (the quantum size and the hysteresis window parameters can be adjusted for simulation). Finally, we show outputs in both Flow and Effort ports for Capacitor C, which matches the expected behavior of the circuit.

## 7.7 VHDL

Digital designers have often relied on M&S for the design of circuits, as simulated studies allow reducing the number of design bugs and integration errors, while easing product maintenance and reducing the overall cost. Design and simulation of digital logic with HDLs (Hardware Descriptor Languages) is a well-proven methodology. Mixed signal HDL simulators allow combining discrete time digital and continuous time analog models. Here, we discuss a method for simulating mixed signal HDLs by converting the designs to DEVS (Mehta and Wainer 2005). The discrete-event nature of DEVS is well suited to model digital logic, and signal quantization with QSS allows modeling the continuous systems components.

VHDL-AMS is targeted toward register transfer level modeling of digital circuits with limited behavioral modeling and analog constructs. The main construct used by the language is the *Entity*, which describes the interface to a VHDL-AMS design or design unit. The entity declaration contains a list of *ports*, each of which is assigned a *type* and an optional *mode*. Ports of type *std_logic* or *std_logic_vector* (a standardized type for digital logic) are used for digital signals while electrical ports are used for analog signals. In the case of digital signals, ports will have mode *in*, *out*, *inout* or *buffer*. Analog ports do not require a mode. The syntax of an entity declaration is as follows:

```
entity  entity_name is
     port ( [signal |  terminal | quantity] identifier {,
            identifier}: [mode | ] signal_type | electrical
        {; [signal | terminal | quantity ] identifier {,
            identifier}: [mode | ] signal_type | electrical});
         ) ;
end [entity] [entity_name] ;
```

For instance, Fig. 7.13 shows the declaration for an analog low pass filter entity.

```
entity LPF is
   port (
   terminal tout, tin,
        tgnd: electrical
   );
end entity LFP;

architecture top of LPF is
  signal clk : std_logic;
  signal vin : std_logic;
  quantity vout across tout to tgnd;

begin
  vout'dot = (1/(R*C))*(vin-vout);
```

```
clk: entity clk
       port map (clk=>clk);
       vin<=clk;
end architecture top;
```

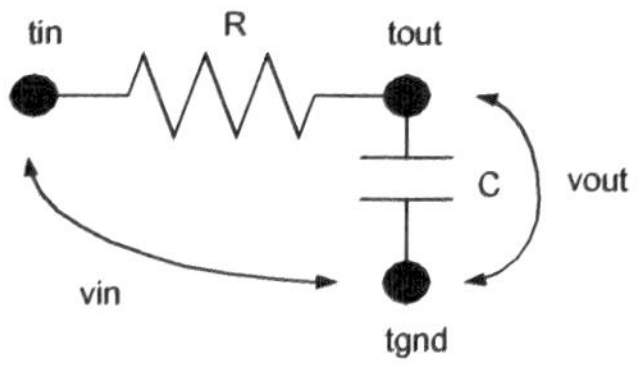


**Fig. 7.13** Low-pass filter

A design *architecture* describes the functionality of a design or design unit; this may be a structural, dataflow or behavioral description. A single architecture is associated with exactly one entity.

```
architecture architecture_name of entity_name is
     signal_declaration
     | constant_declaration
     | component_declaration
begin
     {process_statement
     | concurrent_signal_assignment_statement
     | component_instantiation_statement
     | simultaneous_statement}
end [architecture] [architecture_name] ;
```

The body of an architecture consists of statements that may be categorized as concurrent, sequential or simultaneous. These statements operate on signals and quantities that are declared within the scope of the architecture, and on ports that are declared in the entity.

```
  signal   signal_name   :   std_logic_vector   (upper_bound   downto
lower_bound) | std_logic ;
  quantity identifier : REAL | Voltage | Current | Charge ;
  quantity identifier {, identifier} across identifier {, identifier}
through free_terminal to reference_terminal ;
```

Signals and quantities are defined in the declarative region. These belong to the scope of the architecture in which they are declared and may only be referenced within that architecture. Quantities may also be declared relative to terminals in an entity declaration. These may be either *across* or *through* quantities. Across quantities, represent the voltage at the free terminal relative to the reference terminal. Through quantities represent the current from the free terminal into the reference terminal. *Process*, *Simultaneous*, *Concurrent Assignment* and *Conditional Concurrent Assignment* Statements execute concurrently within an architecture. The conditional concurrent assignment statement assigns the target signal the value of an expression if the condition is true and the value of a different expression otherwise. A *process* executes the statements between begin and end process when an event occurs on a signal in its sensitivity list. No signals modified by the process are updated until the process body is completed. The statements between *begin* and *end* clauses are referred to as *sequential statements*, and they are executed in sequence.

```
[process_name:]
process (sensitivity_list)
   { type_declaration      }
begin
  {signal_assignment_statement
  | if_statement
  | case_statement
  end process [process_name] ;
```

The *if* statement has identical semantics to that of an if-then-else statement in C/C++; the *case-when* statement runs the sequence of statements that are listed under the when clause whose expression matches that of the expression in the case statement. Simultaneous statements are generally used for describing Differential Algebraic Equations, and may consist of quantities or signal, for instance:

```
x1'dot'dot == -f*(x1 – x2) / m1;
x2'dot'dot == -f*(x2 – x1) / m2;
...
```

The *'dot* notation denotes the derivative with respect to time of the quantity listed before the *'dot*. For example *signal'dot* is the first derivative with respect to time of *signal*, while *signal'dot'dot* is the second derivative.

Figure 7.14 shows the software architecture of the tools used to build VHDL-AMS models. We start with a syntax check phase, in order to ensure that the VHDL-AMS model is syntactically correct. During the elaboration phase, each component description is assigned to a structure in the VHDL-AMS design hierarchy. The description of the architecture and entity for each component in the design is parsed, and a Netlist is produced, which includes interconnected integrators, algebraic operators, processes, signals and sub-component instances. We then generate DEVS models in CD++, using a library for each components of the design and a coupled model to define the architecture. The CD++ models are then compiled, after which the Netlist and model library are used by the model file generation, after which we obtain a DEVS coupled that can be used for simulation. In order to convert VHDL-AMS models to CD++ coupled models (done during the Model Code and Netlist Generation phases above) we first need to identify the components that constitute the design hierarchy. Basic components do not contain

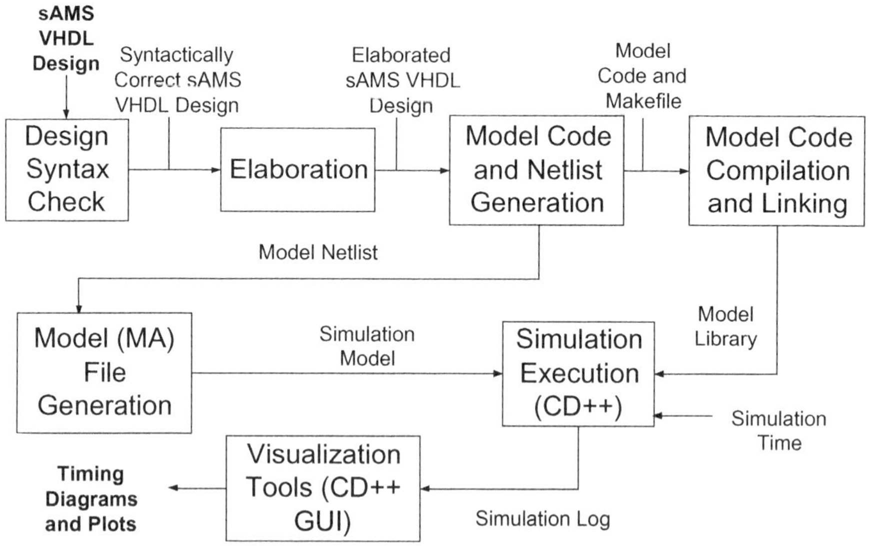


**Fig. 7.14** Architecture of the VHDL to DEVS compiler

sub-component instances in their architectures, while aggregate components do. We generate a dependency tree must in which the leaves are basic components, while branches are aggregate components, and the root is the top-level model. VHDL-AMS sub-component instances are connected to the architecture in which they are instantiated as defined by the port map clause in their component instantiation statement.

The process body is implemented within the external transition function in CD++. Buffering is also done within the external transition function for each signal referenced in a rising_edge or falling_edge operation. Buffers are also created for each output port on the model in CD++. The output ports on the model represent all the signals that are driven from within the VHDL-AMS process. The values that are assigned to these buffers will be output on their respective ports when the model schedules an output event. Expressions and case statements are converted directly to equivalent ones in C++. The CD++ signal model is used to implement transport delay on messages sent between process model ports. The implementation of the process model in DEVS does not allow assignment statements to have transport delays, since the output events for all driven signals must occur simultaneously. Therefore, we implemented transport delays in the CD++ signal model, which receives and buffers data on its input port, enters the active state for the time specified by the assignment statement transport delay, and outputs the buffered data on its output port.

As discussed earlier, simultaneous statements in VHDL-AMS allow the definition of continuous time systems through differential algebraic equations (DAE); in our case, we approximated this by solving ordinary differential equation systems

Fig. 7.15 Euler integration method

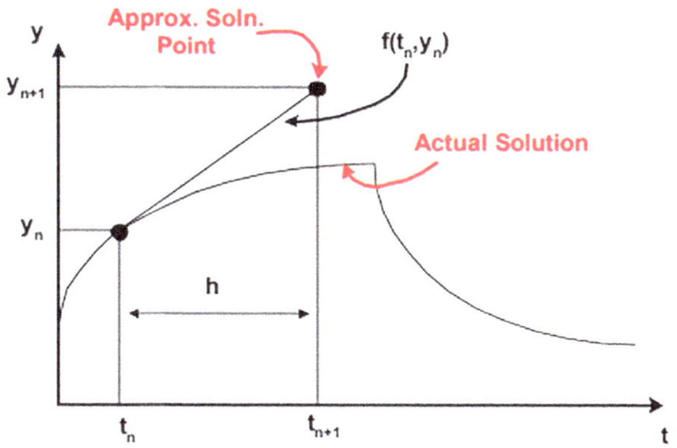


with initial conditions, and combining them with a DEVS approximation of Runge-Kutta and Euler integration methods. For instance, in the case of Euler, we compute $y_{n+1} = y_n + h.f(t_n, y_n)$. The method extrapolates the solution over the interval using an approximation to the derivative at the beginning of the interval (Fig. 7.15).

We implemented both Euler and a fourth-order Runge-Kutta Method Integration method (which is more accurate and stable than Euler for a given step size, as it uses the derivative at the beginning of the interval, the derivative at two trial midpoints and the derivative at a trial end point). These two methods were defined

```
[top]
components : int@rkIntegModel clock
out : clk y
Link : y@int y              Link : y@int dydt@int
Link : out@clock clk        Link : out@clock vin@int

[int]
y0 : 0       dydt0 : 0      C : 1.0E-6      R : 1000
[clock]
components : inv@Process_Inv sig1@Signal qm@QuantumMultiply
out : out
Link : out@sig1 in@inv      Link : out@inv in@sig1
Link : out@sig1 in@qm       Link : out@qm out

[sig1]
Transport_Delay : 00:00:1:000

[qm]
Transport_Delay : 00:00:000Attenuation : 100
```

Fig. 7.16 DEVS implementation of the low pas filter in Fig. 7.13

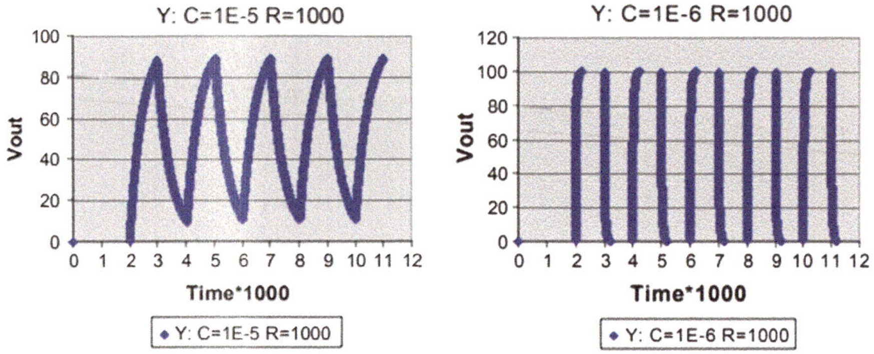


**Fig. 7.17** Simulation results of the low-pass filter

using the QSS method (order two), providing accurate results for a small quantum size. To do so, we inverted the equations used by the numerical methods to determine at what time (relative to the present time) the integral of the first order differential equation will enter the quantum state above or below the current quantum state. We decompose the ODE into a set of first order differential equations, and convert them into a Quantized Integrator model.

The following figure shows a CD++ definition of the DEVS model derived from the low pass filter presented in Fig. 7.13, and simulation results of the model (Figs. 7.16 and 7.17).

## 7.8 The ATLAS Traffic Modelling Language

Urban traffic analysis is a problem whose complexity makes the analysis with traditional analytical methods difficult. The use of simulation is now the tool of choice urban traffic analysis. ATLAS (Advanced Traffic Language Specifications) is devoted to build models of city sections using microsimulation (Davidson and Wainer 2006). The basic language constructions allow defining a static topology of the section to be studied. The dynamic behavior of the section can be modified by including traffic lights, traffic signs, etc. Once the urban section is outlined, models are converted into cell spaces and the traffic flow is automatically set up. Language constructions were mapped into DEVS and Cell-DEVS models. The models were formally specified, which made easier the verification of the language constructions.

ATLAS allows to represent the structure of a city section defined by a set of streets connected by crossings. The language constructions define a static view of the model, and they are considered as cellular models. The main constructions are *Segments*, *Crossings*, *Parking segments*, *Traffic lights*, *Railways*, *Construction sites*, and *Traffic Signs*. Different *Experimental frameworks* can be used to conduct experimentation and analysis.

ATLAS formal language defines the structure and behavior of these constructs, and we used the formal specification with validation purposes. Based on ATLAS, we built TSC, the Traffic Simulation Compiler, which understands ATLAS models and translates them into DEVS models. In order to make the model definition easier, we defined two different tools for model definition and simulation visualization. MAPS allows the user to build maps and decorations (traffic signs, speed, etc.), and the results can be animated using CD++ (or Google Maps) to visualize the simulation results. Figure 7.18 shows a workflow and the software stack used for ATLAS.

Following, we will summarize the constructions of ATLAS, their formal specifications, and their implementation in the TSC simulation language (the reader interested in further details about the formal language and the compiler tools can refer to (Lo Tártaro et al. 2001)).

The main construction is the *segment,* which represent a section between two intersections. Each lane in a given segment has the same direction (i.e., they are one-way) and they have a maximum speed. They are formally specified as:

$$Segment = \{(p1, p2, n, a, dir, max)\}$$

where *p1* and *p2* represent the boundaries of each segment, n is the number of lanes, and *dir* represents the vehicle direction, 0 represents the traffic direction is from *p2* to *p1* and 1 represents the traffic direction is from *p1* to *p2*. The *a* parameter defines the shape of the segment (0-straight or 1-curve, allowing to define the city shape precisely, and to include the exact number of cells), and max is the maximum speed allowed in this segment (Fig. 7.19).

The syntax of TSC (Lo Tártaro et al. 2001) allows defining the segments by delimiting them using the sentences *begin segments* and *end segments*. At least one segment must be defined, using the following syntax:

```
id = p1, p2, lanes, shape, direction, speed, delay,
parkType
```

where,

`p1:` (x, y), integers; the start of the segment.
`p2:` (x, y), integers; the end of the segment.
`lanes:` integer, the number of lanes in the segment.
`shape: [curve | straight]`, the shape of the segment.
`direction: [go | back]`, the direction of the segment.
`speed:` integer, maximum speed in the segment.
`delay:` integer, defines a delay value used for parking lanes.
`parkType: [parkNone | parkLeft | parkRight | parkBoth]`, defines if parking is allowed in the segment, and where.

The following example shows the definition of the segments section; in this case there is only one segment with start/end points at cells (4, 1) and (4, 3), two lanes wide. It is a straight segment, and traffic moves from origin (cell (4, 3)) to destination (cell (4, 1)). The maximum speed is 60 km/h. You can park on the right lane, and the simulation will receive a parameter of 1100 s as the average delay

```
begin segments
    S1 = (4,3), (4,1), 2, straight, go, 60, 1100, parkRight
end segments
```

As we can see in Fig. 7.18, these TSC specifications are translated into CD++, as a set of rules based on the Cell-DEVS formalism. We defined models for segments from 1 to 5 lanes (unidirectional), and we formally verified the correctness of the rules. We will now discuss this mapping for 2-lane segments.

A segment s t = (p1, p2, 2, a, dir, max) is defined as a 2D Cell-DEVS model, using transport delays, with the structure presented in Fig. 7.20.

Each row has a different specification, based on the asymmetry of the cell space. The vehicles in the first row can move straight or to the right (and the ones in the second row, in the opposite direction). The first row is defined as:

$$C0j = \langle I, X, S, Y, N, d_{int}, d_{ext}, delay, d, t, l, D \rangle$$

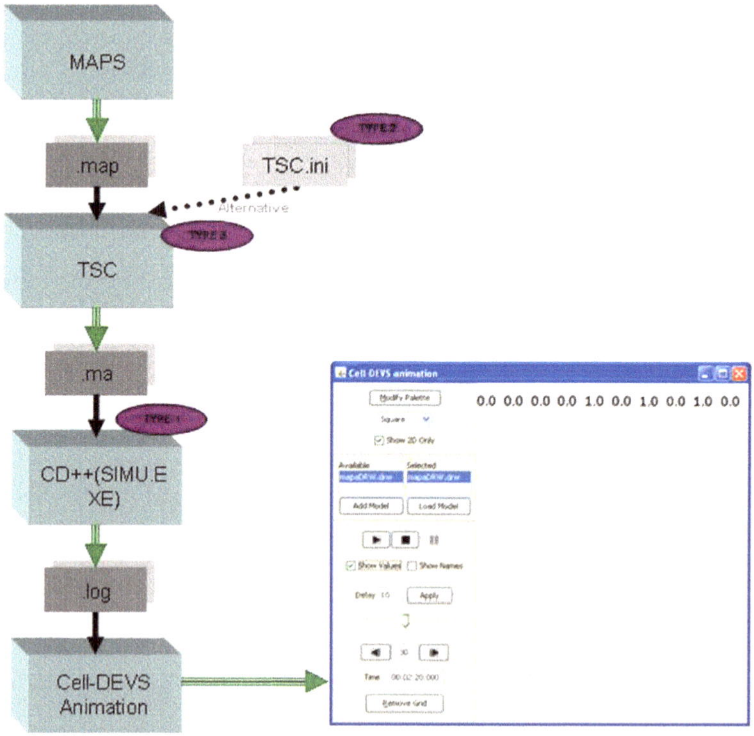


**Fig. 7.18** ATLAS Tool stack

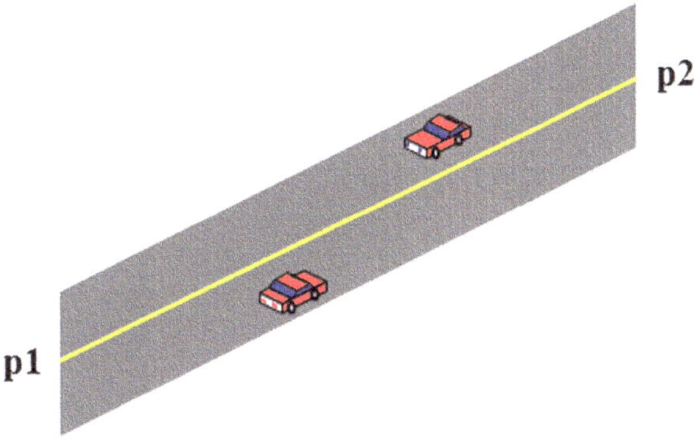


**Fig. 7.19** A road segment

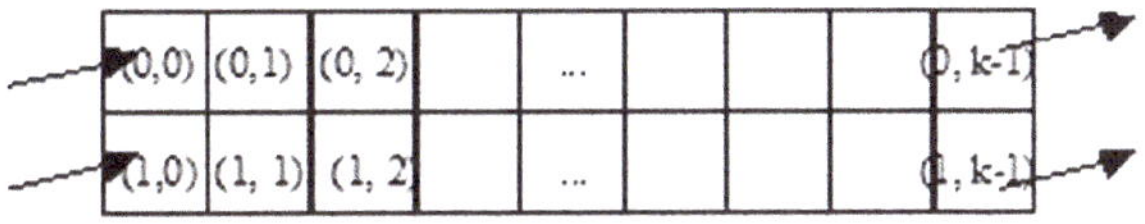


**Fig. 7.20** 2-lane segment

**I** = < η, $P^x$, $P^y$>, with η = 6; and $P^x$ = { ($X_1$, boolean), ($X_2$, boolean), ($X_3$, boolean), ($X_4$, boolean), ($X_5$, boolean), ($X_6$, boolean) }; $P^y$ = { ($Y_1$, boolean), ($Y_2$, boolean), ($Y_3$, boolean), ($Y_4$, boolean), ($Y_5$, boolean), ($Y_6$, boolean) }.

X = Y = boolean;

S:

$$s = \begin{cases} 1 \text{ if there is a vehicle;} \\ 0 \text{ otherwise.} \end{cases}$$

delay = transport;

d = convert_to_delay(speed(max)); where *speed* is a random function that uses using a probabilistic distribution based on vehicle traffic. One expects a few vehicles with maximum and minimum speed, and a majority between them. Based on the maximum speed, we compute the mean $\bar{x} = \frac{2}{3} * \max(km/h)$ and the standard deviation, $\sigma = \frac{1}{3} * \max(km/h)$, which are passed to the function, which returns a natural number representing the random speed in km/h for the vehicle. Based on this, we compute the delay to cross a cell, which is 7.5 m (the size needed for a vehicle (Chopard et al. 1996)), dividing by the speed in km/h, and multiplying by $60^2$ (to convert the delay in km/h into seconds).

N = { (0, 0), (0, 1), (−1, 0), (−1, 1), (0, −1), (−1, −1) };
$\tau$ is:

| New State | Preconditions | Rule name |
|---|---|---|
| 1 | ((0, 0) = 0 and (0, −1) = 1) or | From_Behind |
| | ((0, 0) = 0 and (0, −1) = 0 and (−1, −1) = 1 and (−1, 0) = 1) | From_Right |
| 0 | ((0, 0) = 1 and (0, 1) = 0) or | Move_Forward |
| | (0, 0) = 1 and (−1, 1) = 0 and (−1, 0) = 0 | Right_of_Way_Right |
| (0, 0) | True | Default |

The vehicles can only arrive from the cell behind, or from the right (due to the neighborhood definition). For diagonal movements, we need to consider the right of way. In this way, we avoid collisions. The rules in the second lane are symmetrical to these ones, but we also need to consider that the vehicles from the right have the right of way (so, we evaluate those rules first) (Fig. 7.21).

The coupled model corresponding to the segment is defined as:

S2L(k, max) = < Xlist, Ylist, I, *X*, *Y*, n, $\{t_1,\ldots,t_n\}$, $\eta$, N, C, B, Z, select >

Ylist = { (0, 0), (1, 0), (0, k–1), (1, k−1) }
Xlist = { (0, 0), (1, 0), (0, k−1), (1, k−1) }
I = <$P^x$, $P^y$>, with $P^x$ = {x-c-vehicle, boolean>, x-c-vehicle, boolean>, <x-c-free, boolean>, <x-c-free, boolean>} and $P^y$ = {<y-c-free, boolean>, <y-c-free, boolean>, <y-c-vehicle, boolean>, <y-c-vehicle, boolean>}
X = Y = Boolean;
n = 2; t1 = 2; t2 = k; $\eta$ = 6 and N is described for each lane.
C = { $C_{ij}$ / i ∈ [0, 1] ∧ j ∈ [0, k−1] }, with Cij a Cell-DEVS atomic model
B = { (0, k−1), (1, k−1), (0, 0), (1, 0) }
select = { (0, 1), (1, 1), (0, 0), (1, 0), (0, −1), (1, −1) }

In this case, S2L(k, max) is a segment of 2 lanes, k cells long, and maximum sped *max* km/h. The number of cells *k* is computed automatically as the distance between start/end of the segment divided by 7.5 m, using the inclination angle as in Fig. 7.21.

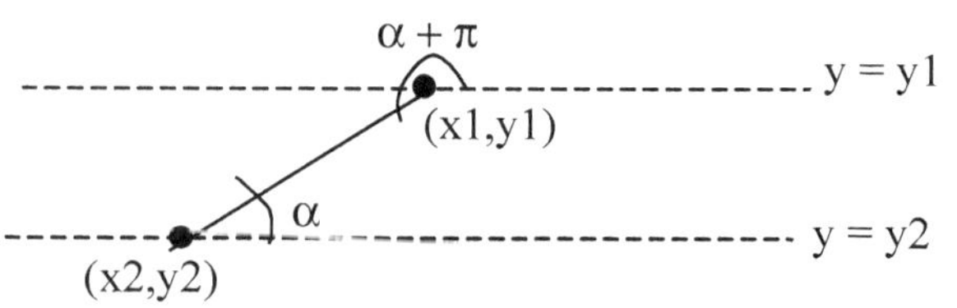


**Fig. 7.21** Computing the length based on inclination angles

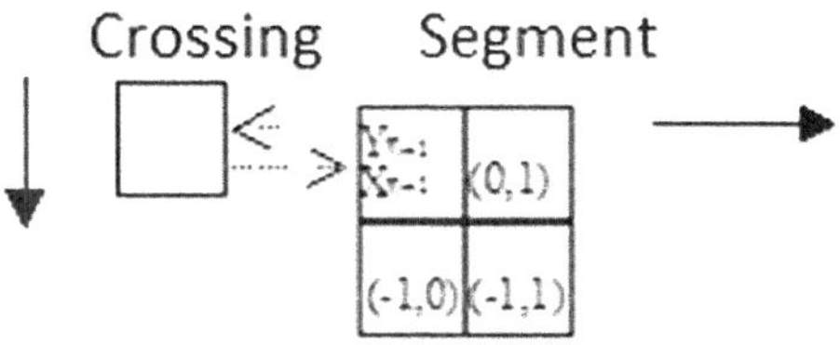


**Fig. 7.22** Neighborhood definition and ports for border cell (0, 0)

The model's interface includes the cells of the first and last column of the segment; these ones can interchange vehicles from/to the crossings. Therefore, the behavior of these cells is different to the rest of the segment, and the formal definitions of the borders change as follows, for cell (0, 0) (Fig. 7.22):

$\eta = 4$; $\mathbf{N} = \{ (0, 0), (0, 1), (-1, 0), (-1, 1) \}$

$\tau$ is:

| New State | Preconditions | Rule name |
|---|---|---|
| 1 | (0, 0) = 0 and portvalue(x-c- vehicle) = 1 | From_Crossing |
| 0 | ((0, 0) = 1 and (0, 1) = 0) or | Move_Forward |
| | (0, 0) = 1 and (−1, 1) = 0 and (−1, 0) = 0 | Right_of_Way_Right |
| (0, 0) | True | Default |

This cell only receives vehicles that are coming from a crossing, while the rules to advance are similar to the other cells. We define a similar set of rules for cell (1, 0), the second input border. Similarly, cells (k, 0) and (k, 1) need a modification of neighborhood and input/output ports to allow vehicles to leave the segment in the direction of a crossing. In this case, $\tau$ is:

| New State | Preconditions | Rule name |
|---|---|---|
| 1<br>send(0, y-c-vehicle) | (0, 0) = 0 and (0, −1) = 1 | From_Behind |
| | ((0, 0) = 0 and (0, −1) = 0 and (−1, −1) = 1 and (−1, 0) = 1) | From_Right |
| 0<br>send(1, y-c- vehicle) | (0, 0) = 1 and portvalue(x-c-free) = 0 | To_Crossing |
| (0, 0) | True | Default |

This cell only receives vehicles coming from a crossing, while the rules to advance are similar for lane one and the rest of the model do not change. We have symmetric rules for cell (1, 0). Likewise, cells (0, k−1) and (1, k−1) must generate outputs to the crossings, and the behavior generated is similar.

*Parking* defines different behavior in the border cells in a segment, as these can be used for parking, as seen in Fig. 7.2. They are formally defined as:

*Parking={(s, n1)}*

Every pair (s1, n1) identifies the segment and the lane where car parking is allowed. If n1= 0, the cars park on the left; if n1 = 1, the cars park on the right. As seen in the *segments* definition, the parking has been added as an attribute of the segment for each of them in the TSC compiler.

*Crossings* are defined in those points in the plane where several segments intersect. They are formally defined as:

*Crossings={(c, max)}*

which represents the points where the crossings are located, and the maximum speed to cross a segment (this allows defining a different speed for the crossing and the segments they are connected to, and defining specialized behavior for countries where speed in crossings should be lower than in the crossing streets. Likewise, it allows defining reduced speeds for roundabouts. Finally, if there are segments with different speeds entering a given crossing – for instance, a main avenue crossing a residential street – this lets us defining the desired speed for the crossing). Crossings are translated to a cellular model built as a ring of cells with moving vehicles. A vehicle in a cell of the crossing has higher priority to obtain the next position in the ring than the cars outside the crossing (see Fig. 7.23). The number of cells in the ring is calculated as the number of input and output segments connected to the crossing, as shown in Fig. 7.23.

The TSC definitions for crossings are delimited by the separators `begin crossings` and `end crossings`. Each sentence defines a crossing using the following syntax:

```
id = p, speed, tLight, crossHole, pout
```

Parameters `p` and `speed` represent the coordinate of the crossing `(p1, p2)` and max `speed` for the crossings. `Pout` defines the probability of a vehicle to leave the crossing, used to simulate random routing of different vehicles.

Crossings with *traffic lights* are defined as:

$$TLCrossing = \{(c|c \in Crossings)\}$$

Here, $c \in TLCrossings$ defines a set of models representing the traffic lights in a corner and the corresponding controller. Each of these models is associated with a crossing input. The model sends a value representing the color of the traffic light to a cell in the intersection corresponding to the input segment affected by the traffic light. The following qualifier is added to a standard crossing definition in TSC when a crossing must include traffic lights:

```
tLight : [withTL | withoutTL]
```

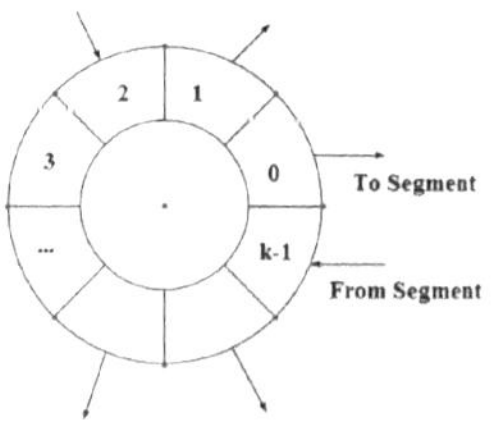


$$k = \sum_{\substack{t \in T \wedge t=(c1,c2,n,a,dir) \\ \wedge (c1=p \vee c2=p)}} n$$

**Fig. 7.23** Crossing definition and calculation of number of cells in the crossing

The crossings C1 and C2 in the following definition show two crossings, one at cell (4, 3), and another at cell (4, 14). Both have a maximum speed of 40 km/h and a probability to leave the cell of 70%.

```
begin crossings
   C1 = (4,3), 40, withTL, withoutHole, 0.7
   C2 = (4,14), 40, withTL, withoutHole, 0.7
end crossings
```

*Railways* are defined as a set of level crossings overlapped with the road segments (see Fig. 7.5). The railway network is defined by:

$$RailNet = \{(Station, Rail)$$

Where *Station* is a model and $Rail \in RailTrack$ and *RailTrack* is defined as:

$$RailTrack = \{(s, \delta, seq)\}$$

Railtrack associates a level crossing with other existing constructions in the city section. Each element identifies the segment that is crossed ($s$) and the distance to the railway from the beginning of the section ($\delta$). Finally, a sequence number ($seq$)

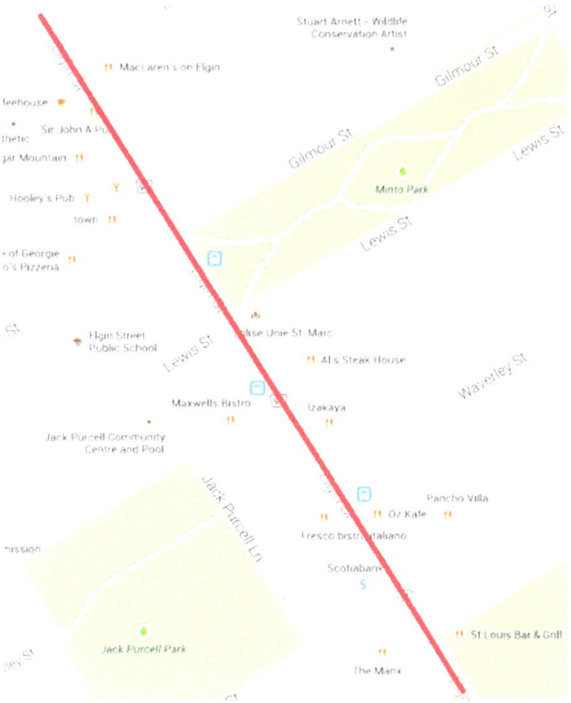


**Fig. 7.24** Traffic section for case study: Elgin Street in Ottawa

is assigned to each level crossing, defining its position in the RailTrack. *RailNet* represents a set of stations connected to railways. When a railway is defined in TSC, we use the *begin railnets* and *end railnets* clauses. Each *RailNet* is defined using the following syntax:

`id =(s1, d1) (s2, d2) (s3, d3)...`

where `si` defines an identifier of a segment crossed by the railway, and `di` defines the distance between the beginning of the segment `si` and the railway. The compiler automatically generates the sequence number.

Similarly, we have defined language constructions for potholes, traffic signs, construction zones, etc.

The following picture shows a main section in downtown Ottawa. The traffic in this area is usually crowded, and congestion occurs frequently. The area includes two parks, several one-way streets and avenues. In several of these streets, parking is allowed, while in others it is forbidden. There are traffic lights in several of the crossings.

Figure 7.24 shows the ATLAS definition for this area, labeling the segments and crossings. Using MAPS, we built a model of this section, which is then translated to ATLAS TSC as follows:

```
begin segments
     S1 = (4,1), (4,3), 2, straight, back, 60, 1100, parkRight
     S2 = (5,1), (4,3), 2, straight, go, 60, 1100, parkRight
     S3 = (4,3), (4,14), 2, straight, back, 60, 1100, parkRight
     S4 = (4,3), (5,14), 2, curve, go, 60, 1100, parkRight
   ...

      S28 = (4,36), (10,36), 1, straight, go, 40, 1100, parkNone
      S29 = (10,36), (13,36), 1, straight, go, 40, 1100, parkNone
      S30 = (4,39), (1,40), 1, straight, go, 50, 1100, parkNone
      S31 = (4,39), (1,39), 1, straight, back, 50, 1100, parkNone
end segments

begin crossings
   C1 = (4,3), 40, withTL, withoutHole, 220, 110
   C2 = (4,14), 40, withTL, withoutHole, 221, 111
   ...
   C8 = (10,28), 20, withoutTL, withHole, 229, 119
   C9 = (10, 36), 20, withoutTL, withHole, 230, 120
end crossings

begin ctrElements
   in S21 : pedestrian crossing, 1, 110
   in S22 : pedestrian crossing, 1, 111
   in S22 : school, 2, 112
   in S23 : stop, 7, 113
   in S24 : stop, 6, 114
   ...
   in S30 : pedestrian crossing, 1, 117
end ctlElements
```

For instance, we can see, that segment S1 is a two-lane road, which includes all the parameters as discussed earlier. Following the method to compute the cells in the cell space, this model must be mapped into a two-dimensional Cell-DEVS with transport delays. Similarly, the TSC definition of segment S29 represents is a one way street starting at (10, 36) and ending at (13, 36). The traffic direction on this road is from the source to the destination. The maximum speed is 40 km/h and parking is not permitted. The crossing C1 is at position (4, 3) and the maximum speed in this intersection is 40 km/h. The crossing has traffic lights, and there are no potholes. We can also see that in road segment 22, there is a school located 2 cells away from the beginning point, and that in segment 28, there is a pedestrian crossing located 1 cell away from the beginning point.

The compiler generates a CD++ file, using the TSC template rules, and the model generated is as follows, following the rules for 2-lane models described above:

```
[TOP]
components : S1  S2  S3 ... S31   ; segments
       C1   C2   ... C7           ; crossings
       S17tl@TrafficLight S17Gen@Generator S15Cons@Consumer ...
       C1stl@SincroTrafficLight C2stl@SincroTrafficLight
       C21stl@SincroTrafficLight C4stl@SincroTrafficLight
       C41stl@SincroTrafficLight
...

link : y-t-car0@S17Gen x-ge-car00@S17
link : y-t-car1@S17Gen x-ge-car10@S17
...

[S17]
type : cell
width : 3    height : 2      delay : transport    border : nowrapped
neighbors : (1,-1) (1,0) (1,1) (0,-1)(0,0)(0,1) (-1,-1) (-1,0) (-1,1)
in : x-ge-car00 x-ge-car10 x-c-space02 x-c-space12
out: y-c-car02 y-c-car12
link : x-ge-car00 x-ge-car@S17(0,0)
link : x-ge-car10 x-ge-car@S17(1,0)
...
localtransition : S17-segment2-lane0-rule
[S17-segment2-lane0-rule]
#Macro(S17-From_Behind)
#Macro(S17-From_Right_Lane)
#Macro(S17-Move_Forward)
#Macro(S17-To_Right_with_Right_of_Way)
#Macro(S17-Default)

  ...
#BeginMacro(S27-From_Behind)
rule : 1         40 { (0,0) = 0 and (0,-1) = 1 }
```

As we can see, the segments definitions are translated into CD++ coupled models using the definitions discussed earlier. For instance, we can see the

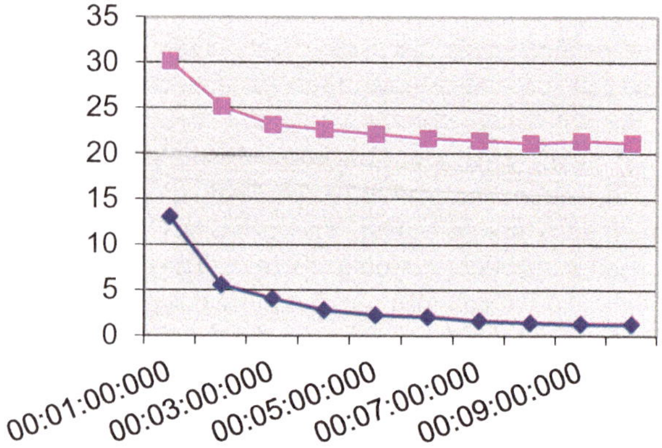


**Fig. 7.25** Simulation scenario with traffic lights and without traffic lights

definition of segment S17, a small Cell-DEVS model, of 2 lanes (3 cells long), connected to a *generator* model (which feeds the section with vehicles through the ports *x-ge*. The local transition rule is defined using the formal specifications discussed above, and in particular, we show a definition of the macro used to define the rule "From_Behind", which represents movement to a cell from the vehicle behind it. When we simulate the model, we can change initial conditions, for instance, the traffic lights, number of vehicles generated, etc. Figure 7.25 shows the simulation results at different times in segment S17, in which we injected a similar amount of vehicles, and we can see the influence of traffic lights (lower line) that help regulating the traffic in the section (Fig. 7.25).

## 7.9 Conclusion

The last 50 years of SummerSim have brought a wide range of advances to the field of Modeling and Simulation. Here, we have discussed our lab's research in the area, focusing on using DEVS as a common formalism for simulating a diverse variety of models. DEVS provides the means of building complex models evolving incrementally from simple subcomponents in incremental, hierarchical fashion. This view enables the reuse of simulations and components, where the integration of simulations and components is seamless.

The experiments were carried out using CD++, a DEVS tool that has been built following the formal definitions of DEVS and Cell-DEVS. The modeler can then to focus on the modeling formalism of interest, and to use all the advantages of DEVS (in terms of integration with other models, multimodeling, sharing of repositories, proving the validity of simulation code, etc.) for model composition and simulation.

The use of formal modeling techniques enhances model verification. Specifically, automated rule verification, based on meeting basic logical properties in cellular models and coupled model definitions, can be provided. Our tools also provide mechanisms for automating the verification of multicomponent model coupling. In the same sense, we showed how to provide multimodeling and hybrid modeling. The models are able to include both continuous and discrete event model components. For instance, the behavior governing the physics of a vehicle can be described with Bond Graphs (continuous modeling technique), while vehicle cruise control system might be better modeled using a discrete event formalism. We showed how to integrate these model views in a seamless fashion, using DEVS and the CD++, combined with the QSS method and model transformations. Spatial notions can provide extra facilities for understanding and visualizing the resulting simulation. For example, it allows incorporating geographical data obtained in GIS software.

**Acknowledgements** This work was partially funded by NSERC. The research has been conducted by a large number of students in our lab in the last 20 years, including: M. Lo Tártaro, C. Torres, T. Zheng, M. D'Abreu, L. Checiu, C. Jacques, H. Saadawi, D. Vicino, R. Castro, S. Mehta, A. González and numerous others, who have spent long hours in this research.

## References

Alur, R., & Dill, D. (1994). Theory of timed automata. *Theoretical Computer Science, 126,* 183–235.

Bowman, H., & Gomez, R. (2006). *Concurrency theory: Calculi and automata for modelling untimed and timed concurrent systems*. London: Springer.

Castro, R., Kofman, E., & Wainer, G. (2010). A formal framework for stochastic DEVS modeling and simulation. *SIMULATION: Transactions of the Society for Modeling and Simulation International, 86*(10), 587–611.

Chechiu, L., & Wainer. G. (2005). Experimental results on the use of modelica/CD++. In *Proceedings of the 2005 SCS Summer Computer Simulation Conference*. Philadelphia, PA.

Chopard, B., Queloz, P. A., & Luthi, P. (1996). Cellular automata model of car traffic in two-dimensional street networks. *J. Phys., A.vol.29,* 2325–2336.

D'Abreu, M., & Wainer, G. (2006) A Bond-Graph mapping mechanism for M/CD++. In *Proceedings of the 2006 SCS Summer Computer Simulation Conference*. Calgary, AB, Canada.

Davidson, A., & Wainer, G. (2006). ATLAS: A specification language for traffic modelling and simulation. *Simulation, Practice and Experience*. Elsevier. *14*(3), 317–337.

Fritzson, P. (2004). *Principles of object-oriented modelling and simulation with Madelia 2.1.* Wiley-IEEE Press.

Hopcroft, J., & Ullman, J. (1979). *Introduction to automata theory, languages, and computation.* Reading/MA: Addison-Wesley.

Jacques, C., & Wainer, G. (2002). Using the CD++ DEVS toolkit to develop Petri Nets. In *Proceedings of the 2002 Summer Computer Simulation Conference*, San Diego, CA. USA.

Karnopp, D., Margolis, D., & Rosenberg, R. (1990). *System Dynamics: A unified approach.* Wiley.

Kofman, E., & Junco, S. (2001). Quantized state systems. *A DEVS Approach for Continuous System simulation Transactions of the SCS, 18*(3), 123–132.

Liu, F., & Zhang, H. (2018). A class of extended time Petri nets for modelling and simulation of discrete event systems. *SIMULATION, 94*(8), 753–762.

Lo Tártaro, M., Torres, C., & Wainer, G. (2001). TSC: A compiler for the ATLAS language. In *Proceedings of 2001 Winter Simulation Conference. Arlington*, VA. U.S.A. IEEE Press.

Mehta, S., & Wainer, G. (2005). DEVS for mixed-signal Modeling based on VHDL. In *Proceedings of 2005 DEVS Integrative M&S Symposium, Spring Simulation Conference*. San Diego, CA.

Peterson, J. L. (1977). Petri nets. *ACM Computing Surveys, 3*(5), 221–252.

Pietro, F. D., Migoni, G., & Kofman, E. (2018). Improving linearly implicit quantized state system methods. *SIMULATION*. https://doi.org/10.1177/0037549718766689.

Saadawi, H., & Wainer, G. (2010). Rational time-advance DEVS (RTA-DEVS). In *Proceedings of DEVS Symposium 2010*, Orlando, FL., April 11–15.

Villa-Villaseñor, N., & Rico-Melgoza, J. J. (2018). Complementarity framework formulation from bond graphs to model a class of nonlinear systems and hybrid systems with fixed causality. *SIMULATION, 94*(9), 783–795.

Wainer, G. A. (2009). *Discrete-event modeling and simulation: A practitioner's approach*. Boca Raton, FL, USA: CRC Press.

Zeigler, B. P., Kim, T., & Praehofer, H. (2000). *Theory of modeling and simulation*. San Diego, CA: Academic Press.

Zheng, T., & Wainer, G. (2003). Implementing finite state machines using the CD++ toolkit. In *Proceedings of the 2003 Summer Computer Simulation Conference*. Montreal, QC. Canada.

**Gabriel A. Wainer,** FSCS, SMIEEE, received the M.Sc. (1993) at the University of Buenos Aires, Argentina, and the Ph.D. (1998, with highest honors) at UBA/Université d'Aix-Marseille III, France. In July 2000, he joined the Department of Systems and Computer Engineering at Carleton University (Ottawa, ON, Canada), where he is now Full Professor and Associate Chair for Graduate Studies. He has held visiting positions at the University of Arizona; LSIS (CNRS), Université Paul Cézanne, University of Nice, INRIA Sophia-Antipolis, Université de Bordeaux (France); UCM, UPC (Spain), University of Buenos Aires, National University of Rosario (Argentina) and others. He is one of the founders of the Symposium on Theory of Modeling and Simulation, SIMUTools and SimAUD. Prof. Wainer was Vice-President Conferences and Vice-President Publications, and is a member of the Board of Directors of the SCS. Prof. Wainer is the Special Issues Editor of SIMULATION, member of the Editorial Board of IEEE Computing in Science and Engineering, Wireless Networks (Elsevier), Journal of Defense Modeling and Simulation (SCS). He is the head of the Advanced Real-Time Simulation lab, located at Carleton University's Centre for advanced Simulation and Visualization (V-Sim). He has been the recipient of various awards, including the IBM Eclipse Innovation Award, SCS Leadership Award, and various Best Paper awards. He has been awarded Carleton University's Research Achievement Award (2005, 2014), the First Bernard P. Zeigler DEVS Modeling and Simulation Award, the SCS Outstanding Professional Award (2011), Carleton University's Mentorship Award (2013), the SCS Distinguished Professional Award (2013), and the SCS Distinguished Service Award (2015). He is a Fellow of SCS.

# Chapter 8
# A Profiling Study of the Summer Computer Simulation Conference

**Navonil Mustafee and Korina Katsaliaki**

**Abstract** The Summer Computer Simulation Conference (SCSC) is part of the Summer Simulation Multi-conference that is organized by the Society of Modelling and Simulation International. SCSC celebrated its 50th Anniversary in 2018! We therefore thought it timely to present a profiling study of SCSC conferences and report on metrics associated with publications, such as, contribution of authors and institutions, county-specific affiliation data, funding organizations, scholarly outlets (journals and conferences) and authors being cited by SCSC authors, frequency of citations by years, and citations count of papers published in SCSC proceedings. The period of review was from 2005 to 2017 and our dataset consisted of 911 papers. Our findings confirm the international character of this conference, with leading SCSC authors and institutions based in Canada, Europe and North America. Our citation-based analysis reveals the breath of literature being cited by the authors, and which includes Operations Research, industrial and process engineering, and applied computing. This book chapter adds to the profiling studies that have been undertaken by the authors in the field of M&S, and which includes, a profile of the journals Simulation and the Journal of Defence Modelling and Simulation, and the Winter Simulation Conference.

## 8.1 Introduction

The *Summer Computer Simulation Conference* is a leading international conference in the field of Modelling and Simulation (M&S). It is a part of the *Summer Simulation Multi-conference*, SummerSim for short, and is organized by the *Society*

N. Mustafee (✉)
The Centre for Simulation, Analytics and Modelling (CSAM),
University of Exeter Business School, Building:ONE, Rennes Drive, Exeter EX4 4PU, UK
e-mail: N.Mustafee@exeter.ac.uk

K. Katsaliaki
School of Economics, Business Administration and Legal Studies,
International Hellenic University, 57001 Thessaloniki, Greece

 
J. Sokolowski et al. (eds.), *Summer of Simulation*, Simulation Foundations, Methods and Applications, https://doi.org/10.1007/978-3-030-17164-3_8

*of Modelling and Simulation International (SCS).* SCS is a technical society that is devoted to furthering the field of Modeling and Simulation (M&S) (SCS 2018). Since its inception in 1952, the Society has widely disseminated the advancements in this field through its peer-reviewed journals—*SIMULATION: Transactions of The SCS* and *Journal of Defense Modeling and Simulation: Applications, Methodology, Technology*, and conferences such as the *Spring Simulation Multi-conference (SpringSim), Power Plant Simulation (PowerPlantSim),* SummerSim. Summer is a multi-conference, and as such, it includes several other international symposia, and which may change from year to year, for example, *International Symposium on Performance Evaluation of Computer and Telecommunication Systems (SPECTS), International Conference on Modeling and Simulation Methodology Tools Software Applications (M&SMTSA), Grand Challenges in Modeling and Simulation Symposium (GCMS)* and *International Conference on Bond Graph Modeling and Simulation (ICBGM).* The annual conference generally takes place in USA, Canada and Europe. In some cases, SummerSim has also been co-located with other international conferences, for example, in 2011, SummerSim was co-located with the *2011 SISO European Simulation Interoperability Workshop (EURO SIW).* The SummerSim conferences generally have a theme. For example, the theme for the 2019 conference (which will be held in the Technical University of Berlin, Germany; July 22–24) is 'M&S for Smart and Connected World'. The usual policy of the conference is that the conference proceedings are archived in both SCS digital library and the ACM Digital Library; further, the papers are indexed in bibliographical databases like DBLP and Scopus. For this study we were mostly reliant on Scopus and the SCS digital library.

Our book chapter is a profiling study of the Summer Computer Simulation Conference (SCSC), which in 2018, celebrated its 50th Anniversary. SCSC marked its golden jubilee year as part of the 2017 SummerSim, which was held in Bellevue, Washington from July 9–12, 2017. The theme for the conference was 'Pervasive Simulation for Enhanced Decision Making'. The authors have written this chapter, drawing inspiration from their previous work, done to mark the 60th anniversary year of the *SCS*, where they prepared a profiling study of literature published in the Society's journal—*Simulation: The Transactions of the SCS* (Mustafee et al. 2012)—and presented a co-citation analysis for the same journal (Mustafee et al. 2014). One of the co-authors has also published a profiling study to mark the 50th Anniversary of the Winter Simulation Conference which was held Dec 3–6, 2017 in Las Vegas, Nevada (Mustafee and Fishwick 2017). Similar to their previous work, which they considered to be a *fitting tribute* to those "scientists and engineers, who had actively shaped and influenced the growth and development of SCS and continue to contribute to the theory, methodology, and applications of simulation science" (Yilmaz 2011), the book chapter acknowledges the contributions of the SpringSim authors and their affiliated institutions in the development of the field of M&S.

The remainder of the book chapter is organized as follows. Section 8.2 presents an overview of profiling studies associated with scholarly publications, and the motivation for undertaking such an endeavor. It further describes the objectives of the current study. Section 8.3 describes the dataset that is used for the analysis. Section 8.4 presents the results of the analysis and discusses its findings. This is followed by the conclusion section.

## 8.2 Objectives of the Profiling Study

Our profiling study will enable the readers to better understand the publication trends associated with the SCSC papers published from 2005–2017, and identify the contributions of key authors, institutions, seminal articles, etc. Previous profiling studies have mainly been done for journals, for example, Palvia et al. (2007) presented a profiling study of the journal *Information and Management (I&M);* authors have conducted similar analysis for the *European Journal of Information Systems (EJIS)* (Dwivedi and Kuljis 2008), *Information Systems Frontiers* (Dwivedi et al. 2009), *Journal of the Operational Research Society* (Katsaliaki et al. 2010), *Simulation: Transactions* (Mustafee et al. 2012) and *Journal of Defense Modeling and Simulation* (Mustafee et al. 2017). Further, there have been several studies that have compared between journals, for example, *Management Information Systems Quarterly (MISQ)* and *I&M* (Claver et al. 2000), *MISQ* and *EJIS* (Mustafee 2011), and *I&M, EJIS* and *MISQ* (Palvia et al. 2017). Mustafee and Fishwick (2017) present a profiling of papers published in the 2017 Winter Simulation Conference.

Having discussed profiling studies both M&S and other fields, we now list the objectives which will define the variables for data collection and its subsequent analysis. Our objectives are, (a) to analyze authorship and identify authors with the most number of publications in the period considered in this study, (b) to determine the authors' geographical locations, (c) to identify institutions associated with the most number of publications (analysis based on authors' affiliation), (d) to perform an analysis of organizations providing funding for research published in SCSC, (e) to analyze authors' keywords with the objective of identifying the most frequently used keywords, (f) to identify scholarly outlets that are cited by SCSC papers, the frequency of citation (year-based analysis) and the most authors that are most cited, and (g) to determine the citation count of paper published from 2005–2017. The findings of the study will thus present a ranking of the most productive authors, institutions, etc.; however, we would like to voice a note of caution to the readers with regard to interpreting this data. Being a conference-specific profiling exercise, such findings should be regarded as indicative only of SCSC activity and that too within the timeframe of analysis.

## 8.3 Dataset for the Analysis

Our first task was identify SCSC papers using Scopus and the SCS digital library. Compiling the dataset was a challenge, since we had to extract the relevant SCSC papers from the other SummerSim symposia (SPECTS, IGBGM etc.). There was also some inconsistency in the outlet through which the papers were published, for example, from 2012 to 2017 they were included as *Simulation Series* with volume numbers 44 (2012) through to 49 (2017) respectively. However, prior to 2012, these were not specifically labelled as 'Simulation Series' but only as proceedings. The exception to this was in the year 2009, when the proceedings of SummerSim'2009 were published as volume 41 (issue 3). In year 2007, SCSC papers were labelled only as 'Volume 2' and in 2010 it was referred to as 'Issue 1 Book'. Table 8.1 lists the SCSC conferences that have been included in the analysis, together with the venue of the conference, the volume and issue numbers and page count (where available). The table also lists, in addition to SCSC, the symposia/conferences that were organized as part of the Summer Simulation Multi-Conference (please refer to Sect. 8.1 for the abbreviations).

The period of review was from 2005 to 2017, both years inclusive. Thus, we undertook the review of the conference proceedings for the last 13 years, excluding editorial comments, governing board introductions or papers with no abstract. A total of 911 papers were included in the analysis for the selected period, with an average of around 70 paper per year. The number of papers varied from a minimum of 19 in 2012 to a maximum of 186 articles in 2007 (Fig. 8.1). Having created the final data set comprising of a total of 911 papers, we exported the results for cross-checks (e.g., identification different name variants of authors) and analysis. Further, we have used several features already present in Scopus to analyze this data.

## 8.4 Findings

In this section we present the findings of our analysis based on the objectives that have been defined. The mapping of the objectives and the sub-sections are as follows:

(a) To analyze authorship and identify authors with the most number of publications in the period considered in this study—Sect. 8.4.1
(b) To determine the authors' geographical locations—Sect. 8.4.2
(c) To identify institutions associated with the most number of publications—Sect. 8.4.3
(d) To perform an analysis of organizations providing funding for research—Sect. 8.4.4

**Table 8.1** List of SCSC conferences included in the study

| Year | Conference | Location/Dates | Volume/Issue/ Pages |
|---|---|---|---|
| 2005 | SCSC'2005 *(Part of SummerSim'05—with SPECTS'05)* | Philadelphia, USA, July 24–28, 2005 | Proceedings, 520 pages |
| 2006 | SCSC2006 *(Part of SummerSim'06—with SPECTS'06 and M&SMTSA'06)* | Calgary, Canada; July 31–August 2, 2006 | Proceedings, 525 pages |
| 2007 | SCSC'2007 *(Part of SummerSim'07 with SPECTS'07)* | San Diego, USA; July 16–19, 2007 | Proceedings, 1492 pages |
| 2008 | SCSC'2008 *(Part of SummerSim'08—with SPECTS'08 and GCMS'08)* | Edinburgh (Scotland), UK; June 16–19, 2008 | Proceedings, 569 pages |
| 2009 | SCSC'2009 *(Part of SummerSim'09—with SPECTS'09)* | Istanbul, Turkey; July 13–16, 2009 | Proceedings, 485 pages (Volume 41, Issue 3) |
| 2010 | SCSC'2010 *(Part of SummerSim'2010—with GCMS 2010)* | Ottawa, Canada; July 11–15, 2010 | Proceedings, 626 pages (Issue 1 Book) |
| 2011 | SCSC'2011 *(Co-located with The 2011 SISO European simulation interoperability workshop)* | The Hague, Netherlands; June 27–30, 2011 | Proceedings, 225 pages (Issue 1 Book) |
| 2012 | SCSC'2012 *(Part of SummerSim'2012—with GCMS 2012, ICBGM'2012 and SPECTS'2012)* | Genoa, Italy; 8–11 July 2012 | Simulation series, Vol 44 #10 (152 pages) |
| 2013 | SCSC'2013 *(Part of SummerSim'2013—with GCMS 2013 and SPECTS'2013)* | Toronto, Canada; July 7–10, 2013 | Simulation series, Vol 45 #11 |
| 2014 | SCSC'2014 *(Part of SummerSim'2014—with GCMS 2014, ICBGM'2014)* | Monterey, US; July 6–10, 2014 | Simulation series, Vol 46 #10 (540 pages) |
| 2015 | SCSC'2015 *(Part of SummerSim'2015—with SPECTS'2015)* | Chicago, US; July 26–29, 2015 | Simulation series, Vol 47 #10 (547 pages) |
| 2016 | SCSC'2016 *(Part of SummerSim'2016—with SPECTS'2016 and ICBGM' 2016)* | Montreal, Canada; July 24–27, 2016 | Simulation series, Vol 48 #9 |
| 2017 | SCSC'2017 *(Part of SummerSim'2017—with SPECTS'2017)* | Bellevue, USA; 9–12 July 2017 | Simulation series, Vol 49 #9 (449 pages) |

**Fig. 8.1** Number of papers published in SCSC proceedings per year

(e) To analyze authors' keywords—Sect. 8.4.5
(f) To identify scholarly outlets that are cited by SCSC papers, the frequency of citation (year-based analysis) and the authors most cited—Sect. 8.4.6
(g) To determine the citations received by the SCSC paper published from 2005–2017—Sect. 8.4.7

### *8.4.1 Analysis Based on Authorship and Authors' Publications*

Our analysis revealed that 154 authors have contributed to at least two SCSC articles (either as first author or co-author) during the period 2005–2017. Of these, 27 have contributed to two papers, 61 authors have contributed to three papers and the remaining 66 authors have four and above contributions. For further assessing research published in the proceedings of the conference, we analyzed the number of publications from each author/co-author. Table 8.2 lists the nine most published authors, with 10 plus contributions, along with their affiliations and geographical locations, sorted by the number of publications as well as alphabetically for authors sharing the same number of publications.

Table 8.2 shows that, in total, the nine authors have contributed to 133 scholarly publications, which represents 15% of all selected publications. *Alessandro A G Bruzzone* from University of Genoa has the highest number of contributions to the conference, with 21 papers as the first author and the remaining as co-author.

**Table 8.2** List of authors with ten or more publications, with affiliation and the order of authorship

| Author | Institution | Country | Total papers | First author | Co-author |
|---|---|---|---|---|---|
| Bruzzone, A.G. | University of Genoa, Department of Engineering | Italy | 26 | 21 | 5 |
| Wainer, G. | Carleton University, Department of Systems and Computer Engineering | Canada | 20 | 0 | 20 |
| Longo, F. | University of Calabria, Department of Mechanical, Energy & Management Engineering | Italy | 16 | 2 | 14 |
| Massei, M. | University of Genoa, Department of Engineering | Italy | 14 | 0 | 14 |
| Bocca, E. | University of Genoa, Department of Engineering & MAST srl | Italy | 13 | 4 | 9 |
| Barjis, I. | New York City College of Technology, Department of Biological Sciences | USA | 12 | 11 | 1 |
| Yilmaz, L. | Auburn University, Department of Computer Engineering | USA | 12 | 7 | 5 |
| Azadeh, A. | University of Tehran, Department of Industrial Engineering | Iran | 10 | 10 | 0 |
| Piera, M.A. | Autonomous University of Barcelona, Telecommunications and Systems Engineering | Spain | 10 | 4 | 6 |

Second in the list is *Gabriel Wainer* from Carleton University in Canada and who has co-authored 20 papers. Third is *Francesco Longo* from University of Calabria, Italy, followed by *Marina Massei* from University of Genoa (Italy). In the top list of authors we see many authors who come from Italy, although the conference has been based mostly in Canada and USA (on nine occasions during the period of analysis).

### 8.4.2 Analysis Based on Authors' Geographic Location

Our analysis of the authors' affiliations revealed that contributors came from 52 different countries, with *US* (33.4%) dominating. The second largest category comprised of authors affiliated to *Canadian* institutions; this was followed by authors with *Italian* affiliations (10.5% and 7.5% respectively). *France* and *Germany* are next with over 50 contributions. Table 8.3 lists the top 14 countries in terms of the total region-specific contributions of the authors taking into consideration the fact that authors could have contributed to more than one paper; also, a paper could have been co-authored with authors from other countries. The total number of contributions is 1036.

**Table 8.3** Countries with the highest number of author contributions

| Country | Author contributions | Total % | Country | Author contributions | Total % |
|---|---|---|---|---|---|
| USA | 346 | 33.4 | Spain | 39 | 3.8 |
| Canada | 109 | 10.5 | Iran | 25 | 2.4 |
| Italy | 78 | 7.5 | Turkey | 23 | 2.2 |
| France | 56 | 5.4 | South Korea | 21 | 2.0 |
| Germany | 53 | 5.1 | Brazil | 14 | 1.4 |
| United Kingdom | 45 | 4.3 | Mexico | 13 | 1.3 |
| China | 39 | 3.8 | Netherlands | 13 | 1.3 |

It is perhaps expected that the largest volume of contribution is from authors affiliated to *USA-based* institutions. This may be because the *Summer Simulation Multi-Conference* is a SCS conference and was established in the US. Further, SpringSim/SCSC was organized in the US on five of the 13 occasions; on four occasions the conference has taken place in *Canada*. The other countries where the conference has been organized are *Italy*, *UK (Scotland)*, *Turkey* and *The Netherlands* and more recently in *France* (2018). *Germany* will be the host for the 2019 edition of SummerSim/SCSC. The level of contributions appear to be in line with the location of the conference (and the frequency of hosting the conference, this is particularly true in case of USA and Canada). The large representation of other countries (162; 15.6%—not included in Table 8.3), however, indicates that the conference has an international appeal. It is also important to note that *Italy* is the third leading country in terms of contributions and is a further evidence of the vibrant M&S research community in that country.

### 8.4.3 Analysis Based on Authors' Institutional Affiliations

The data for this analysis is derived from the institutional affiliation of the contributing authors that are reported at the time of publication. 160 institutions were identified with more than one contribution. For every contribution, only unique author affiliations are taken into consideration. For example, if a paper has been jointly co-authored by researchers belonging to different universities, then the respective institutional count is incremented for each institution. If, however, the co-authors are from the same university, then this will be counted as only one instance of institutional contribution. This is different to the analysis presented in Table 8.2 (authorship), wherein the count for every author (first author or co-author) is incremented, irrespective of whether or not the researchers are based in the same institution. Table 8.4 lists only those institutions with ten or more contributions. The total contributions approach results in the combined count of all authors/affiliations being greater than the total number of articles.

From Table 8.4 we see that two Italian Universities occupy the first two positions, *University of Genoa* (*Universita degli Studi di Genova*) and *University of Calabria (Universita della Calabria)* with 29 and 25 contributions respectively. Following this, we have three academic institutions with the same score of contribution (24 instances), with two from Canada (*Carleton University* and *University of Ottawa*) and one from USA (*Old Dominion University*). The top-19 institutions publishing in SCSC come from 10 different countries. This is further validation of the international appeal of this conference. Table 8.4 does not include any practitioner organizations (institutions not engaged in teaching). However, this should not be taken to mean that SCSC is solely an academic-focused conference, since there are several organizations that appear later-on in the list, for example, *Defence Research and Development* (Canada; 9 contributions), *NASA* (USA; 7), *MITRE Corporation* (USA; 8), *National Defence and the Canadian Forces* (Canada; 4), *Space and Naval Warfare Systems Center* (USA; 4), *U.S. Army Research Laboratory* (USA; 4), *National Institute of Standards and Technology* (USA; 3), and *Wright-Patterson Air Force Base* (USA; 3). Two consortiums also prominently

**Table 8.4** Institutions with ten or more contributions

| Institution | # Total contributions |
|---|---|
| University of Genoa, Italy | 29 |
| University of Calabria, Italy | 25 |
| Carleton University, Canada | 24 |
| University of Ottawa, Canada | 24 |
| Old Dominion University, USA | 24 |
| Auburn University, USA | 17 |
| University of Arizona, USA | 16 |
| Complutense University of Madrid, Spain | 15 |
| University of Central Florida, USA | 14 |
| Korea Advanced Institute of Science & Technology (KAIST), South Korea | 13 |
| Autonomous University of Barcelona, Spain | 13 |
| Laboratory of Information Science and Systems (LSIS), Marseille, France | 12 |
| New York City College of Technology, USA | 11 |
| Beihang University, China | 10 |
| Florida State University, USA | 10 |
| National University of Defense Technology, China | 10 |
| University of Hamburg, Germany | 10 |
| University of South Carolina, USA | 10 |
| University of Tehran, Iran | 10 |

appear in our affiliation-based analysis—*Liophant Simulation* and the *Consortium for the Development of Innovative* with nine and eight instances respectively.

### 8.4.4 Analysis Based on Funding Body

In our dataset, 50 papers report the source of funding. The data is based on indexing information included in Scopus and does not take into account the acknowledgements that may have been included by the authors in the full-text. Our analysis shows that 5.5% of the total papers report a source of funding. Altogether 30 different funding bodies are acknowledged. Table 8.5 presents the list of the top 10 institutions that have funded two or more studies. The *National Science Foundation* comes at the top with 6 studies, followed by the *Erzincan Üniversitesi in* Turkey, which has probably supported research in simulation through internal research grants. The Spanish funding bodies have a good representation in our list with 3 different institutions (*FEDER*, *MINECO* and *Santander*). From Table 8.5 we observe that SCSC authors secured funding from specialist grant providing institutions like the *NSF* and *NSFC*, research labs/government agencies like the *DOE* and *MINECO*, commercial businesses like *Banco Santander*, universities like *Erzincan Üniversitesi* and Charities like *Bill and Melinda Gates Foundation.*

### 8.4.5 Keywords Analysis

In this section, we present an analysis of the most frequently used keywords that are included by SCSC authors and which were indexed by Scopus. The 20 most popular keywords for the whole time-frame is presented in Table 8.6. *Computer Simulation* with 569 occurrences is the most frequently used keyword. The second

**Table 8.5** List of top 10 organizations providing funding for research published in SCSC

| Funding organization | # Count |
|---|---|
| National Science Foundation (NSF), US | 6 |
| Erzincan Üniversitesi (EU), Turkey | 4 |
| Federación Española de Enfermedades Raras (FEDER), Spain | 4 |
| Ministerio de Economía y Competitividad (MINECO), Spain | 3 |
| Natural Sciences and Engineering Research Council of Canada (NSFRC), Canada | 3 |
| Banco Santander, Spain | 2 |
| Bill and Melinda Gates Foundation, US | 2 |
| CRC Health Group (CRC), US | 2 |
| National Natural Science Foundation of China (NSFC), China | 2 |
| U.S. Department of Energy (DOE), US | 2 |

**Table 8.6** The top 20 keywords in our dataset

| Keywords | Frequency | Keywords | Frequency |
|---|---|---|---|
| Computer simulation | 569 | Decision making | 37 |
| Discrete event simulation | 116 | Computational methods | 35 |
| Modeling and simulation | 68 | Virtual reality | 35 |
| Simulation | 61 | Agent-based model | 32 |
| Software engineering | 59 | Decision support systems | 32 |
| Algorithms | 50 | Computer software | 31 |
| Circuit simulation | 49 | Discrete events | 30 |
| Optimization | 49 | Three dimensional computer graphics | 30 |
| Simulation model | 47 | Behavioral research | 29 |
| DEVS | 43 | Petri nets | 29 |

most popular keyword is *Discrete Event Simulation* with 116 occurrences (and a further 30 instances of the keyword *Discrete Events* and 43 instances of *DEVS*), followed by *Modeling and Simulation* with 68 occurrences. Overall, we observe that many keywords include the word Simulation in the top 10 list, and as we go further down the list, we find a mix of methods, systems and models, all closely related to simulation.

Our analysis reveals DEVS to be an important community in SCSC. It also confirms that the conference is predominantly on discrete modelling methods (*Discrete Event Simulation*, *DEVS*, *Petri Nets*, etc.). There is a mix of applied papers (keywords *Decision Making* and *Decision Support Systems*) and those that focus on algorithm development and computational methods. In terms of application area, only *Circuit Simulation* is included in the top-20 list. There are also some surprises, for example, *Virtual Reality* and *Behavioral Research* have 35 and 29 occurrences respectively. Behavioral Operations Research (BOR) is a growing area of research in the Operations Research (OR) community (*European Journal of Operational Research* special issue on BOR—Franco and Hamalainen 2015), and *Behavioral Research* being identified as the 19th most popular SCSC keyword demonstrates the existence of behavioral literature in the simulation community; this work is perhaps mostly associated with *Agent-based Model* (32 instances) and agent behavior.

### 8.4.6 Analysis of Articles Cited by SCSC Authors

Table 8.7 presents the first 20 source titles which have cited by the 911 papers of our dataset. Altogether 13,886 references (these are not unique articles) were cited by the 911 selected documents published in the SCSC proceedings. Hoverer, in our analysis, we included only the 12,353 references published between 1980 and 2018 (Oct). It is apparent that most bibliography comes from papers published in the

**Table 8.7** The top 20 sources of references cited by the papers in the dataset

| Source title | Citations |
|---|---|
| Proceedings Winter Simulation Conference (WSC) | 291 |
| Lecture notes in computer science—*Including subseries lecture notes in artificial intelligence and lecture notes in bioinformatics* | 194 |
| European Journal of Operational Research (EJOR) | 89 |
| Simulation: Transactions of the SCS | 81 |
| Nature | 41 |
| International Journal of Production Economics (IJPE) | 40 |
| Simulation Modelling Practice and Theory (SMTP) | 39 |
| Computers and industrial engineering | 37 |
| International Journal of Production Research (IJPR) | 37 |
| Journal of the Operational Research Society (JORS) | 35 |
| Proceedings IEEE INFOCOM | 33 |
| Proceedings of the national academy of sciences of The United States of America | 29 |
| ACM Transactions On Modeling And Computer Simulation (TOMACS) | 28 |
| Computer | 28 |
| IEEE transactions on power electronics | 28 |
| Science | 28 |
| IEEE transactions on computers | 26 |
| Proceedings of the IEEE | 26 |
| Communications of the ACM (CACM) | 25 |
| Operations research | 25 |

*Winter Simulation Conference (WSC)* proceedings whose technical sponsors include the leading simulation societies and simulation interest groups, for example, The Society for Modeling & Simulation International (SCS), ACM/SIGSIM, IEEE/SMC, and Institute for Operations Research and the Management Sciences: Simulation Society (INFORMS-SIM) and other societies. The second source of bibliography are the conference proceedings that are published by Springer as *Lecture Notes in Computer Science* (together with its subseries). Moreover, many of the renowned Operations Research (OR) journals are listed as sources of references, e.g. *Operations Research, European Journal of Operational Research, Journal of the Operational Research Society*. It does not come as a surprise that most of the simulation journals feature in the top-20 (*Simulation: Transactions of the SCS, ACM Transactions on Modeling and Computer Simulation, Simulation Modelling Practice and Theory*). Notable exception to these are the *Journal of Simulation* (published by the UK OR Society) and *The Journal of Defense Modeling and Simulation: Applications, Methodology, Technology* (published by SCS). *International Journal of Production Economics* (published by Elsevier) and *International Journal of Production Research* (Taylor & Francis) publish research

**Table 8.8** #Num references cited by SCSC authors, shown by year of publication of the cited article

| Year | Count | Year | Count | Year | Count |
|---|---|---|---|---|---|
| 2018 | 4 | 2005 | 815 | 1992 | 141 |
| 2017 | 44 | 2004 | 830 | 1991 | 136 |
| 2016 | 157 | 2003 | 750 | 1990 | 118 |
| 2015 | 216 | 2002 | 646 | 1989 | 82 |
| 2014 | 311 | 2001 | 517 | 1988 | 96 |
| 2013 | 381 | 2000 | 493 | 1987 | 73 |
| 2012 | 470 | 1999 | 424 | 1986 | 86 |
| 2011 | 467 | 1998 | 425 | 1985 | 64 |
| 2010 | 501 | 1997 | 313 | 1984 | 48 |
| 2009 | 588 | 1996 | 291 | 1983 | 48 |
| 2008 | 540 | 1995 | 260 | 1982 | 47 |
| 2007 | 735 | 1994 | 217 | 1981 | 48 |
| 2006 | 752 | 1993 | 173 | 1980 | 46 |

on manufacturing, engineering and management, process management, operations management and allied topic. The inclusion of both these journals in the top-20 list demonstrate the importance of simulation in making decisions pertaining to both manufacturing and services process. Another finding from this analysis is that two of the most renowned research journals of all times, namely *Nature* and *Science* are listed as sources of references and in particular *Nature* is in the top-5 list.

Table 8.8 shows the distribution of the above-mentioned references by year of publication. As mentioned, we only considered 12,353 references (these are not unique articles, but rather, unique articles multiplied by total number of occurrences/instances = total number of references) originating from articles published between 1980 and 2018 (Oct) and which were referenced by authors that published in SCSC proceedings from 2005 to 2017. The fact that the table identifies four references in 2018 is because some authors of the SCSC proceedings have referenced papers first published online or accepted for publication. From the table we see that fewer articles are generally cited from the 1980s and 1990s. We see a significant increase in the number of references cited from original articles published in the period 1998–2012. Paper published from 2013 onwards show a decreasing trend in terms of citations; this is expected as only five SCSC conferences (2013–2017) could have cited articles from 2013 onwards.

Our final analysis of this section reports on authors that have received the most number of citations. The source of these citations are the 911 papers in our dataset. Table 8.9 lists the shows the top-16 most cited authors based on the above selection of references. *Bruzzone* comes first with 223 references. With a total of 36 first and co-authored papers, *Bruzzone* is also the author with the highest number of papers in the SCSC proceeding (refer to Table 8.2). Other names that are repeated among the two lists are *Wainer* (second highest in terms of SCSC publications and sixth highest in terms of citations), *Longo* (reported as third highest in both the lists, with

16 SCSC papers and 54 citations respectively) and *Massei* (fourth in the list of authors with 14 co-authored publications; also fourth in terms of citations with 54 instances). Our top-16 list includes Bernie Ziegler, Richard Fujimoto, Andreas Tolk and Tuncer Ören.

### 8.4.7 Citation Analysis of the SCSC Papers

In our last analysis, Table 8.10 lists the total citations received by the 911 papers in our dataset; the data is shown by the year of the SCSC conference. The column *Citation Count* presents the total number of citations received, whereas the last column *Citation Count (no self-citation)* lists only the citations received from papers not authored or co-authored by any of the authors of the cited paper. Therefore, compared to the second column, there is a reduction of the number of citations in the last column.

As can be seen from Table 8.10, the number of citations received decline from 2012 onwards. This is to be expected since the most recently published papers in the SCSC conferences have been around only for a few years and is in line with the general trend in citations for academic literature. SCSC papers from 2007 have received over 700 citations, and this data point appears as an outlier. What could be the reason for this? One contributing factor for this spike is the number of papers that were published in the 2007 SCSC conference. From Fig. 8.1 we see that a total of 186 papers were included in the proceedings for that year, and for all other years this varied from the low of 19 papers (2012) to 82 papers (2010). Again, it comes as no surprise that papers published at the 2012 SCSC conference received only 35 citations (19 papers) and is lower than those received for the 2013 and 2014 conferences (with 59 and 72 papers published in the proceedings respectively).

**Table 8.9** List of the top 16 most cited authors

| Author | Citations received | Author | Citations received |
|---|---|---|---|
| Bruzzone, A.G. | 223 | Kim, T.G | 38 |
| Zeigler, B.P | 65 | Fujimoto, R.M | 37 |
| Longo, F | 54 | Revetria, R | 37 |
| Massei, M | 54 | Tolk, A | 36 |
| Mosca, R | 51 | Ören, T.I | 35 |
| Wainer, G | 45 | Jịvor, A | 33 |
| Giambiasi, N | 41 | Frydman, C | 31 |
| Mittal, S | 40 | Giribone, P | 30 |

**Table 8.10** Total citations received by SCSC papers published 2005–2017

| Year | Citation count | Citation count (no self-citation) |
|---|---|---|
| 2005 | 214 | 201 |
| 2006 | 132 | 121 |
| 2007 | 705 | 673 |
| 2008 | 84 | 74 |
| 2009 | 150 | 145 |
| 2010 | 200 | 188 |
| 2011 | 100 | 96 |
| 2012 | 35 | 35 |
| 2013 | 50 | 48 |
| 2014 | 56 | 54 |
| 2015 | 31 | 28 |
| 2016 | 19 | 16 |
| 2017 | 2 | 2 |
| **Total** | **1778** | **1681** |

## 8.5 Conclusion

The objective of this profiling paper was to present an overall picture of research published in the proceedings of the *Summer Computer Simulation Conference (SCSC)* from 2005 to 2017. Our thirteen years of analysis included a total of 911 papers. In this chapter we outline the structured approach that we followed for constituting the underlying dataset. We have presented seven different analysis, including, metrics associated with authorship and affiliations, sources of funding, analysis of keywords and citation analysis. Unlike our other journal profiling papers (Katsaliaki et al. 2010; Mustafee et al. 2012, 2017) where we read the full-text to capture variables directly from an article, for this work, we relied largely on analysis that was made available through Scopus. The challenge for us was to develop the dataset and which allowed for such an analysis. This could be considered as a limitation of this study since the results could be affected by unintended errors from the indexing editors at Scopus. However, it is also true that such errors are unlikely to have any significant effect on the summary results that have been presented in this book chapter. We hope that the results of the profiling study will be useful for the simulation community and, read along with several other chapters of this book, will highlight the contribution of the SCSC authors in advancing simulation methodology and practice in its 50th year!

## References

Claver, E., Gonzalez, R., & Llopis, J. (2000). An analysis of research in information systems (1981–1997). *Information & Management, 37*(4), 181–195.

Dwivedi, Y. K., & Kuljis, J. (2008). Profile of IS research published in the European journal of information systems. *European Journal of Information Systems, 17*(6), 678–693.

Dwivedi, Y. K., Lal, B., Mustafee, N., & Williams, M. D. (2009). Profiling a decade of information systems frontiers' research. *Information Systems Frontiers, 11*(1), 87–102.

Franco, L. A., & Hamalainen, R. P. (2015). Behavioural operational research: returning to the roots of the OR profession (Editorial to the special issue on BOR). *European Journal of Operational Research, 249*(3), 791–795.

Katsaliaki, K., Mustafee, N., Dwivedi, Y. K., Williams, T., & Wilson, J. M. (2010). A profile of OR research and practice published in the journal of the operational research society. *Journal of the Operational Research Society, 61*(1), 82–94.

Mustafee, N. (2011). Evolution of IS research based on literature published in two leading IS journals—EJIS and MISQ. In *Proceedings of the 19th European Conference on Information Systems (ECIS 2011)*. Helsinki, Finland, Paper 228.

Mustafee, N., & Fishwick, P. (2017). Analysis of M&S literature published in the proceeding of the winter simulation conference from 1981 to 2016. In A. Tolk, J. Fowler, G. Shao & E. Yücesan (Eds.), *Advances in Modeling and Simulation Seminal Research from 50 Years of Winter Simulation Conferences*. Springer.

Mustafee, N., Katsaliaki, K., Fishwick, P., & Williams, M. D. (2012). SCS—60 years and counting! A time to reflect on the Society's scholarly contribution to M&S from the turn of the millennium. *Simulation: Transactions of the Society for Modeling and Simulation International, 88*(9), 1047–1071.

Mustafee, N., Katsaliaki, K., & Fishwick, P. (2014). Exploring the modelling & simulation knowledge base through journal co-citation analysis. *Scientometrics, 98*(3), 2145–2159.

Mustafee, N., Katsaliaki, K., Le Forsonney, T., & Diallio, S. (2017). Profile of modeling and simulation research published in the journal of defense modeling and simulation. *The Journal of Defense Modeling and Simulation: Applications, Methodology, Technology, 15*(3), 337–349.

Palvia, P., Pinjani, P., & Sibley, E. H. (2007). A profile of information systems research published in Information & Management. *Information & Management 44*(1), 1–11.

Palvia, P., YK, C. P., Kakhki, M. D., Ghoshal, T., Uppala, V., & Wang, W. (2017). A decade plus long introspection of research published in information & management. *Information & Management, 54*(2), 218–227.

SCS. (2018). *World premier professional society: Society for modelling & simulation international*. http://scs.org/about/. Accessed November, 2018.

Yilmaz, L. (2011). Reflections on the 60th year anniversary of SCS. *Simulation: Transactions of the Society of SCS, 87*(1–2), 3–4.

**Navonil Mustafee** is Associate Professor of Operations Management and Analytics in University of Exeter Business School, UK. Nav has research interests in Hybrid Modelling, Simulation Methodologies, Healthcare Simulation and Bibliometric analysis. He is an Associate Editor for *Simulation: The Transactions of the SCS* and the *Journal of Simulation*.

**Korina Katsaliaki** is Assistant Professor in Operations and Technology Management at the School of Economics and Business Administration of International Hellenic University (IHU), Thessaloniki, Greece. She is also a collaborating teaching member of the Open University in Greece and Cyprus in the area of Healthcare Management.

# Chapter 9
# New Frontiers in Modeling and Simulation in Complex Systems Engineering: The Case of Synthetic Emergence

**Saurabh Mittal**

**Abstract** Complex systems are everywhere, in both natural and artificial world. One of the characteristic properties of complex systems is the manifestation of emergent behavior that continuously keeps the system evolving. Many times, this emergent behavior is the very reason for the systems' survival. To perform complex systems engineering for artificial systems, Modeling and Simulation (M&S) provides a harness to analyze, experiment and test the system before it goes into production. Developing an M&S environment that can manifest emergent behavior is a challenge in itself. Part of the problem is the elusive nature of emergence. This chapter will provide an overview of the evolutionary path the M&S of emergent behavior has taken in the last few years. It defines Synthetic Emergence as emergence engineering in artificial environments. The chapter provides the pre-, post-conditions and the engineering process to achieve synthetic emergence.

## 9.1 Introduction

Complex systems are everywhere, in both natural and artificial (man-made) world. Complexity is a universal problem that is being tackled across many disciplines including those of science and engineering. Man-made or artificial systems are generally top-down design. Natural systems are mostly bottom up with numerous feedback cycles that keep them both sustainable and evolving. Modeling and Simulation (M&S) has been applied to the study of complex systems from the 1950s and the quest hasn't ended yet. M&S, along with Big Data technologies are at the forefront of such exploration and investigation (Mittal et al. 2017; Fujimoto et al. 2017; Tolk and Oren 2017).

In a recent National Science Foundation (NSF) Workshop on Research Challenges in Modeling and Simulation for Engineering Complex Systems

S. Mittal (✉)
The MITRE Corporation, 7515 Colshire Dr, VA 22102 Mclean, USA
e-mail: smittal@mitre.org



J. Sokolowski et al. (eds.), *Summer of Simulation*, Simulation Foundations, Methods and Applications, https://doi.org/10.1007/978-3-030-17164-3_9

(Fujimoto et al. 2017), the following four key challenges were identified and explored:

- *Conceptual Modeling*: A conceptual model aligns various stakeholders and is the foundation for model-based engineering (MBE). When aligned for a specific domain, it provides the motivation to have a Domain-Specific Model (DSM). A valid conceptual model must be supported by a potent strategy to move to a computational model in a collaborative development environment. The collaborative environment ensures that the shared understanding stays close to the objectives at-hand. The computational modeling environment requires the development of modeling language, the selection of a modeling paradigm (discrete, continuous, agent-based, system dynamics) and mechanisms to facilitate conceptual exchange.
- *Computational Challenges*: This aspect refers to the computational technology (including High Performance Computing) available for both modeling and simulation activities. A valid computational model must lend itself to a verifiable simulation implementation. Advanced model transformation technologies are required that maintain the model-simulator relationship. Incorporating new age technologies like Big Data and necessary modalities like real-time Live, Virtual and Constructive (LVC) simulation environments present an engineering challenge.
- *Uncertainty*: A model is an abstraction of reality and a digital simulation implementation discretizes a continuous phenomenon. The former defines the knowledge boundary while the latter uses approximation and introduces error in the manifested computational behavior. Experimentation using stochastic processes provide a window to bringing uncertainty to light. When multiple engineering disciplines (including human-in-the loop) are integrated through the available models, the challenge becomes enormous. New technologies are needed that allow better control of abstraction and limit the error propagation. Accurate visualization technologies to instrument at both the modeling and simulation levels facilitate uncertainty control.
- *Reuse of models and simulations*: Models and the supporting simulation environments are often created for a specific purpose. Advanced mechanisms to repurpose the models to address new objectives are becoming a necessity as a valid model is an asset and should be reused to the maximum extent possible. The simulation environment that supports a model must be able to integrate with other simulation environments through the co-simulation methodology to provide the needed integrated simulation substrate.

All complex systems have two fundamental characteristic properties: emergent behavior and nonlinearity. Emergent behavior, a macro behavior that is more than the sum of its parts is sometimes traceable to individual component(s), and sometimes this behavior is just "novel". Complexity in these systems is multi-variable, multi-resolution and multi-dimensional in space-time continuum. Many artificial systems take the form of complex systems replete with emergent behavior when the manifested behavior deviates from the intended design considerations. Such artificial complex systems are known with different labels as system of systems, or complex adaptive system of systems. Complexity is further amplified by the various instituted control structures: centralized, decentralized or anarchy, in such artificial systems. Large scale enterprise systems, mission command, missile defense systems, healthcare, etc. belong to this category of artificial complex systems. Getting a handle on complexity and managing the emergent behavior is required to ensure predictable, reliable and repeatable behavior. Currently, there are a lack of tools to understand the inherent complexity in such complex systems so that they are useful in steering the evolution of these systems.

In the current taxonomy of Emergent Behavior across different schools of thought such as Complexity Science, Physics, Mathematics, etc., the following two fundamental definitions are prevalent (Wolf and Holvoet 2005):

1. *Weak*: The emergent behavior is a macro-level behavior that is reducible to agent/system composition and is not causal at micro-levels.
2. *Strong*: The emergent behavior is novel, causal to lower levels and is irreducible at lower levels. This behavior has to be proven consistent for adoption.

There are two processes by which emergent behavior is categorized in the above two categories:

1. *Epistemological*: The emergent behavior is described using existing science. This points out to the explanation process, various laws and governance procedures that will eventually put the emergent behavior in the above four categories.
2. *Ontological*: This emergent behavior questions the processes of the existing Science itself. The emergent behavior expands the scientific nomenclature or taps into an orthogonal domain ontology not relevant to the current ontology of the system under study. It attends to the pure novelty aspect as the observed phenomena is outside the boundary of current understanding (captured as the ontology of the modeled system).

While the above presented four challenges in Fujimoto et al. (2017) that are universal to M&S discipline, when it comes to emergent behavior in M&S, they have a specific interpretation (Table 9.1).

This chapter provides an overview on how the application and incorporation of emergent behavior in M&S discipline has shaped in the recent years and where the new opportunities are. Section 9.2 provides a chronological chart of evolution of emergence research in M&S. Section 9.3 introduces the concept of Synthetic Emergence and explores the notion of Synthetic Emergence in the context of Cyber Physical Systems (CPS). It describes the principles behind engineering Synthetic emergence and a framework by which it can be achieved in M&S environments. Section 9.4 discuss the challenges, new opportunities and summarizes the chapter.

**Table 9.1** M&S for engineering complex system challenges and their interpretation for Emergent Behavior

| Challenge | Emergent behavior interpretation |
|---|---|
| Conceptual modeling | Epistemological emergent behavior provides foundation for the expected emergent behavior. Ontological emergence ensures that ontological boundary is accurately established |
| Computational challenges | Epistemological emergence provides the guidance to ensure the model stays valid and simulation verifiable. Any deviation from the epistemological behavior provides the evidence to perform model and simulator correction |
| Uncertainty | Presence of uncertainty in any behavioral trajectory opens a window to ontological emergence that expands the ontological boundary of the model |
| Reuse of models and simulations | The epistemological emergent behavior can be reproduced in an M&S environment |

## 9.2 Evolution of Emergent Behavior Research in M&S

This section provides an overview of published work in emergent behavior and its application to M&S discipline.

**1985**

In *Complexity and Emergence*, Foo and Zeigler (1985) introduce the subject of emergent behavior in M&S through holistic effects. They define holism as:

Holism = reductionism + computation + higher-order-effects

While the computation is strictly the algorithmic complexity of the system, the higher-order effects are the emergent behaviors that are considered in the holistic representation.

**2013**

Nearly three decades later, In *Emergence in Stigmergic and Complex Adaptive Systems: A Formal Discrete Event Systems Perspective*, Mittal (2013) analyzed two classes of complex systems: stigmergic systems and complex adaptive system. A stigmergic system is a multi-agent system where agents interact through a persistent environment. In a complex adaptive system, the agents adapt and learn as well. The paper provides 29 characteristics of these system and how the Discrete Event Systems (DEVS) formalism (Zeigler et al. 2000) can be used to ensure weak emergent behavior in M&S environments. Mittal stated that DEVS formalism with its closure-under-coupling property guarantees weak emergent behavior and multi-level DEVS (ML-DEVS) with variable structure capability can support strong emergence in simulation environments, subject to validation. These concepts were further explored in netcentric complex adaptive systems by Mittal and Martin (2013).

**2014**

In *Attention-Focusing in Activity-based Intelligent Systems*, Mittal (2014) introduced the idea of a cyclical relationship between weak and strong emergence. Figure 9.1 shows the concept how it is related to *Open* system and *Closed* system concepts. This is also related to the *closure-under-composition* principle from Systems Theory (Mittal 2013). For M&S, a closed solution provides a guarantee that the system behavior trajectories are predictable.

This concept was then explored in two papers: *Modeling Attention-Switching in Resource-constrained Complex Intelligent Dynamic Systems (RCIDS), and Context and Attention Switching in Activity-based Intelligent Systems*, both by Mittal and Zeigler (2014a, b) that explored the behavior of resource-constrained system and the cyclical loop in reference to Control Theory as described by Ashby (1956) and Systems Theory as described by Wymore (1967). They also described how attention-switching is an emergent property of such a system. They developed a DEVS architecture based on *Winner-take-all* algorithm from the Cognitive Science domain that directs the attention and eventually the resources to the subject of interest (area of high activity). They also explored the concept of quantized context to discretize run-time phenomena and how it can be incorporated as a new system state.

**2015**

This year is marked with three important works. The first work is a book titled: *Modeling and Simulation Support to System of Systems Engineering* by Rainey and Tolk (2015). The chapter titled *The Role of Modeling and Simulation in System of Systems Engineering* (Maier 2015) related the taxonomy of emergent behavior with M&S explicitly as:

- *Simple*: Emergent property is readily predicted in a reduced abstracted model of the system
- *Weak*: Emergent property is consistently reproduced in simulations but not consistently predicted in advance. It can be understood through reduced complexity models.

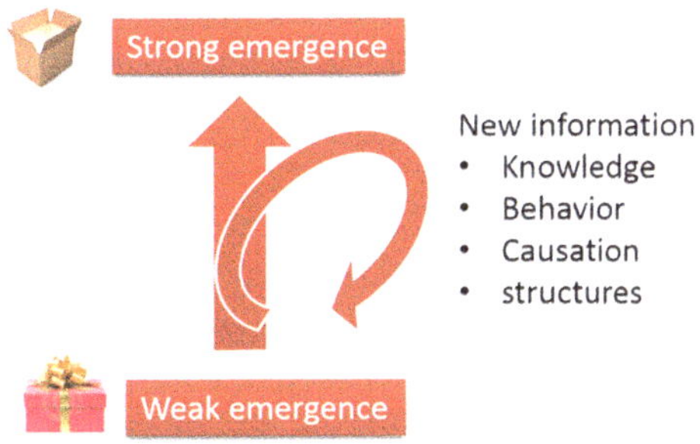


**Fig. 9.1** Weak and Strong emergent behavior in design of complex adaptive systems (Mittal 2014)

- *Strong*: Emergent property is consistent with the known properties of the system but is not reproduced in any simplified model of the system. Reduced complexity models or simulations do not reliably predict where the property will occur.
- *Spooky*: Emergent property is inconsistent with the known properties of the system and cannot be reproduced even with a model of the same complexity as the real system.

These definitions support the cyclical loop (Fig. 9.1) that weak emergent behavior can be consistently reproduced, while the strong emergent behavior requires new information (guidance and constraints) to be made predictable. Once it becomes predictable, it is no longer strong, no longer novel but is an expected, predictable weak emergent behavior (Mittal 2013). Maier conceptualized the spooky emergent behavior, which as we shall see ahead, is the emergent behavior we must strive to avoid or carefully incorporate in the new ontology.

The second work is in the same book (Rainey and Tolk 2015) titled *Human in the Loop in System of Systems (SoS) Modeling and Simulation: Application to Live, Virtual and Constructive (LVC) Distributed Mission Operations (DMO) Training* (Mittal et al. 2015). This chapter used the emergent behavior to describe intelligent systems, especially of artificial agents for pilot training. They brought the concept of affordance in relation to emergent behavior. Affordances can be seen as opportunities for action as they align agent's goals with observed emergent properties. While in weak emergence, such affordances are not available to the agent, in strong emergence, a *situated* agent displays a proactive role by taking advantage of emergent affordances. They developed the concept of Environment Abstraction (EA) to situate the agent in the LVC environment and bring the concept of ontology harmonization for handling epistemological emergent behavior in DMO Training.

The third work is the Best Paper for Summersim'15 titled: *Harnessing Emergence: The Control and Design of Emergent Behavior in System of Systems Engineering* (Mittal and Rainey 2015). The authors proposed Emergent Behavior Observer (EBO) snapshot to capture these macro-level states that can be attributed to holistic SoS states. Emergent behavior is a cross-cutting aspect of SoS. This aspect cuts though SoS characteristics, to name a few: operations, technology, processes, techniques, procedures, autonomy, structure. Figure 9.2 shows these aspects ($S_A$–$S_Z$) in the x-y plane and their relative impact with the length of these lines. The aspects enumerated above can correspond to $S_i$ in Fig. 9.2a. For example, $S_A$ is less impactful than $S_G$. Alternatively, small changes in $S_G$ can impact the overall system's behavior. The red outlined area in (a) depicts the cross-cut that defines a complex state of the system. Time is on z-axis and the blue shaded region signifies the duration for which the red region is sustainable. This complex state and the specified duration on z-axis is then put on a time-series with complex states on y-axis and time on x-axis, as shown in Fig. 9.2b. This gives the state-transition for emergent behavior that leads to a detector for emergent behavior. They introduced a methodology for SoS analytics. These transition systems can very well be developed using formal DEVS specifications. They also related an SoS with various

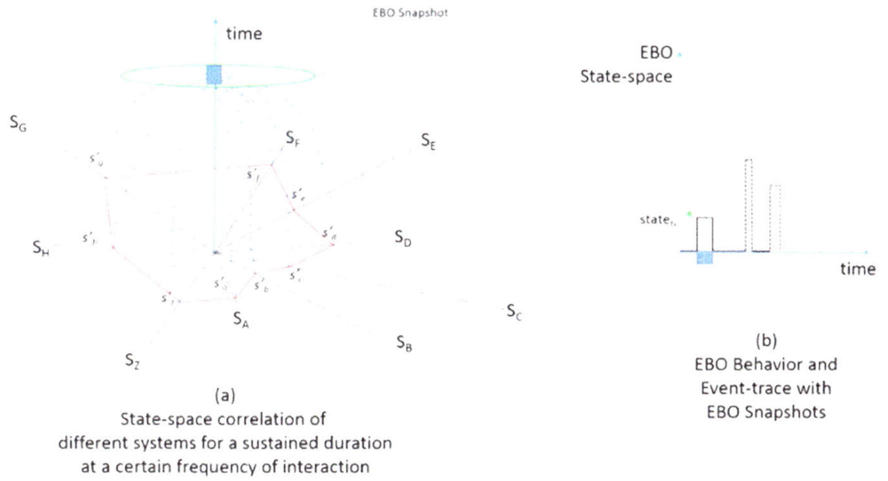


**Fig. 9.2** Emergent Behavior Observer spatiotemporal snapshots and EBO time-series behavior (Mittal and Rainey 2015)

branches of study. They brought in the Complexity perspective into the M&S of emergent behavior through Emergence Cone and the concepts of second order cybernetics to forefront. Figure 9.3 shows Emergence Complexity Cone linking emergent behavior taxonomy in increasing complexity on the y-axis, dividing the Cone into stochastic and deterministic search-spaces on the x-axis.

The deterministic domain is supported by established theories, like Cybernetics, Systems Theory, Control Theory and Network theory. The stochastic domain is supported by domains such as Estimation Theory. Cone volume depicts the variety (as conceptualized by Ashby 1956). Cone perimeter depicts constraints, and the knowledge boundary as a cylinder that addresses the variety and constraints. Knowledge cylinder around simple and weak emergence in the deterministic domain signifies ample knowledge available to develop abstractions. A diverging cone reflects the increasing complexity as constraints are loosened in the stochastic domain leading to an increase in variety and lack of theoretical constructs to understand the overall complex behavior.

**2016**

This year helped develop both formal and engineering approaches to model emergent behavior in M&S environments. Architecture and application were developed that incorporated the subject of emergence explicitly in their undertaking.

In *A note on promoting positive emergence and managing negative emergence in systems of systems*, Zeigler (2016a) expanded on the work of Mittal and Rainey (2015) to introduce a tri-layered architecture (Fig. 9.4) of positive emergence: an emergent behavior that sustains the overall purpose of SoS. The three layers (from the bottom) are: (1) *SoS ecology*: the composition layer where constituent systems

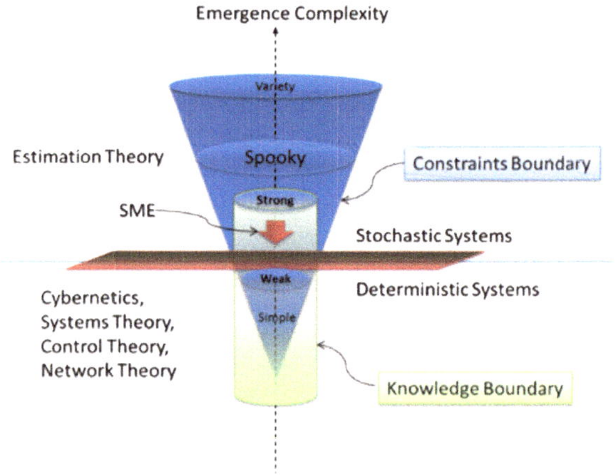


**Fig. 9.3** Emergence Complexity Cone (Mittal and Rainey 2015)

**Fig. 9.4** Tri-layered architecture for positive emergence (Zeigler 2016a)

| |
|---|
| Coordination Economics |
| Pragmatic Level Communication Network |
| System of Systems Ecology |

are deliberately considered for an SoS. (2) *Network supporting pragmatic level of communication*: the communication layer that assumes syntactic and semantic interoperability to provide pragmatic interoperability. (3) *Coordination economics*: a coordination layer that establishes control and management to create economic conditions for sustainment of emergent behavior. The tri-layered architecture was applied to emergence of topics in social networks with Twitter as an example (Zeigler 2016b).

In *Formalization of Weak Emergence in Multiagent Systems*, Szabo and Teo (2016) formalized the description of a weak emergent behavior for multi-agent systems. They introduced a grammar-based approach to formalize and identify the existence and the size of emergent property state sets without the need for prior knowledge of emergent properties. Their formalism permits the modeling of open systems and they acknowledge the state-space explosion in determining global "emergent" states. This explosion is controlled using degree-of-interaction metric and is applicable only to systems where emergent behavior is a result of direct interaction and does not consider indirect interaction.

Building on the work of Mittal (2013); Zeigler and Muzy (2016) in *Iterative Specification of Input/Output Dynamic Systems: Emergence at the Fundamental Level*, advanced the DEVS formalism and laid the conceptual groundwork for Iterative systems that are well-defined (static and deterministic) and progressive (temporal dynamic) i.e. do not suffer from halting problem, when autonomous systems are composed as coupled systems, and more specifically, as DEVS systems. They provide proofs on DEVS supporting fundamental emergence modeling as it operationalizes closure-under-coupling conditions that form the basis of well-defined resultants of system composition, especially where feedback coupling prevails.

In *Contextualizing Emergence in System of Systems Engineering with Gap Analysis*, Mittal and Cane (2016) used gap analysis to identify and contextualize the gap between the SoS' emergent behavior purpose and actual observed behavior with an objective of identifying SoS factors that contribute to the existence of such a gap, as depicted in Fig. 9.5. Various qualitative methods such as Analysis of Alternative (AoA), capability-based assessments, crown jewel analysis, dependency analysis, functional dependency network analysis, model-assisted experimentation, including analyses of risk, scenario, visualization, temporal behaviors and data, were presented.

**2017**

This year is marked with two works on computational infrastructure and applications. In *Simulation-based Complex Adaptive Systems*, Mittal and Martin (2017) presented the implementation issues with regards to simulation of a complex adaptive systems model. They advocate simulation infrastructure be free of any unknown approximations and accumulated error because of computational implementation. They also incorporate the elements of Machine Learning, Feature Engineering, Deep Learning and Hyper-heuristics supported by Big Data technology for a data-driven M&S in Complex Adaptive Systems (CAS) engineering (Fig. 9.6). They emphasized the importance of verification and validation of CAS models and simulation infrastructure.

In *Emergence of Human Language: a DEVS-Based Systems Approach*, Zeigler (2017) developed a DEVS model using Iterative Systems approach to formulate the emergence of language. This work was later refined as a book chapter (Zeigler and

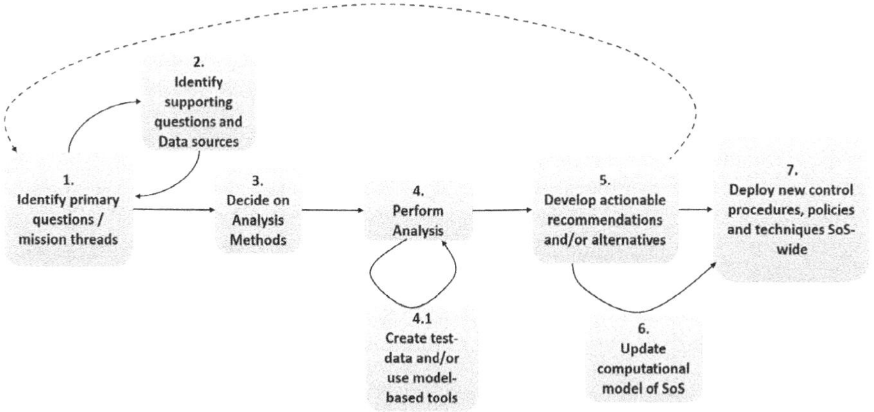


**Fig. 9.5** Gap Analysis Methodology for SoS Emergent Behavior contextualization

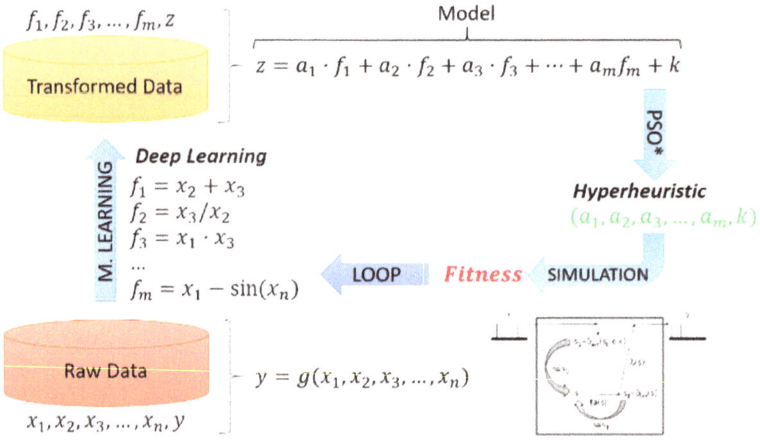


**Fig. 9.6** Specific view of the CAS modeling part incorporating Machine Learning in M&S (Mittal and Martin 2017)

Mittal 2018) using the RCIDS construct that supports attention-switching as a necessary condition for emergence of language between two hominid agents.

**2018**

With a relatively silent 2017, the year 2018 arrived with the release of two books: one about emergent behavior in M&S inspired from multiple disciplines and another on engineering emergent behavior, specifically in SoS.

In *Emergent Behavior in Complex Systems Engineering: A Modeling and Simulation Approach* (Mittal et al. 2018), the editors brought together leading experts in various disciplines that are struggling with emergent behaviors in their M&S application. Herein, the authors present the theoretical considerations and the tools required to enable the study of emergent behaviors in artificial systems. The text examines the underlying complex system model utilized to address a specific problem and discusses the emergent behavior obtained through the application of M&S. Table 9.2 shows a summary of this multi-disciplinary undertaking.

The text offers several simulation-based methods, technologies, and approaches that are designed to encourage the reader to incorporate simulation technologies to further their understanding of emergent behavior in complex systems across multiple disciplines. The authors present a resource for those designing, developing, managing, operating, and maintaining complex systems. The text is designed to help better detect, analyze, understand, and manage the emergent behavior inherent in complex systems engineering (CSE) in order to reap the benefits of innovations and avoid the dangers of unforeseen consequences. This vital resource also contains information on the next generation of CSE.

**Table 9.2** Summary of undertakings in Emergent Behavior in Complex Systems Engineering

| Discipline | Complex system | Emergent behavior tackled | Author(s) |
|---|---|---|---|
| Linguistics | Resource-constrained complex intelligent dynamical system | Language between two hominid agents | Mittal and Zeigler (2018) |
| Science | Generative systems | Creative cognition | Yilmaz (2018) |
| Engineering | System of system | Depends on operational use | Tolk et al. (2018c) |
| Enterprises | System of system, complex adaptive system | Organizational synergy | Rouse (2018) |
| Defense (Unmanned aerial vehicles) | Multi-agent system | Swarming behavior | Norman et al. (2018) |
| Sociology | Computational social system | Multi-model, role/identity | Oren et al. (2018) |
| Software | Complex software system | Failure | Gore (2018) |
| Anthropology | Lossy information networks | Culture | Lane (2018) |
| Ecology (Birds) | Multi-agent system | Grouping behavior detectability | Szabo and Birdsey (2018) |
| Transportation | Supply-chain distribution system | Economic losses | Ojha et al. (2018) |
| Chemistry | System of system | Reduction in entropy | Johnson et al. (2018) |
| Information economy | Multi-agent system | Technological lock-in as a social phenomenon | Frydenlund and Earnest (2018) |

The chapter titled: *Complex Systems Engineering and the Challenge of Emergence* (Tolk et al. 2018a) discussed the difference between complicated and complex systems. While complicated systems have an established engineering process, the complex systems are replete with emergent behavior. Going back to Fig. 9.1, this chapter brings insight that complex system need to be transitioned back to complicated systems, thereby bringing strong emergent behavior into the weak emergent category and making it predictable. Figure 9.7 illustrates how the emergence taxonomy is related to various types of systems.

In the second effort of 2018, in *Engineering Emergence: A Modeling and Simulation Approach*, the editors Rainey and Jamshidi (2018) brought in various authors that explored the subject of emergence in SoS engineering. It investigates emergence to interrogate or explore the domain space from an M&S perspective to facilitate understanding, detection, classification, prediction, control, and visualization of the phenomenon.

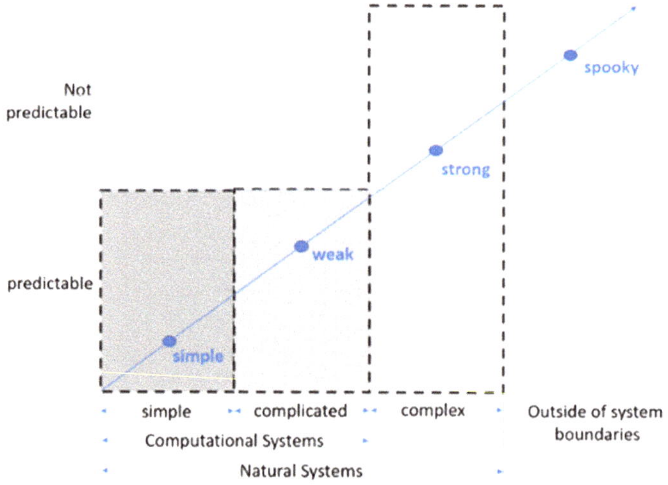


**Fig. 9.7** Systems and Emergence (Tolk et al. 2018a)

### *9.2.1 Summary*

Table 9.3 summarizes the key takeaways from the emergence research as applicable to the M&S discipline. The research incorporated in this section represents significant ideas that contributed to the advancement of the subject and is not an exhaustive list.

## 9.3 Synthetic Emergence

As stated earlier, epistemological emergent behavior deals with the explanation of the manifested behavior and how such behavior can be arrived at, and ontological emergent behavior deals with identifying what is manifested. The former relates to the process while the latter relates to an ontological label.

Arriving at synthetic emergent behavior in an experimental environment such as in simulation environment, is the objective of harnessing emergence in complex systems engineering. Therefore, Synthetic Emergence is defined as:

> Emergent Behavior created in artificial environments that include both computational and physical manifestations in controlled settings.

**Table 9.3** Summary of Emergence research in M&S domain

| Year | Emergent behavior M&S evolution | Author(s) |
|---|---|---|
| 1985 | Holism | Foo and Zeigler |
| 2013 | Discrete perspective, closure-under-coupling guarantee, Incorporation of DEVS Systems Theory and Network Theory for weak and strong emergent behavior | Mittal |
| | Application to netcentric complex adaptive system | Mittal and Martin |
| 2014 | Cyclical relationship between weak and strong emergent behavior. | Mittal |
| | Incorporation of systems theory, control theory and cybernetics | Mittal and Zeigler |
| | Application to context and activity in intelligent systems | Mittal and Zeigler |
| 2015 | Revised emergent behavior Taxonomy for M&S: simple, weak, strong, spooky | Maier |
| | Application to SoS Engineering, LVC and DMO Training | Mittal, Doyle and Portrey |
| | Incorporation of estimation theory, network theory, control theory, stochastic and deterministic systems. introduction of emergence behavior observer (EBO) snapshots, positive and negative emergence | Mittal and Rainey |
| 2016 | Tri-layered architecture for positive emergence in SoS | Zeigler |
| | Application to social networking sites such as Twitter | Zeigler |
| | Formalization of weak emergent behavior using context-free grammars | Szabo and Teo |
| | Iterative system specification for DEVS-based emergent behaviors | Zeigler and Muzy |
| | Contextualization of emergent behavior in SoS using gap analysis | Mittal and Cane |
| 2017 | Simulation infrastructure for eliminating computational emergent behavior and incorporating Machine Learning, Big Data and hyper-heuristics for data-driven M&S | Mittal and Martin |
| | Application to linguistics | Zeigler |
| 2018 | Application to Complex systems engineering | Mittal, Diallo and Tolk |
| | Application to SoS Engineering | Rainey and Jamshidi |

Artificial environments include purely man-made systems and hybrid systems. A purely man-made system can be complicated in terms of engineering that goes in developing it. The macro behavior obtained from such a system is tractable and reproducible. Hybrid system, on the other hand, are man-made systems that interface with organic systems. For example, in the near future, Smart City will be such a hybrid system.

### 9.3.1 Use Case for Synthetic Emergence

The computational world is a digital world, i.e. it is discrete. Conversely, the physical world is a continuous world, i.e. it is continuous. The discrete and continuous systems have been integrated in the past to engineer complex hybrid systems using both the systems engineering and M&S technology when the hybrid system is a closed system. However, in modern world, when these two class of system are separated by a communication medium such as Internet, which makes it an open system. Developing a hybrid system such that both the discrete computational system and the continuous physical system interact robustly is a challenge in CSE. Establishing robust and secure communications between the two components is a subject of ongoing research in a new class of systems now known CPS. Developing a corresponding hybrid M&S environment is fraught with its own challenges (Tolk et al. 2018b).

Performing a CPS simulation requires that a CPS model be first built. While CPS modeling is not the focus of this chapter, a recent panel explored the state of the art of CPS modeling and the complexity associated in engineering intelligence, adaptation and autonomy through M&S. The literature survey conducted in (Tolk et al. 2018b) enumerate the following active research areas and the associated technologies for CPS modeling and concluded that the need for a common formalism that can be applied by practitioners in the field is not yet fulfilled:

- *DEVS formalism*: Strong mathematical foundation that support multi-paradigm modeling (Vangheluwe 2000), multi-perspective modeling (Traoré et al. 2018), and complex adaptive systems modeling (Mittal 2013) to handle emergent behaviors.
- *Process algebra*: Provides hybrid processes using multi-paradigm modeling. Models combine behavior on a continuous time scale with discrete state transition behavior at given points in time.
- *Hybrid automata*: Combines finite state machines with Ordinary Differential Equations (ODE) to account for non-deterministic finite states. Bond graphs are used to govern changes.
- Simulation languages: Combines discrete event and continuous system simulation languages. Involves modular design of hybrid languages, multiple abstraction levels combining different formalisms.
- *Business processes*: Use of standardized notational languages like Business Process Modeling Notation (BPMN) provide value in securing buy-in from the stakeholders in an efficient manner.
- *Interface design for co-modeling*: Functional Mock-up Interface (FMI) as a means of integration of various CPS components. DEVS can also be used as a common denominator in a vendor neutral manner.
- *Model-driven approaches*: Model transformation chains to arrive at a single formal model. Governance is required to develop such automation.

- *Agent-based modeling*: Paradigm to employ component models at scale with individual behaviors, to study ensemble effects.

The above-mentioned approaches and technologies allow the development of CPS models, albeit in a piecewise manner. These model pieces and their definitions and specifications are dictated by the cross-domain CPS operational use-case. Assuming we now have a validated model (i.e. a model that has been deemed valid by the stakeholders), next comes the task of executing it on a computational platform i.e. simulation. The piece-wise model composition sometime does not directly translate into a monolithic simulation environment due to the confluence of both the continuous and discrete system in the hybrid system. In the literature survey (Tolk et al. 2018d) as well as in many discussions with the experts (Tolk et al. 2018b), the use of co-simulation was identified as the best way to go in support of CPS for development, testing, and eventually training. According to (Gomes et al. 2018), co-simulation is the co-existence of two or more simulators, to produce a behavioral trace by coupled simulators. It is initiated by a co-simulation scenario and requires a top-level coordinator. Mittal and Zeigler (2017) elaborated on the definition as:

> The task of integrating various simulators to perform together as a composite simulation is termed as co-simulation. This involves weaving the time series behavior and data exchanges accurately, failure of which, will yield inaccurate simulation results. Every such hybrid system would require a dedicated effort to build a co-simulation environment.

So, given the complexity of such a hybrid M&S system, limiting the emergent behavior out of simulation infrastructure and ensuring that it is robust, and close to the modeling layer, engineering synthetic emergent behavior is the next topic in progression.

## *9.3.2 Principles of Synthetic Emergence*

To engineer synthetic emergence, we must achieve three goals:

1. There is no emergent behavior as a result of simulation platform (Mittal and Martin 2017)
2. There is explicit communication protocol between the component systems (Zeigler and Mittal 2018)
3. The observed emergent behavior as useful (validated) or eliminated (pursuant to further investigation and resolution)

To achieve these three goals, the system must have the necessary pre-conditions to arrive at the synthetic emergent behavior.

**Pre-conditions**

1. There are at least two systems engaged in communication.
2. The systems are autonomous and maintain their individuality.

3. The systems need to be on the same time-base i.e. their clocks need to be synchronized if they are on different time-base.
4. The propagation delays in communication are accounted for abstract time execution.
5. They share a common ontology or data dictionary.
6. The systems do not exhibit *zenoness* property i.e. not advancing time or running an infinite loop calculation in zero virtual time with real-time consequences.
7. The systems are intended to be connected and share semantic context.

Once the observed resulting behavior, due to the semantic-temporal interactions, is detected and deemed *novel* and *consistent* in presence of various experimental conditions, such as: random seeds, temporal characteristics, quantization of context, time-base and other domain-dependent parameters, etc., the next step is to arrive at post-conditions for such emergent behavior.

**Post-conditions**

1. The new sematic-temporal signature can be recorded for a finite duration, as in EBO Snapshot described earlier.`
2. The system components share a semantic context.
3. The system falls out of emergent behavior as context is switched.
4. The granularity of time advancement in hybrid system is explicitly handled i.e. the emergent behavior should be reproducible at higher granularity but the system may fall out of emergence at coarse granularity.

**Process to Achieve Synthetic Emergence**

Having met the pre-conditions and post-conditions for obtaining synthetic emergent behavior, the next step is to bring it back into the system model, understand it (epistemology and ontology), thereby, executing the cyclical loop in Fig. 9.1.

Following is the stepwise process for executing the cycle for the model:

1. Discover emergent behavior.
2. Perform *consistency* check.
3. Perform *novelty* check. Label, if novel. Create nomenclature if outside the bounds of current ontology.
4. Develop instrumentation, analytics and visualization of system components states at multiple resolutions.
5. Develop heuristics and perform experimentation. Identify operating boundaries.
6. Acknowledge reproduction. Instrument precursors, transients and state vectors.
7. Develop controls to assure, test and track system components.
8. Incorporate control mechanisms and interaction protocols in system components.
9. Develop standards, if needed.
10. Document epistemology

This sequence increases the abstraction level of the M&S effort incrementally, to arrive at a place where M&S provides a model that is useful, and its behavior supported by the needed hardware.

**Verification and Validation**
As established in the pre-conditions, emergence is a multi-body phenomenon, beginning with a two-body system. In a larger context, such as a System of System, we may not have access to any or all of systems' internal architecture. Because we are dealing with a lot of black boxes, i.e. systems behind interfaces, proprietary technologies, etc., the best we can do is validation. Verification implies we have access to the black boxes and can instrument and setup testing procedures to make it a white box and then verify. In one study at Air Force Research Lab (AFRL) (Mittal et al. 2015), the developed system integrated virtual simulators (e.g. cockpit simulators) and constructive elements (cognitive architectures). It focused on the validation as verification was just hard due to both the "business" nature and the technical scope of the problem. Deliberate efforts were made in making sure the validation mechanisms were sound (through SME input).

The situation is same in CPS M&S. At some point in system specification level, we end up with a black-box in the model and its behavior has to be assumed and ascertained by how it responds in the larger SoS. If we are pursuing verification, we must adhere to a domain Reference Architecture which is unfortunately, not the de facto method of developing system across multiple vendors/systems. Co-simulation for hybrid models is the preferred way (Tolk et al. 2018c).

## *9.3.3 LVC M&S Environments for Synthetic Emergence*

Exploration of emergent behavior in conjunction with Model-based Systems Engineering (MBSE) in LVC environment was demonstrated at AFRL study (Mittal et al. 2015). In the chapter, *Research Agenda for Next-Generation Complex Systems Engineering*, Diallo et al. (2018) described a mixed method approach using LVC environments to engineer artificial environments required for complex systems engineering.

Ontological (spooky) emergence marks the upper bound of systems thinking and epistemology can help bring strong emergent behavior back into the weak emergent behavior, thereby executing the cyclical relationship (Mittal et al. 2018). The mixed method approach puts the model in situations and environments that allow generation of new knowledge and categories, thereby setting the model on an evolutionary path of continuous improvement and usefulness. As it has been discussed by Tolk (2019) in his chapter in this book, a computable model alone would not be able to achieve this feat alone, the model must be put in an environment that allows its augmentation and evolution.

Simulation of a model affords two things: experience and experimentation with the model (Mittal et al. 2018). LVC environment provides ample opportunities to experience the model in experimental settings. An LVC environment has the following characteristics (Diallo et al. 2018):

1. *Open-Ended*: The participants can draw from their personal and professional experiences. The interaction between the live (people and environments) and computational elements allows development of shared mental models that facilitate evolution of the computational model (Yilmaz 2018) which further allows model validation and model orientation of the entire exercise.
2. *Rich Data*: The LVC environment is heavily instrumented such that the experiment can be post-analyzed for generating additional insights that aid model developments. The instrumentation mechanisms may be multi-modal, i.e. the instrumented output can be expressed physically, verbally, physiologically, and mentally.
3. *Delay and misunderstanding*: The LVC is designed to be non-linear, to put model in situations that involve open systems such as live players. The exchange between the elements of LVC will test model assumptions, limitations and capabilities. Consequently, there is an acceptable delay before the experienced simulation generates insights that inform model evolution. During this process of adaptation, misunderstandings may occur and is acknowledged as an acceptable part of the discovery process.
4. *Transdisciplinarity*: LVC environment allows inclusion of participants from different perspective and professional backgrounds to experience and challenge the model. To achieve ontological emergence, we must embrace the differences between artists, engineers, humanists, and neuro-atypical thinkers. This inclusion of diversity will only strengthen the usefulness of model. Models created by one Subject Matter Expert (SME) for one purpose may not serve him eternally, but the same model may become useful to another domain when given an opportunity to experience in LVC environment.

The Simulation Experience Approach (SEA) framework to engineering LVC environment (Column 1) and its associated source discipline (Column 2) is summarized in Table 9.4.

The SEA framework opens the M&S discipline to multiple disciplines for incorporation of novel perspectives that help validate the model in the original context and further aid its evolution and possible application to other disciplines. While the modeling aspect of SEA framework undoubtedly increases its usefulness, the simulation aspect is a challenging one. As described in Sect. 3.1 the simulation infrastructure, which now takes the shape of co-simulation infrastructure requires dedicated effort involving computer, software and system engineers. As Mittal and Martin (2017) point out, the emergent behavior arising out of bad engineering of simulation infrastructure must be eliminated or made aware of before the true emergent behavior at the modeling level is to be accepted and integrated.

Synthetic emergence in an artificial environment must be a modeling phenomenon only that is experienced in a run-time LVC environment as an immersive experience. It is my recommendation that the process of achieving Synthetic emergence as laid out in Sect. 3.2 be adopted by the SEA framework once all the steps in SEA framework (Table 9.4) have been executed.

**Table 9.4** SEA framework as a transdisciplinary approach (Diallo et al. 2018)

| Step | Discipline |
|---|---|
| Formulate modeling question | Psychology, anthropology, social studies, performing arts, visual arts |
| Design live environment | Engineering, theater, scenography, education, special education |
| Design virtual environment | Film, language and literature, mathematics, computer science |
| Design constructive environment | Computer engineering, social sciences, history, geography, physics, biology, geology |
| Create Interfaces | Neuroscience, humanities, physiology, linguistics, systems engineering |
| Identify data collection | Statistics, psychology, sociology |
| Execute experiment | Sociology, philosophy |
| Debrief | Ergonomics, communication, criminal justice |

## 9.4 Discussion and Conclusion

The four grand challenges of complex systems engineering from the NSF workshop, namely, conceptual modeling, computational implementation, uncertainty and reuse of models and simulations, are interpreted for emergent behavior research in M&S discipline (Table 9.1). The interpretation requires the process of developing epistemology and ontology of the emergent behavior. This chapter provided an evolution of M&S research geared to development of both the epistemological processes (Table 9.3) and the ontological process through SEA framework (Table 9.4).

This chapter has made the following additional contributions:

1. Introduced the concept of Synthetic Emergence and the principles to engineering it in artificial environments such as LVC.
2. Summarized the SEA framework that provides step-wise process to engineer LVC environment that involve multi-disciplinary participation.
3. Introduced the Synthetic emergence process and applied with the SEA framework step-wise process by which emergence can be achieved in artificial environment

This chapter further reinforces the fact that a model is never complete and is continuously evolving. Giving adequate infrastructure to experience the model in a simulated artificial environment provides opportunities to deem the model useful to participants across multiple disciplines. Engineering artificial simulation environments such as LVC is expensive and challenging. This comes to be the economic cost of generating new knowledge, either epistemology or ontology. Once engineered and assured, the resulting synthetic emergence is worth paying the price, for it not only advances Science but allows transdisciplinarity and knowledge sharing across diverse use cases, making M&S a worthy cause in itself (Tolk and Oren 2017, Mittal and Tolk 2019).

**Disclaimer**

The author's affiliation with The MITRE Corporation is provided for identification purposes only and is not intended to convey or imply MITRE's concurrence with, or support for, the positions, opinions, or viewpoints expressed by the author. This paper has been approved for Public Release; Distribution Unlimited; Case Number PR_18-2996-2.

## References


Ashby, W. (1956). *An introduction of cybernetics*. Chapman and Hall.

Diallo, S., Mittal, S., & Tolk, A. (2018). Research agenda for next generation complex systems engineering. In S. Mittal, S. Diallo & A. Tolk (Eds.), *Emergent behavior in complex systems engineering: A modeling and simulation approach*. Wiley and Sons.

Foo, N., & Zeigler, B. P. (1985). Complexity and emergence. *International Journal of General Systems, 10*(2–3), 163–168.

Frydenlund, E., & Earnest, D. C. (2018). Emergence in information economies: An agent-based modeling perspective. In S. Mittal, S. Diallo, & A. Tolk (Eds.), *Emergent behavior in complex systems engineering: A modeling and simulation approach*. Wiley and Sons.

Fujimoto, R., Bock, C., Chen, W., Page, E., & Panchal, J. H. (Eds.). (2017). *Research challenges in modeling and simulation for engineering complex systems*. Springer AG.

Gore, R. (2018). Isolating the causes of emergent failures in computer software. In S. Mittal, S. Diallo, & A. Tolk (Eds.), *Emergent behavior in complex systems engineering: A modeling and simulation approach*. Wiley and Sons.

Gomes, C., Thule, C., Broman, D., Larcen, P. G., & Vangheluwe, H. (2018). Co-simulation: A survey. *ACM Computing Surveys, 51,* 3. https://doi.org/10.1145/3179993.

Johnson, J. J., Padilla, J. J., & Sousa-Poze, A. (2018). Modeling emergence in system of systems engineering using thermodynamic concepts. In S. Mittal, S. Diallo, & A. Tolk (Eds.), *Emergent behavior in complex systems engineering: A modeling and simulation approach*. Wiley and Sons.

Lane, J. E. (2018). The emergence of social schemas and lossy conceptual information networks: How information transmission can lead to the apparent "emergence" of culture. In S. Mittal, S. Diallo, & A. Tolk (Eds.), *Emergent behavior in complex systems engineering: A modeling and simulation approach*. Wiley and Sons.

Maier, M. (2015). The role of modeling and simulation in system of systems development. In L. B. Rainey, & A. Tolk (Eds.), *Modeling and simulation support for system of systems engineering applications*. Hoboken, NJ: John Wiley and Sons.

Mittal, S. (2013). Emergence in stigmergic and complex adaptive systems: A formal discrete event systems perspective. *Cognitive Systems Research, 21,* 22–39. https://doi.org/10.1016/j.cogsys.2012.06.003.

Mittal, S., & Martin, J. L. R. (2013). *Netcentric system of systems engineering with DEVS unified process*. Boca Raton, FL: CRC Press.

Mittal, S. (2014). *Attention-focusing in activity-based intelligent systems*. Activity-based Modeling and Simulation. Zurich, Switzerland.

Mittal, S., & Zeigler, B.P. (2014a). Modeling attention-switching in resource-constrained complex intelligent dynamical system (RCIDS). In *Symposium on Theory of M&S/DEVS*. Springsim, Tampa, FL.

Mittal, S., & Zeigler, B. P. (2014b). Context and attention in activity-based intelligent systems. In *Proceedings of Activity-based Modeling and Simulation* (ACTIMS'14), *ITM Web of Conferences,* Vol. 3. http://dx.doi.org/10.1051/itmconf/20140303001.

Mittal, S., Doyle, M., & Portrey, A. (2015). Human in the loop in system of system (SoS) modeling and simulation: Applications to live, virtual and constructive (LVC) distributed mission operations (DMO) training. In L. Rainey, & A. Tolk (Eds.), *Modeling and simulation support to system of systems engineering*. Wiley and Sons.

Mittal, S., & Rainey, L. B. (2015). Harnessing emergent behavior: The design and control of emergent behavior in system of systems engineering. In *Proceedings of Summer Computer Simulation Conference*. Chicago, IL.

Mittal, S., & Cane, S. (2016). Contextualizing emergence in system of systems engineering with gap analysis. Springsim'16.

Mittal, S., & Zeigler, B. P. (2017). Theory and practice of modeling and simulation in cyber environments. In A. Tolk, & T. Oren (Eds.), *The profession of modeling and simulation: Education, vocation, discipline and ethics*, pp. 223–263. Hoboken, NJ: John Wiley & Sons.

Mittal, S., & Martin, J. L. R. (2017). Simulation-based complex adaptive systems. In S. Mittal, U. Durak, & T. Oren (Eds), *Guide to simulation-based disciplines: Advancing our computational future*. Springer AG.

Mittal, S., Durak, U., & Oren, T. (2017). *Guide to simulation-based disciplines: Advancing our computational future*. Springer AG.

Mittal, S., Diallo, S., & Tolk, A. (2018). *Emergent behavior in complex systems engineering: A modeling and simulation approach*. Wiley & Sons.

Mittal, S., & Tolk, A. (2019). *Complexity challenges in cyber physical systems: Using modeling and simulation to support intelligence, adaptation and autonomy*. Wiley (in press).

Norman, M., Koehler, M. T., & Pitsko, R. (2018). Applied complexity science: Enabling emergence through heuristics and simulations. In S. Mittal, S. Diallo, & A. Tolk (Eds.), *Emergent behavior in complex systems engineering: A modeling and simulation approach*. Wiley and Sons.

Ojha, A., Corns, S., Shoberg, T., Qin, R., & Long, S. (2018). Modeling and simulation of emergent behavior in transportation infrastructure restoration. In S. Mittal, S. Diallo, & A. Tolk (Eds.), *Emergent behavior in complex systems engineering: A modeling and simulation approach*. Wiley and Sons.

Oren, T., Mittal, S., Durak, U. (2018). Induced emergence in computational social systems engineering: Multimodels and dynamic couplings as methodological basis. In S. Mittal, S. Diallo, & A. Tolk (Eds.), *Emergent behavior in complex systems engineering: A modeling and simulation approach*. Wiley and Sons.

Rainey, L. B., & Tolk, A. (2015). *Modeling and simulation support to system of systems engineering*. Wiley & Sons.

Rainey, L. B., & Jamshidi, M. (2018). *Engineering emergence: A modeling and simulation approach*. CRC Press.

Rouse, W. (2018) Emergence in complex enterprises. In S. Mittal, S. Diallo, & A. Tolk (Eds.), *Emergent behavior in complex systems engineering: A modeling and simulation approach*. Wiley and Sons.

Szabo, C., & Birdsey, L. (2018). Towards the automated detection of emergent behavior. In S. Mittal, S. Diallo, & A. Tolk (Eds.), *Emergent behavior in complex systems engineering: A modeling and simulation approach*. Wiley and Sons.

Szabo, C., & Teo, Y. M. (2016). Formalization of weak emergence in multiagent systems. *ACM Transactions on Modeling and Computer Simulation (TOMACS), 26*(1), 6.

Tolk, A., Diallo, S., & Mittal, S. (2018a). Complex systems engineering and the challenge of emergence. In S. Mittal, S. Diallo, & A. Tolk (Eds.), *Emergent behavior in complex systems engineering: A modeling and simulation approach*. Wiley and Sons.

Tolk, A., Page, E, & Mittal, S. (2018b). Hybrid simulation for cyber physical systems–state of the art and a literature review. In *Proceedings of Annual Simulation Symposium, Spring Simulation Multi-Conference*, pp. 122–133. Baltimore, MD: SCS.

Tolk, A., Mittal, S., & Page, E. (2018c) Hybrid simulation: Framework, methods and foundations with focus on cyber physical systems, MITRE Technical Report 180341.

Tolk, A., Barros, F., D'Ambrogio, A., Rajhans, A., Mosterman, P., Shetty, S., et al. (2018d). Hybrid simulation for cyber physical systems–a panel on where are we going regarding complexity, intelligence, and adaptability of CPS using simulation. In S.Mittal, J. L. R. Martin, & M. Luetzenberger (Eds.), *Proceedings of MSCIAAS–Spring Simulation Multi-Conference*, pp. 681–698. Baltimore, MD: SCS.

Tolk, A. (2019). Limitations and usefulness of computer simulations for complex adaptive systems research. In U. Durak, N. Mustafee, J. Sokolowski, & A. Tolk (Eds.), *50 years of Summer Simulation Conference*. Springer AG.

Traoré, M. K., Zacharewicz, G., Duboz, R., & Zeigler, B. P. (2018). Modeling and simulation framework for value-based healthcare systems. *Simulation*. https://doi.org/10.1177/0037549718776765.

Vangheluwe, H. L. (2000). DEVS as a common denominator for multi-formalism hybrid systems modelling. In *IEEE International Symposium on Computer-Aided Control System Design*. IEEE, pp 129–134.

Wolf, T., & Holvoet, T. (2005). Emergence versus self-organization: Different concepts but promising when combined. *Lecture Notes in Artificial Intelligence, 3464*, 1–15.

Wymore, W. (1967). *A mathematical theory of systems engineering: The elements*. Huntington, NY: Krieger.

Yilmaz, Y. (2018). Generative parallax simulation: Creative cognition models of emergence for simulation-driven model discovery. In S. Mittal, S. Diallo, & A. Tolk (Eds.), *Emergent behavior in complex systems engineering: A modeling and simulation approach*. Wiley and Sons.

Zeigler, B. P., Praehofer, H., & Kim, T. G. (2000). *Theory of modeling and simulation: Integrating discrete event and continuous complex dynamic systems*. Academic Press.

Zeigler, B. P. (2016a). A note on promoting positive emergence and managing negative emergence in systems of systems. *The Journal of Defense Modeling and Simulation: Applications, Methodology, Technology* 133–136. https://doi.org/10.1177/1548512915620580,13,1.

Zeigler, B. P. (2016b). Contrasting emergence: In systems of systems and in social networks. *The Journal of Defense Modeling and Simulation: Applications, Methodology, Technology*. Retrieved from March 10, 2016, https://doi.org/10.1177/1548512916636934.

Zeigler, B. P. and Muzy, A. (2016). Iterative specification of input/output dynamic systems: Emergence at the fundamental level. *Systems 4*(4), 34. Special Issue "Second generation general system theory: Perspectives in philosophy and approaches in complex systems" https://doi.org/10.3390/systems4040034.

Zeigler, B. P. (2017). Emergence of human language: A DEVS-based systems approach. SpringSim, Virginia Beach, VA, April 5, 2017.

Zeigler, B. P., & Mittal, S. (2018). System theoretic foundations for emergent behavior modeling: The case of emergence of human language in a resource-constrained complex intelligent dynamical system. In S. Mittal, S. Diallo, & A. Tolk (Eds.), *Emergent behavior in complex systems engineering: A modeling and simulation approach*. Wiley and Sons.

**Saurabh Mittal** is Chief Scientist with Simulation, Experimentation and Gaming Department at the MITRE Corporation. He has a Ph.D. in Electrical and Computer Engineering from the University of Arizona, Tucson. He serves on the Board of Directors for Society of Modeling and Simulation (SCS) International and on the editorial boards of the Transactions of the SCS and Journal of Defense Modeling and Simulation (JDMS).

# Chapter 10
# An Introduction to Pedestrian Modeling Using Spatial Discrete-Event Modeling and Simulation

**Gabriel A. Wainer and Ala'a Al-Habashna**

**Abstract** We discuss an approach for spatial discrete-event Modeling and Simulation (M&S) of crowds using the Cell-DEVS formalism, which provides some advantages over existing crowd modeling approaches, as it provides a trade-off between the simulated granularity level and the computational requirements. We explain through simple one- and two-dimensional models how Cell-DEVS is used to build pedestrian models. Furthermore, the usability of the approach is verified by employing it in real-life case studies. We discuss different case studies, employing the Cell-DEVS approach to build a model for a general building evacuation, and a fire evacuation model.

## 10.1 Introduction

Modeling and Simulation (M&S) is an important tool that has been employed to study the behavior of many complex systems. One of the emerging applications for M&S is studying crowd dynamics. The simulation of virtual crowds has found a certain appeal as an illustrative tool in urban and architectural projects. It allows designers and engineers to visualize the utility of their project's space and facilitates the feedback, discussion, and decision-making among all stakeholders. This is becoming increasingly important considering that the world's population is increasing at a high rate, and moving towards urban areas, which increased the occurrence of the crowd phenomena (Zhan et al. 2008).

Crowd analysis can be used for developing crowd management strategies to increase safety in highly crowded situations (like concerts, convocations, demonstrations, public celebrations, etc.). Crowd analysis can also be used in building design, in order to provide a more efficient use of spaces. Crowd analysis is also important in virtual environments as it leads to better simulation in such artificial

G. A. Wainer (✉) · A. Al-Habashna
Department of Systems and Computer Engineering, Carleton University, 1125 Colonel by Dr., Ottawa, ON K12 5B6, Canada
e-mail: gwainer@sce.carleton.ca



J. Sokolowski et al. (eds.), *Summer of Simulation*, Simulation Foundations, Methods and Applications, https://doi.org/10.1007/978-3-030-17164-3_10

settings. Crowd M&S is now used in many areas such as safety, architectural design, computer games, transportation, etc. (Zhou et al. 2010), as in most situations, it can be very difficult (or costly) to study the behavior of crowds using simulacra. On the other hand, crowds exhibit highly complex dynamics, which makes it difficult to characterize its behavior using pure analytical models. When modeling a small crowd (i.e., one with a few dozen individuals), it is easy to investigate the behavior of each individual. However, in large crowds (i.e., with thousands of individuals), the interest is in the overall emergent behavior.

In this chapter, we present an approach for modeling and simulation of crowds based on Cell-DEVS. As Cell-DEVS is an entity-based approach, it provides a higher level of details of the behavior of the individuals in the crowd than fluid dynamic models. At the same time, the Cell-DEVS models require less computational demands than agent-based models.

## 10.2 Related Work

There are many models have been proposed recently for M&S of crowds. These models can be categorized based on the modeling approach into fluid dynamics-based models, social force-based models, agent-based models, Cellular Automata (CA)-based models, and others. In this section, we will discuss each of these methods.

### *10.2.1 Fluid Dynamic Models*

Fluid dynamics models study the crowd as a continuum using coupled nonlinear, partial differential equations that can be solved for simple geometries. It has been proposed that crowds move in a similar manner as fluid flows. Based on that, early work in this area (Bradley 1993) suggested that the Navier-Stokes equations could be applied to pedestrian flows. However, this does not take into account many factors that affect the crowd behavior such as various physiological, psychological, and social factors. As such, recent models do not use the Navier-Stokes equations in their entirety; rather the concepts of fluid dynamics are combined with consultation from behavioral scientists.

In Hughes (2003), the author extended the work in this area, where the crowd is modeled using classical fluid dynamics, but with the additional assumption that human flows "think". Bradley's model was based on a sociological view of crowds which considered that non-orchestrated crowds are rational and can therefore be expected to abide by scientific rules of behavior (McPhail 1991). Hence, the nonlinear, time-dependent, simultaneous equations representing a crowd are conformably mappable. This property makes many interesting applications analytically tractable. The theory has been used to study the Hajj (annual Islamic pilgrimage to

Mecca), in an attempt to improve the flow of pilgrims over the Jamarat Bridge near Mecca. For further and more up to date examples and applications in the literature on Fluid dynamics models, the reader is referred to Treuille et al. (2006), Xia et al. (2009), and Dogbe (2010). The problem with fluid dynamics is that the models do not provide high-resolution details of the behavior of the individuals.

### 10.2.2 Social-Force Models

Another method, proposed in Helbing and Molnár (1995) is called *social force*. In this method, the human motion is viewed as a complex behavior subject to a self-driving force and repulsive forces from the environment (other pedestrians and obstacles). Each pedestrian is assumed to be affected by four major factors; the destination to be reached, the distance to be kept from other people, the distance to be kept from borders and other obstacles, and other people or objects that might attract the pedestrian. The completed social force model can be found from these factors, and the paths taken by pedestrians can be predicted.

In Helbing et al. (2000) the social force model was extended to exhibit numerous phenomena to study the buildup of pressure observed during escape panics. The model considers a mixture of socio-psychological and physical forces influencing the behavior in the crowd. For further date examples, the reader is referred to Hoogendoorn and Bovy (2004) and Campanella et al. (2009).

### 10.2.3 Agent-Based Models

In agent-based models, each individual in the crowd is modeled as separate agent that takes its decisions independently. The local phenomena can affect the decision-making process of each individual, while the entire crowd can produce emerging patterns deducted by the social and physical aspects of each individual.

In Klügl and Rindsfüser (2007), the authors presented an agent-based simulation of pedestrian traffic of the complete railway station of Bern during rush hour. In their simulations, more than 40,000 agents pass through the station during 1.5 h. Furthermore, in their simulations, pedestrians are not only capable of avoiding collisions, but also able to flexibly plan their way through the railway station.

In Ronald et al. (2007), the authors investigated the behaviors that pedestrians may exhibit, and the belief-desire-intention (BDI) architecture, presenting the development of a sample model using Prometheus, an agent-oriented design methodology, and JACK, an agent-oriented programming language.

Agent-based models usually try to simulate crowd at fine scale, which makes them computationally demanding and more suitable for short-term simulations with small-sized crowds. Further examples in the literature on agent-based models can be found in Pluchino et al. (2014) and Liu et al. (2014).

### 10.2.4 Cellular Automata Based Models

CA is one of the oldest models of natural computing; it was introduced by John von Neumann in the late 1940s (Neumann and Burks 1966; Burks 1971; Wolfram 1986). In CA, the studied space is represented as a lattice of cells, with each cell being a state machine. The states of a cell come from a finite set of states (Kari 2005). Cells change their states synchronously at discrete time steps. The state of a cell at the next time step depends on its current state, and the current states of the neighboring cells according to an update rule. The neighborhood usually contains some or all the adjacent cells, but more general neighborhoods can be specified.

CA have been used recently for M&S of pedestrian movement. In Burstedde et al. (2001), the authors proposed a 2D CA model to simulate pedestrian traffic, and to simulate the evacuation of a large room with reduced visibility. The model introduced the concept of "floor field", which can be thought of as a second grid of cells underlying the grid of cells occupied by the pedestrians. Floor field holds the probabilities of moving from a cell to other cells. Dynamic floor cells can evolve with time so that probabilities change with time depending for example on the presence of pedestrians. Hence, floor field is used to model a "long-ranged" attractive interaction between the pedestrians.

The Situated Cellular Automata (SCA) model (Bandini et al. 2006) is a particular class of Multilayered Multi-Agent Situated Systems (MMASS). SCA provides explicit spatial representation, and defines adjacency geometries. SCA was used to build a small-sized model to simulate an environment with the crowd defined as a Multi-Agent Systems (MAS). In Tao and Jun (2009), an entity-based model was used to represent bidirectional pedestrian flow using CA. Different behavioral factors were considered such as position exchange and step back. The pedestrian CA in Ji et al. (2013) focused on acceleration and overtaking. They divided pedestrians into two categories: aggressive and conservative. The model was used to simulate the movement of pedestrians in a corridor.

In Masuda et al. (2014), a simple CA reproduces oscillation phenomena due to formation and destabilization of arches in 2D flows. This is used to study the jamming of pedestrian crowds that occurs due to the formation of arches at bottlenecks. The model predicts critical bottleneck sizes for particle flows without congestion, and it determines the dependency of the jamming probability on the system size. In Vihas et al. (2013), the authors define a CA in which pedestrians follow leaders, as this phenomenon is a fundamental driving mechanism. The model provides microscopic simulation of the crowd, as all configurations of the model are triggered by simple rules applied locally to each of the group members. They also study the emergence of qualitative attributes such as collective effects, random to coherent motion due to a common purpose, and transition to incoordination (arching) due to clogging.

The proxemic approach (Was et al. 2012), which is the process of acquisition of space in evacuation modeling, is based on a detailed representation of space and floor fields. The model allows for efficient, real time simulation of evacuation from large facilities using detailed representation of spatial relations. In order to reduce the computation cost of pedestrian models, (Steffen and Chraibi 2014) reduce the simulation time by using a multicast approach that performs fast simulation of probable evacuation scenarios. The work deals with the problem of passing agents from a CA to a force-based model, and they provide a CA that addresses the problem at less computational cost, with some possible loss of accuracy.

### *10.2.5 Cell-DEVS*

The Cell-DEVS formalism (Wainer 2000; Wainer 2009), allows modeling discrete-event cell spaces built as n-dimensional grids of cells. Each cell is defined as a DEVS atomic model, and a procedure to couple cells is defined. Figure 10.1a shows the contents of an atomic cell. A cell is only active when an external event occurs, or when an internal event is scheduled. When there are no further scheduled events, the cell will passivate. When an external event occurs, the external transition function is executed, and the local computing function ($\tau$) is activated. When the cell's state changes, the external function will schedule an internal transition, and the state change is transmitted after a delay of d. The local computing function in a Cell-DEVS model computes the next state of a cell depending on its current state, and the states of a finite set of nearby cells (like in CA).

The internal computing function is defined using a set of rules indicating the output VALUE for the cell's state after some time DELAY, when a PRECONDITION is satisfied. The rule format is denoted as <VALUE> <DELAY> <PRECONDITION>, which means that when the PRECONDITION is satisfied, the state of the cell will change to the assigned VALUE, and this new value will be transmitted to its neighborhood after a period time of DELAY.

After a cell is defined, it can be integrated into a coupled model representing the cell space. The CD++ M&S tool provides a development environment for implementing Cell-DEVS models using a built-in specification language (Wainer 2009).

Cell-DEVS is built on top of the DEVS formalism (Zeigler et al. 2000), which provides a formal framework for modeling generic dynamic systems and includes

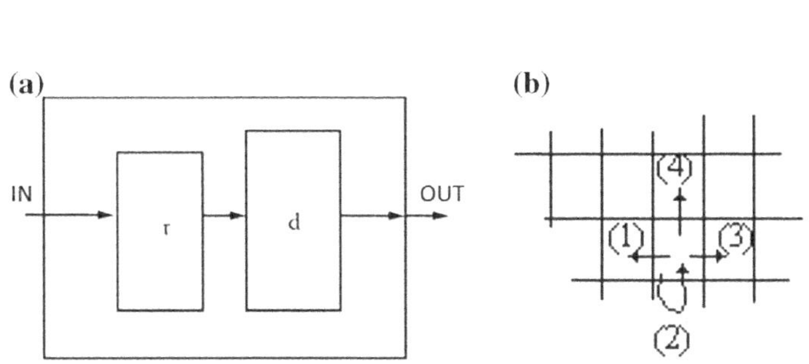


**Fig. 10.1** *Cell-DEVS* model: **a** Atomic cell **b** 2D coupled model

hierarchical, modular, and component-oriented structure, and formal specifications for defining structure and behavior of a discrete event model. Coupled component define the hierarchical structure of the system, while each atomic component is the basic building block of the system, which represents its behavior.

Since Cell-DEVS implement entity-based modeling, it provides a higher level of details than fluid dynamics models, and needs less computation demands than agent-based models (Al-Habashna and Wainer 2015; Zhou et al. 2010). Furthermore, as discussed in Sect. 10.1, Cell-DEVS has multiple advantages over CA, which makes it easier to develop larger models, and allows faster execution of models. In the following section, we discuss how Cell-DEVS can be employed to build pedestrian and crowd models, and present various examples.

### 10.2.6 Centroidal Particle Dynamics

The Centroidal Particle Dynamics (CPD) method is a personal space preserving method, which models pedestrian dynamics based on the concept that a pedestrian, in crowd, tend to maintain its personal space (Hesham et al. 2018). When this concept is employed for modeling pedestrians in highly-dense areas, it produces a very realistic pedestrian behavior. CPD provides a dynamic, autonomous, and adaptive approach that generates realistic pedestrian behavior.

The CPD approach is a variation of the social force approach. With this approach, a pedestrian is modeled as an entity with a personal space, and each pedestrian in the simulation tries to maintain and regain its personal space. This approach very suitable for close-range crowd in areas with high pedestrian density, as the personal space preservation concept produces realistic crowd behavior. The personal space preservation concept was not employed arbitrarily; it is a well-known human behavior in physiology, and a natural reaction that people employ as a mechanism to avoid any uncertainty and unexpected behavior that could be taken by surrounding people (especially strangers).

With the CPD approach, the Personal Space Map (PSM) is first constructed for the pedestrians in the model. This is done by checking the surrounding of each pedestrian within a certain radius (0.8–1.0 m) and calculating the available and violated space for each pedestrian. According to the results obtained in the previous step, the new location for the geometric center (centroid) of a pedestrian with be calculated to regain the full range of personal space. The centroidal force is also calculated, which is a vector pointing from the current location of the pedestrian to the new location.

The movement of a pedestrian can be impacted by various forces such as the movement on a global path to a destination, physical factors such as obstacles, and the different psychological and physiologic factors such as fear or following friends and family. When the centroidal force is calculated, it will be used with the other forces affecting the pedestrian movement to calculate the net force. The net force is then used to calculate the acceleration/declaration experienced by the pedestrian at each time step.

## 10.3 Models of Pedestrian Behavior with Cell-DEVS

In this section, we discuss how Cell-DEVS can be used to develop pedestrian and crowd models. Then, we provide two case studies of 1D and 2D pedestrian models. The models presented in this section are basic models that provide examples and explanation of how Cell-DEVS are used for modeling and simulation of crowd. More complex and realistic models will be provided in the next section.

As discussed in Sect. 10.2, in order to build a Cell-DEVS model we first need to define the following components of the model:

- The dimensions of the model, i.e., the number of cells in each dimension
- The set of possible states for the cells
- The neighborhood shape
- The set of rules that define the behavior of the cells

A Cell-DEVS model can be built in a modular and hierarchical fashion, by integrating various DEVS and Cell-DEVS models together. The cell's states can be used to represent different aspects of the space used by that cell. For instance, certain values can be representing occupied cells while others can represent vacant cells. Depending on the space of each cell, a cell can contain more than one pedestrian. For small cells (e.g., 0.4 $m^2$), a cell can only occupy one pedestrian at a time. In such cases, mechanisms to model collision avoidance should be employed. Moving direction is another important aspect that should be taken into consideration when defining cell states. Different values can be assigned to different movement directions. Pedestrians may have different speeds due to different time constrains. Pedestrians might also walk at a constant speed or have to change their speed. As such, the status of cell can be used to reflect the speed of the pedestrian in the cell.

### *10.3.1 One-Dimensional Movement Model*

We will start with a simple one-dimensional movement model. A cell can be considered to be a single-pedestrian space. This is the case considered in all the Cell-DEVS models in this chapter. Figure 10.2 shows the neighborhood used in

**Fig. 10.2** Neighborhood of the 1-D movement model

| | | |
|---|---|---|
| (0,-1) | (0,0) | (0,1) |

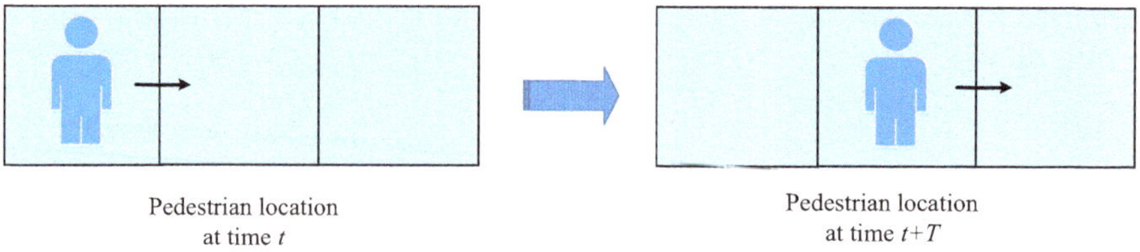


**Fig. 10.3** Pedestrian moving forward in one-direction

such a model. The figure shows a simple, 1×3 neighborhood. There are two states for each cell; a cell can be either empty (state 0) or occupied (state 1).

A simple example is shown in Fig. 10.3; which depicts a pedestrian moving east. A pedestrian moves one cell forward at a time. The movement speed of the pedestrian in the model can be controlled through the time delay.

A simple set of rules can be used for this model. For instance, the pedestrian moves forward if the cell ahead is vacant, which can be represented as follows.

```
Rule 1: 1 400 {(0,0)=0 and (0,-1)=1}
Rule 2: 0 400 {(0,0)=1 and (0,1)=0}
```

The rules are checked sequentially starting from the first rule and continuing until one of the preconditions is satisfied. The first rule states that if the core cell is empty and there is a pedestrian to the west (cell (0,−1)), the next state of the core cell will be 1. The second rule states that if there is a pedestrian in the core cell, and the cell to the east (cell (0,1)) is vacant, the next state of the core cell will be 0. The combination of these two rules represents the movement of the pedestrian. Although we have used a fixed delay of 400 ms to represent the time to cross the 0.4 m distance, we can use a function representing the speed for each cell in each of the rules allowing the modeler to easily represent the time it takes for each individual to cross a cell in different directions. As the cell area is 0.4 $m^2$, which is the average area occupied by a pedestrian, the delay is set to 400 ms. In this way, a constant movement will simulate a walking speed of 1 m/s, which is the average walking speed of pedestrians.

## *10.3.2 Two-Dimensional Movement Model (Crosswalk Model)*

In this section, we explain through an example how to use Cell-DEVS for building a 2-dimensional pedestrian model. The example used is a pedestrian crosswalk model that represents the dynamics of pedestrians crossing a street. Three different scenarios might occur:

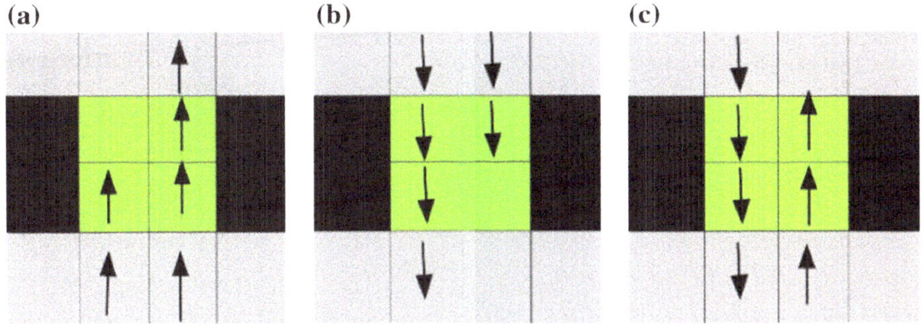


**Fig. 10.4** Crosswalk model scenarios. **a** Upward movement **b** Downward movement **c** Bidirectional movement

1. There are pedestrians standing only at one end of the crosswalk, i.e. only upward movement. This scenario is shown in Fig. 10.4a (arrows represent pedestrians).
2. There are pedestrians standing only at the other end of the crosswalk, i.e. only downward movement. This scenario is shown in Fig. 10.4b.
3. There are pedestrians at both ends of the crosswalk, therefore, during their passing, collisions might occur. This scenario is shown in Fig. 10.4c.

The model represents the movement of pedestrians along the crosswalk, and illustrates collisions that may happen when two pedestrians (one moving upward and another moving downward) are vying for the same cell. In this case one, of the pedestrians must sidestep. If sidestepping cannot be done due to unavailability of cells, then pedestrians must wait until a surrounding cell becomes available.

This model uses two columns of cells to represent the crosswalk. Pedestrians will wait at both ends of the crosswalk until the corresponding cells (i.e. crosswalk) turns green. The crosswalk is only safe for a certain amount of time (e.g., 1 min) and once those cells have changed back to black (representing a street with vehicles) no pedestrian is allowed to cross. The pedestrians are allowed to move three cells ahead (2 green cells representing the crosswalk and 1 gray cell representing the side walk) or one cell sideway (sidestepping).

The Cell-DEVS model will be based on a two-dimensional square grid. The grid dimensions are 4×4. Figure 10.5 shows the cell space. The central 4 cells model the crosswalk. When the traffic light is green, and cars pass the crosswalk, the 4 cells have value of zero. The color of the central cells in this case will be red and pedestrians are not allowed to cross the street. Once the traffic light turns red, the color of these cells will change to green and their corresponding value is set to 2.

The gray cells represent the sidewalks at both sides of the street, and they are occupied by pedestrians. Once the crosswalk cells turn green, the pedestrians will pass the crosswalk and get to the other side of the street. All pedestrians can be standing at one end of the crosswalk and move to the other end, or there could be

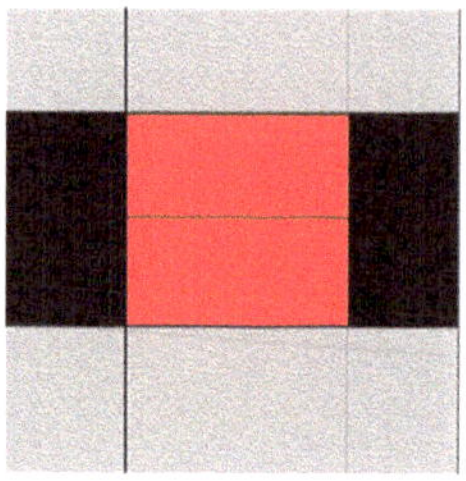

**Fig. 10.5** Cell space representing the crosswalk scene. Red cells represent the crosswalk

bi-directional movement. We use a Moore's neighborhood (the 9 adjacent neighbors) and the cell states in Table 10.1.

More rules are needed in the case of two-dimensional movement to handle collision avoidance. For instance, when two pedestrians walk in opposite directions, the rules governing their movement should consider this, and avoid collisions. We use pedestrians walking at a constant speed.

Following, we define the rules that govern the movement of a pedestrian walking upward. The rules of four different scenarios are listed and explained below. The first rule in each case considers the case where the pedestrian leaves the current cell, while the second rule considers the cell where the pedestrian moves. We show the rules for the first 3 cases. Afterwards, we list rules for the remaining cases.

(a) A crosswalk cell is red, turn it to green after some time delay

```
Rule 1 : 2 400 { (0,0)= 0 }
```

Here, if a crosswalk cell is red (has a value of 0), it will turn green (value 2), and after 400 ms, this change is reported to the neighboring cells, which will make the cell be considered as safe to walk into.

(b) The next rules show the case where there is no pedestrian ahead and the cell is green, then we move upward. This is a case where there is no other pedestrian or an obstacle in the cell ahead (−1,0), and there is no pedestrian in cell (−2,0) walking in the opposite direction. This case is illustrated in Fig. 10.4a.

```
Rule 2: 2 400 {(0,0)=1 and (-1,0)=2 and (-2,0)=2}
Rule 3: 1 400 {(0,0)=0 and (1,0)=1 and (1,0)=2}
```

The first rule checks if the current cell has a pedestrian walking upward. In this case, the pedestrian will move forward, i.e., the current cell will be green. The

**Table 10.1** State values used for the crosswalk model

| State | Value | Color |
|---|---|---|
| Crosswalk is not safe to pass | 0 | Red |
| Crosswalk is safe to pass | 2 | Green |
| Cell is occupied by a pedestrian moving upward | 1 | Blue |
| Cell is occupied by a pedestrian moving downward | −1 | Blue |
| Street | 3 | Black |
| Sidewalk | 5 | Gray |

second rule checks if the core cell is empty and the cell below has a pedestrian walking upward, in which case the pedestrian will move to the core cell.

(c) If there is a pedestrian ahead, move to the right (east) provided that this cell to the right is green. If the next cell forward contains a pedestrian, we try to move to the east if that cell is green and no other pedestrian is going to move into it.

```
Rule 4: 2 400 {(0,0)=1 and (-1,0)!=2 and (0,-1)=2 and
(-1,-1)!=-1 and (1,-1)!=1}
Rule 5: 1 400 {(0,0)=2 and (0,1)=1 and (-1,1)!=2 and
(-1,0)!=-1 and (1,0)!=1}
```

Rule 4 checks if the current cell has a pedestrian walking upward, and the cell forwarded is occupied (or not green), and the cell to the east is available, green, and no one is moving to it (from the bottom or the top). In such case, the current cell will be green, cause the pedestrian will move out of it. Rule 5 checks the same conditions for the cell where the pedestrian is going to move into.

The remaining rules are listed below.

(d) *Pedestrian/obstacle ahead and to the east, move to the left (west) if possible.* In this case, the cell ahead is occupied by an obstacle or a pedestrian moving in the opposite direction, and the cell to the east is occupied by a pedestrian or an obstacle. In such case, the pedestrian will have to move to the cell to the west, given that, it is empty, and that no pedestrian is walking into it

(e) *Two pedestrians vying for the same cell; move to the east if possible.* When there is pedestrian moving upward and another moving downward, and both are moving into the same cell, the one walking upward will move to the cell to the east, given that it is vacant, green, and no other pedestrian is moving into it.

(f) *Two pedestrians vying for the same cell and there is a pedestrian or an obstacle to the east; move to the west.* When there is a pedestrian moving upward and another moving downward, and both are moving into the same cell, and it is not possible for the one moving upward to sidestep to the east, the pedestrian will sidestep to the west, given that it is vacant and no other pedestrian is moving into it.

When none of the cases above is satisfied, the pedestrian simply does not move.

Following, we present simulation results obtained for this model. First, we simulated the case were two pedestrians walking upward. Step by step visualization of the obtained results are shown in Fig. 10.6. Initially, the pedestrians are the bottom of the crosswalk, and the crosswalk is red, which means that the pedestrians cannot cross the street. In the next step, the crosswalk cells turn into green (rule 1) which means that pedestrians can cross the street now. In the next 3 steps the pedestrians move forward, one cell at a time until they cross the street and reach the sidewalk on the other side.

The following scenario shows two pedestrians moving across the street. One of them is moving upward, and the other is moving downward. Each pedestrian was positioned in a different lane so that no collision will be experienced. The results from this case are shown in Fig. 10.7. As in the previous scenario, the crosswalk cells are red in the beginning, and they turn green afterwards. Thereafter, the pedestrians start walking cross the street, each in the opposite direction of the other.

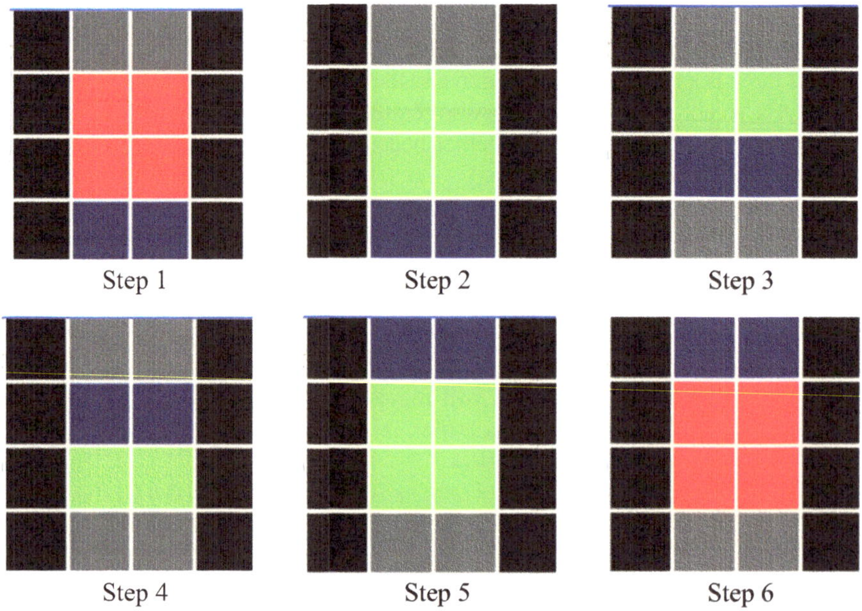


**Fig. 10.6** Simulation results: two pedestrians moving upward

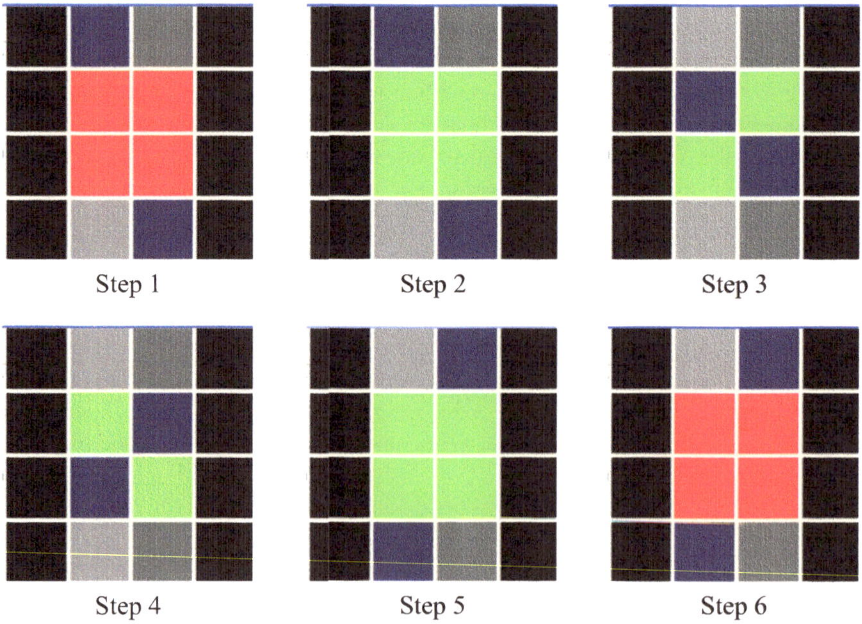


**Fig. 10.7** Simulation results: two pedestrians moving upward (no collision)

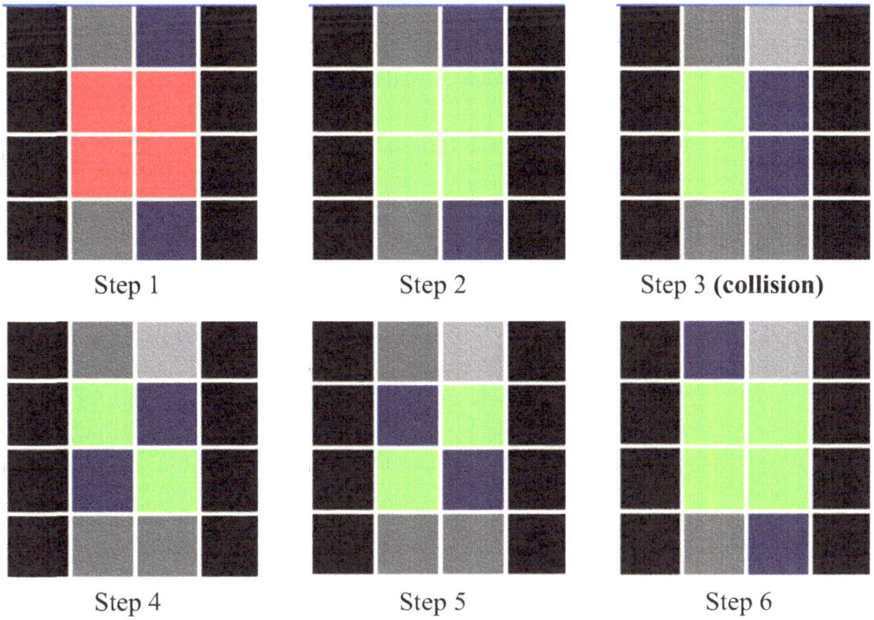


**Fig. 10.8** Simulation results: two pedestrians moving upward (with collision)

Since, each one is walking in a different lane; both reach the sidewalk on the other side of the street without colliding with the other.

In the fourth scenario (shown in Fig. 10.8), we simulated two pedestrians moving across the street. One of them moving upward and one is moving downward. Not like scenario 3, both pedestrians are walking in the same lane, and hence, they will collide at step 3. In step 4, the pedestrian walking upward steps to the west (as the cell to the east is unavailable). This resolves the collision, as the pedestrians are now walking in different lanes. Each pedestrian continue moving ahead until reaching the sidewalk on the other side of the street.

## 10.4 Cell-DEVS Fire-Evacuation Model

Shopping malls, office and school buildings are some examples of buildings that we come across in a day to day basis which include meeting of a large number of people within closed areas. These buildings are made in such ways to maximize the utilization of the limited space. However, safety should be taken into consideration when designing them, especially for emergency evacuation scenarios. One of the most common causes for such evacuations is the occurrence of fires. In such scenarios, a large number of people will have to evacuate the building in a limited time.

In this section, we present a more advanced pedestrian model using Cell-DEVS: a 3-dimensional fire-evacuation model. In the first subsection, we present our evacuation model. We also show some simulation results for the building evacuation model. In the second subsection, we discuss the fire spread model. Thereafter, present fire-evacuation scenarios and simulation results.

## *10.4.1 Building Evacuation*

In our model, the state of each cell will be determined by two main factors. The first factor is the direction of the shortest path to the nearest exit, as illustrated in Fig. 10.9. The figure shows that each cell will have a certain direction the pedestrian (occupying the cell) should take to get to the exit. As we are working with a 3-dimensional model here, some cells will be used as stair-cases that lead from one floor to another.

The second factor that decides the cell sate is whether it is occupied or not. In the beginning of each simulation, the direction of each cell is determined. Then, pedestrians are distributed randomly throughout the building. Depending on the two factors above (direction and occupancy), the state of the cell will be set. In addition to the cells that model the rooms and pathways, other cells are used to represent walls and exits in the building. The cell states are shown in Table 10.2.

As can be seen in Table 10.2, numbers from 3 to 10 represent the pathways. Even numbers in that range represent occupied cells while odd numbers represent vacant cells. For example, a cell with a value of 10 means that the cell is occupied, and the person will move to the cell to the left when it is available. When the person moves, the state of the cell will change to 9. A three-dimensional neighborhood is also used here as pedestrian need to check the availability of the stairwell cells when moving up and down between building floors (in the third dimension).

When pedestrians enter a staircase cell, they "vanish" from the simulation, and they are considered as having left the building. Stairway cells are used to model stairways between the floors. A stairway is modeled with two cells, one to represent the top of the stairway, and another to represent the bottom. In the event that both the top and base of the stairway are occupied, no one can enter until either is vacant.

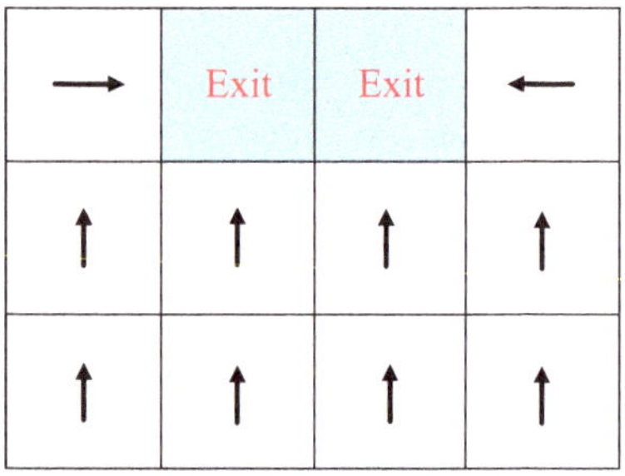


**Fig. 10.9** Movement direction of the cells

**Table 10.2** Cell states of the fire evacuation model

| State | State color | State name | State | State color | State name |
|---|---|---|---|---|---|
| 1 | Black | Wall | 8 | Green | Up occupied |
| 2 | Red | Exit | 9 | White | Left |
| 3 | White | Down | 10 | Green | Left occupied |
| 4 | Green | Down occupied | 11 | Yellow | Top of stairs |
| 5 | White | Right | 12 | Green | Top occupied |
| 6 | Green | Right occupied | 13 | Blue | Bottom of stairs |
| 7 | White | Up | 14 | Green | Bottom occupied |
| 15 | Red | Fire | | Green | |

A pedestrian enters the top of stairwell when the pathway instructs it to and when the top of the stairway has the value of 11 (unoccupied). A pedestrian at the top of the stairway moves down to the bottom of the stairway whenever it is vacant. Afterwards, the pedestrians leave the bottom of the stairway and follow the path to the next stairway or to the exit of the building.

The rules used in the model can be categorized into three sets: initialization, movement, and fire-spread rules. As their name indicates, initialization rules are used to distribute pedestrians in the model, and set the initial states of the pathway cells. A sample of the initialization rules, which are used to initialize the first floor, are shown below:

```
rule : { if (uniform(0,1) < 0.2), 10, 9) } 0 { (0,0,0) = 0 and (0,-1,0) > 1
      and (0,-1,0) < 11}
rule: { if (uniform(0,1) < 0.2), 4, 3) } 0 { (0,0,0) = 0 and (1,0,0) > 1 and
      (1,0,0) < 11}
rule : { if (uniform(0,1) < 0.2), 6, 5) }0 { (0,0,0) = 0 and (0,1,0) > 1 and
      (0,1,0) < 11}
rule : { if (uniform(0,1) < 0.2), 8, 7) } 0 { (0,0,0) = 0 and (-1,0,0) > 1
      and (-1,0,0) < 11}
```

All the pathway cells first have a value of zero. The cells next to the exit will first be initialized to point to the exit. Then, the neighboring cells of the ones that were just initialized will be also initialized to point to the former ones, and so on, until all the pathway cells are initialized. In this way, the direction of the cells will always form the shortest path to the exit. In addition to initializing a cell with direction, each cell will be populated with a pedestrian with a certain probability. The probability determine the density of pedestrians in the building. For instance, the rules above generate pedestrians with a probability of 0.2.

The second set of rules are used to control the movement of pedestrians in the various area of the model, including the pathways, stairwells, and exits. Following, we show as an example a set of rules used to move people into an empty cell within the same floor:

```
rule : 4 400 { (0,0,0) = 3 and ( (0,1,0) = 10 or (-1,0,0) = 4 or
      (0,-1,0) = 6 or (-1,0,0) = 14 or (1,0,0) = 14 or (0,1,0) = 14
      or (0,-1,0) = 14 )}
rule : 6 400 { (0,0,0) = 5 and ( (1,0,0) = 8 or (-1,0,0) = 4 or
      (0,-1,0) = 6 or (-1,0,0) = 14 or (1,0,0) = 14  or (0,1,0) = 14
      or (0,-1,0) = 14 )}
rule : 8 400 { (0,0,0) = 7 and ( (1,0,0) = 8 or (0,1,0) = 10 or
      (0,-1,0) = 6 or (-1,0,0) = 14 or (1,0,0) = 14  or (0,1,0) = 14
      or (0,-1,0) = 14 )}
rule : 10 400 { (0,0,0) = 9 and ( (1,0,0) = 8 or (0,1,0) = 10 or
      (-1,0,0) = 4 or (-1,0,0) = 14 or (1,0,0) = 14  or (0,1,0) = 14
      or (0,-1,0) = 14 )}
```

The first rule states that if the core cell value is 3 (vacant cell with direction of movement pointing down), and any of the cells around it is occupied with direction of movement toward the core cell, the next state of the core cell will be 4 (occupied cell with down direction of movement). The other 3 rules are the same but for cells with different direction of movement, for example, the second rule assumes the core cell state is 5. As another example, we list another set of rules below that are used to move people into an exit or a stairwell:

```
rule : 3 400 { (0,0,0) = 4 and ((1,0,0) = 2 or (1,0,0) = 11) }
rule : 9 400 { (0,0,0) = 10 and ((0,-1,0) = 2 or (0,-1,0) = 11) }
rule : 7 400 { (0,0,0) = 8 and ((-1,0,0) = 2 or (-1,0,0) = 11) }
rule : 5 400 { (0,0,0) = 6 and ((0,1,0) = 2 or (0,1,0) = 11) }
```

The first rule, for example, states that if the current value of the cell is 4, and the cell below (in the same level) is either 2 (an exit) or 11 (empty top of stairs), the next value of the core cell will be 3.

In the following, we show some simulation results for building evacuation with the rules above (without fire). Fire-evacuation is discussed in the next subsection.

We consider a 3-dimensional building model of the dimensions (30, 35, 4). We consider different building designs, and evaluate the evacuation time of each design.

#### 10.4.1.1 Model with One Exit

In first design (shown in Fig. 10.10), there is only one exit/stairwell on every floor on the right-side of the building.

Figure 10.10 shows the 4 floors of the building, with the top floor on right, and the bottom level at the left. The figure shows the building at time $t = 0$, i.e., at the time the building is just initialized with pedestrians. The building was populated with pedestrians with 20% density, i.e., each empty cell is occupied with a probability of 0.2. Figure 10.11 shows a screenshot from the simulation at time $t = 1.14$ min. As

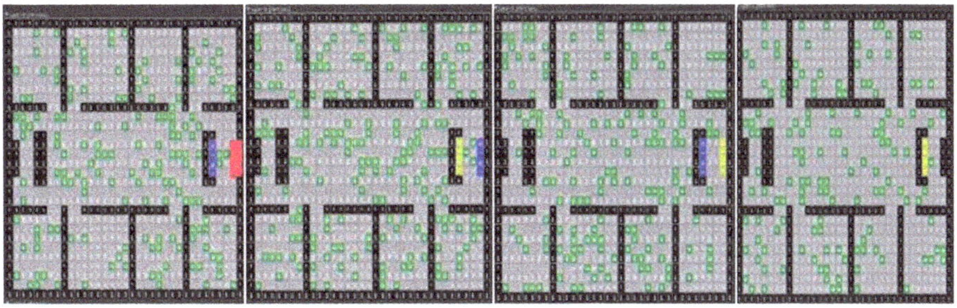

Fig. 10.10 The model of the first 4-floor building at time t = 0

can be seen, pedestrians at each floor are moving towards the entrances of the stairwells to get to the bottom floor and eventually exit the building. With this design, It takes 5:59 m for all the pedestrians to evacuate the building.

#### 10.4.1.2 Model with Two Exits

To improve the above design and reduce the evacuation time, another stairwell has been added to the opposite side of the upper levels, and an exit has been added to the other side of the first level (Fig. 10.12).

From the simulation results we can see that people are evacuating from two exits available on both sides of the building. This should reduce congestion at the stairwells/exits, and hence, reduce evacuation time. With this design, building is evacuated at time $t$ = 3:18 min, which is almost half of the time needed to evacuate the first design. A screenshot of the evacuation process in this design at time, $t = 0.5$ minute, is shown in Fig. 10.13.

#### 10.4.1.3 Model with Obstacles

In previous building designs, there were no structures in each floor. Realistically, buildings usually have structures such as columns. Such structures sometimes

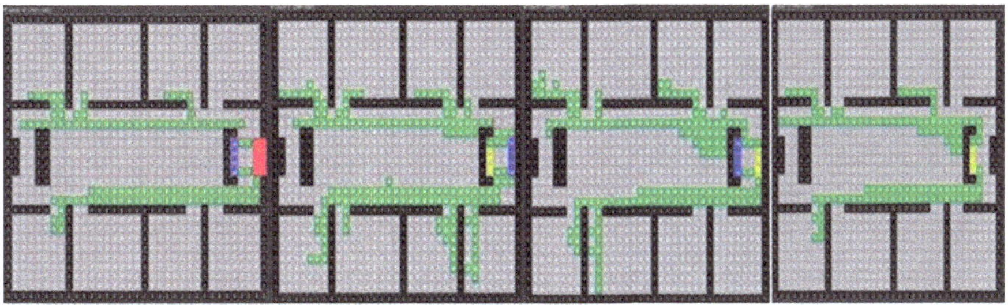

Fig. 10.11 The model of the first 4-floor building at time t = 1.14 min

**Fig. 10.12** The model of the second 4-floor building at time t = 0 min

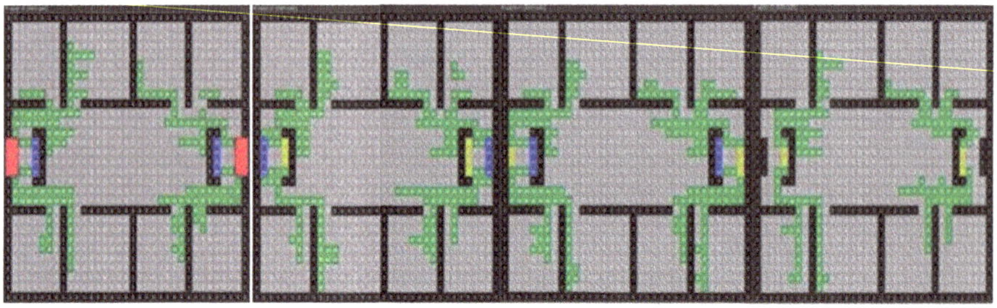

**Fig. 10.13** The model of the second 4-floor building at time t = 0.5 min

present obstructions to pedestrians while evacuating the building. To study the effect of such structures, we added obstacles in the previous building (with two exits).

Figure 10.14 shows the initial state of the simulation. Simulations have shown that when evacuating the rooms of each floor, pedestrians were able to maneuver and avoid the obstacles in the rooms. Hence, such obstacles did not have much impact on the evacuation process. However, the obstacles at the stairwells and exits did have an impact on the evacuation process. These obstacles blocked parts of the exits and stairwell entrances and increases congestion at these locations. Results

**Fig. 10.14** The model of the third 4-floor building at time t = 0 min

show that evacuation of this building took 4:50 m, which is a considerable increase in the evacuation time over the last design (without obstacles).

### 10.4.2 Fire-Evacuation Model

In this model, the initial values of the cells is set so that certain cells can be designated as the origin of fire. Furthermore, a set of rules are defined to govern the spread of fire through the building. The fire spread in all directions and the speed of the fire spread is controlled by time delay of the fire cells. Rules are defined such that everything coming in contact of fire will turn into fire in the simulation except cells with a value of 1, i.e., wall cells. The rules for fire spread are as follows,

```
rule : 15 5000{ (0,0,0) = 3 and ((-1,-1,0)=15 or (-1,0,0)=15 or
      (-1,1,0)=15 or (0,-1,0)=15 or (0,1,0)=15 or (1,-1,0)=15 or
      (1,0,0)=15 or (1,1,0)=15)}
rule : 15 5000{ (0,0,0) = 4 and ((-1,-1,0)=15 or (-1,0,0)=15 or
      (-1,1,0)=15 or (0,-1,0)=15 or (0,1,0)=15 or (1,-1,0)=15 or
      (1,0,0)=15 or (1,1,0)=15)}
...
rule : 15 5000{ (0,0,0) = 13 and ((-1,-1,0)=15 or (-1,0,0)=15 or
      (-1,1,0)=15 or (0,-1,0)=15 or (0,1,0)=15 or (1,-1,0)=15 or
      (1,0,0)=15 or (1,1,0)=15)}
rule : 15 5000{ (0,0,0) = 14 and ((-1,-1,0)=15 or (-1,0,0)=15 or
      (-1,1,0)=15 or (0,-1,0)=15 or (0,1,0)=15 or (1,-1,0)=15 or
      (1,0,0)=15 or (1,1,0)=15)}
```

Here, we adopt the building model with two exits and obstructions. Initially, a fire begins on the top floor in one room, and it spreads across the floor. It destroys everything including people and stairs. Fire starts form the red cell at the room at the bottom-right corner of the top floor, as we can see in Fig. 10.15. Fire spreads across the top floor as per the rules above, until it takes over the whole floor, as depicted in Fig. 10.16. People follow the evacuation rules, and eventually flee the building. Nevertheless, when a person is surrounded by fire, that person dies and gets eliminated from the simulation.

With this model, it takes 2:57 m to evacuate the building, which is less than the evacuation time of model without fire. However, not everyone makes it out of the building safely, which explain the reduction in evacuation time, as there were fewer people leaving the building. The speed of fire spread was 1/5 of the speed of pedestrians in the simulations above. Obviously, the speed of the fire spread has an impact on the number of causalities.

**Fig. 10.15** The model of the third 4-floor building at time t = 0.5 min

We ran simulations for another scenario where the fire starts at the third floor instead. In this case, the fire spreads across the third floor and reaches and destroys the stairs from the fourth floor. As such, pedestrians on the fourth floor will not be able to evacuate. Simulations of this scenarios are shown in Figs. 10.17, 10.18, and 10.19. Only the third and fourth floors are shown in these figures.

Figure 10.17 shows the beginning of the simulations when the fire starts at the bottom-left corner of the third floor. Figure 10.19 shows the simulations at $t = 32$ s. Once can see that the fire took over the whole room and spread outside the room. At $t = 3.05$ min, one can see that the fire took over the whole third floor and destroyed both stairwells.

One can see from the simulations above, that after the fire spread across the whole third floor and destroyed the stairwells, many people are trapped at the fourth floor and are not able to evacuate the building. Similarly, if the fire starts at the second floor, eventually people at both the third and fourth floors will be trapped inside the building and will not be able to evacuate.

This model has many parameters such as time delays which determine the speed of crowd evacuation, fire spread, etc. These parameters can be calibrated to simulate real-life scenarios and build real-life applications. The model can be used to evaluate different building designs in terms of evacuation speed and causalities in fire evacuation under different scenarios (people density, fire-spread speed, etc.).

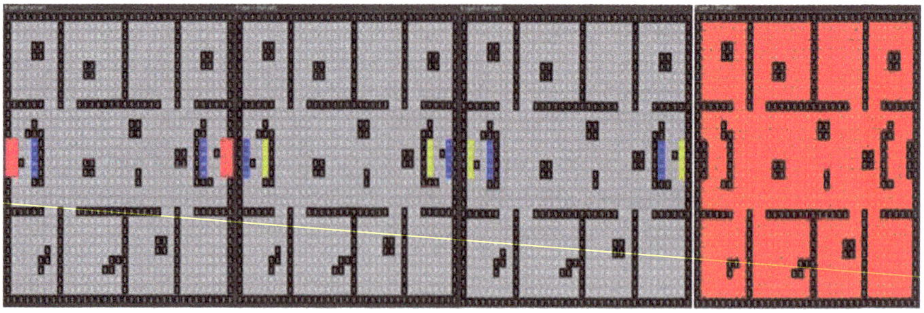

**Fig. 10.16** The model of the third 4-floor building at the end of the simulations

Fig. 10.17 The model of the third-floor fire at time t = 0.0 min

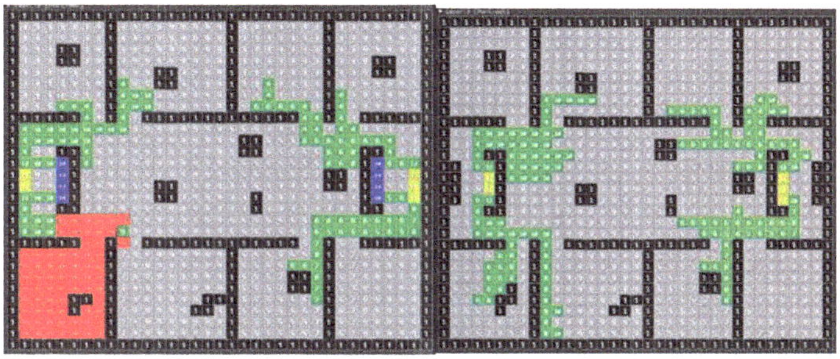

Fig. 10.18 The model of the third-floor fire at time t = 32 s

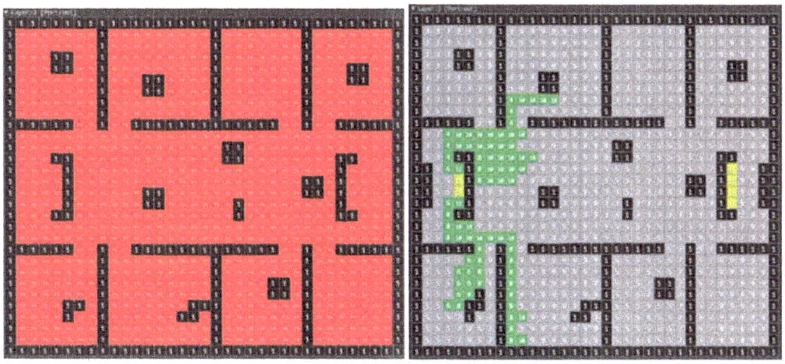

Fig. 10.19 The model of the third-floor fire at time t = 3.05 min

## 10.5 Conclusion

We provided a review of the existing methods for pedestrian modeling in the literature. Furthermore, an approach for spatial discrete-event Modeling and Simulation (M&S) of crowds using the Cell-DEVS formalisms is presented and discussed in detail. Since Cell-DEVS implement entity-based modeling, it provides a higher level of details than fluid dynamics models, and needs less computation demands than agent-based models. Furthermore, Cell-DEVS has multiple advantages over CA-based methods that makes it easier to develop larger models.

We explain through simple one- and two-dimensional models how Cell-DEVS is used to build pedestrian models. Furthermore, the usability of the approach is verified by employing it in real-life case studies. We discussed different case studies, employing the Cell-DEVS approach to build a model for a general building evacuation, and a fire evacuation model.

**Acknowledgements** This research has been partially funded by NSERC. The authors thank S. Wang, R. Farrell, M. Moallemi, W. Xiang, M. Van Schyndel, S. Jafer, H. Sohanpal and R. Chunara who built some of the models presented here.

## References

Al-Habashna, A. A., & Wainer, G. (2015). Modeling pedestrian behavior with Cell-DEVS: Theory and applications. *Simulation: Transactions of the Society for Modeling and Simulation International, 92,* 117–139.

Bandini, S., Manzoni, S., & Vizzari, G. (2006). Crowd modeling and simulation: The role of multi-agent simulation in design support systems. In *Innovations in design and decision support systems in architecture and urban planning* (pp. 105–120).

Bradley, G E. (1993). A proposed mathematical model for computer prediction of crowd movements and their associated risks. In *Proceedings of the International Conference on Engineering for Crowd Safety.* London, UK.

Burks, W. A. (1971). *Essays on cellular automata.* Urbana: University of Illinois Press.

Burstedde, C., Klauck, K., Schadschneider, A., & Zittartz, J. (2001). Simulation of pedestrian dynamics using a two-dimensional cellular automaton. *Physica A: Statistical Mechanics and its Applications, 295,* 507–525.

Campanella, C, Hoogendoorn, S. P., & Daamen, W. (2009). Improving the nomad microscopic walker model. In *12th IFAC Symposium on Control in Transportation Systems.* Fairfield, NJ.

Dogbe, C. (2010). On the cauchy problem for macroscopic model of pedestrian flows. *Journal of Mathematical Analysis and Applications, 372,* 77–85.

Helbing, D., & Molnár, P. (1995). Social force model for pedestrian dynamics. *Physical Review E, 51,* 4282–4286.

Helbing, D., Farkas, I., & Vicsek, T. (2000). Simulating dynamical features of escape panic. *Letters to Nature.*

Hesham, O., Aburime, W., Rabeh, Z., Bhushan, S., & Wainer, G. (2018). Observed behaviours in simulated close-range pedestrian dynamics. In *Proceedings of the Symposium on Simulation for Architecture & Urban Design (SimAUD).* Delft, Netherlands.

Hoogendoorn, S., & Bovy, P. (2004). Pedestrian route-choice and activity scheduling theory and models. *Transportation Research Part B: Methodological, 38,* 169–190.

Hughes, R. L. (2003). The flow of human crowds. *Annual Review on Fluid Mechanics, 35,* 169–182.

Ji, X., Zhou, X., & Ran, B. (2013). A cell-based study on pedestrian acceleration and overtaking in a transfer station corridor. *Physica A: Statistical Mechanics and its Applications, 392,* 1828–1839.

Kari, J. (2005). Theory of cellular automata: A survey. *Theoretical Computer Science, 334,* 3–33.

Klügl, F., & Rindsfüser, G. (2007). Large-scale agent-based pedestrian simulation. *Multiagent System Technologies, 4687,* 145–156.

Liu, S., Lo, S., Ma, J., & Wang, W. (2014). An agent-based microscopic pedestrian flow simulation model for pedestrian traffic problems. *IEEE Transactions on Intelligent Transportation Systems, 15,* 992–1001.

Masuda, T., Nishinari, K., & Schadschneider, A. (2014). Cellular automaton approach to arching in two-dimensional granular media. *Lecture Notes in Computer Science, 8751,* 310–319.

McPhail, C. (1991). *The myth of the madding crowd.* New York: A. de Gruyter.

Pluchino, A., Garofalo, C., Inturri, G., Rapisarda, A., & Ignaccolo, M. (2014). Agent-based simulation of pedestrian behaviour in closed spaces: A museum case study. *Journal of Artificial Societies and Social Simulation, 17,* 1–14.

Ronald, N., Sterling, L., & Kirley, M. (2007). An agent-based approach to modelling pedestrian behaviour. *International Journal of Simulation Systems, 8,* 25–38.

Steffen, B., & Chraibi, M. (2014). Multiscale simulation of pedestrians for faster than real time modeling in large events. *Cellular Automata Lecture Notes in Computer Science, 8751,* 492–500.

Tao, W., & Jun, C. (2009). An improved cellular automaton model for urban walkway bi-directional pedestrian flow. In *Measuring Technology and Mechatronics Automation International Conference.* Zhangjiajie, Hunan.

Treuille, A., Cooper, S., & Popović, Z. (2006). Continuum crowds. In *Proceedings of the 33rd International Conference on Computer Graphics and Interactive Techniques.* Boston, Massachusetts.

Vihas, C., Georgoudas, I., & Sirakoulis, G. (2013). Cellular automata incorporating follow-the-leader principles to model crowd dynamics. *Journal of Cellular Automata, 8,* 333–346.

Von Neumann, J., & Burks, A. (1966). *Theory of self-reproducing automata.* Urbana: University of Illinois Press.

Wainer, G. (2000). Improved cellular models with parallel cell-DEVS. *International Transactions of the Society for Computer Simulation, 17,* 73–88.

Wainer, G. (2009). *Discrete-event modeling and simulation: A practitioner's approach.* Boca Raton: CRC/Taylor & Francis Group.

Was, J., Lubas, R., & Mysliwiec, W. (2012). Proxemics in discrete simulation of evacuation. *Lecture Notes in Computer Science, 7495,* 768–775.

Wolfram, S. (1986). *Theory and applications of cellular automata.* Singapore: World Scientific.

Xia, Y., Wong, S. C., & Shu, C. W. (2009). Dynamic continuum pedestrian flow model with memory effect. *Physical Review E, 79,* 1–20.

Zeigler, B., Praehofer, H., & Kim, T. (2000). *Theory of modeling and simulation.* San Diego: Academic Press.

Zhan, B., Monekosso, D. N., Remagnino, P., Velastin, S. A., & Xu, L. (2008). Crowd analysis: A survey. *Machine Vision and Applications, 19,* 345–357.

Zhou, S., Chen, D., Cai, W., Luo, L., Low, M., Tian, F., Tay V. S., Ong, D. W., & Hamilton B. D. (2010). Crowd modeling and simulation technologies. *ACM Transactions on Modeling and Computer Simulation* 20, 1–35.

**Gabriel A. Wainer** FSCS, SMIEEE, received the M.Sc. (1993) at the University of Buenos Aires, Argentina, and the Ph.D. (1998, with highest honors) at UBA/Université d'Aix-Marseille III, France. In July 2000, he joined the Department of Systems and Computer Engineering at Carleton University (Ottawa, ON, Canada), where he is now Full Professor and Associate Chair for Graduate Studies. He has held visiting positions at the University of Arizona; LSIS (CNRS), Université Paul Cézanne, University of Nice, INRIA Sophia-Antipolis, Université de Bordeaux (France); UCM, UPC (Spain), University of Buenos Aires, National University of Rosario (Argentina) and others. He is one of the founders of the Symposium on Theory of Modeling and

Simulation, SIMUTools and SimAUD. Prof. Wainer was Vice-President Conferences and Vice-President Publications, and is a member of the Board of Directors of the SCS. Prof. Wainer is the Special Issues Editor of SIMULATION, member of the Editorial Board of IEEE Computing in Science and Engineering, Wireless Networks (Elsevier), Journal of Defense Modeling and Simulation (SCS). He is the head of the Advanced Real-Time Simulation lab, located at Carleton University's Centre for advanced Simulation and Visualization (V Sim). He has been the recipient of various awards, including the IBM Eclipse Innovation Award, SCS Leadership Award, and various Best Paper awards. He has been awarded Carleton University's Research Achievement Award (2005, 2014), the First Bernard P. Zeigler DEVS Modeling and Simulation Award, the SCS Outstanding Professional Award (2011), Carleton University's Mentorship Award (2013), the SCS Distinguished Professional Award (2013), and the SCS Distinguished Service Award (2015). He is a Fellow of SCS.

**Ala'a Al-Habashna** received his Master of Engineering degree from Memorial University of Newfoundland in 2010, and his Ph.D. degree from Carleton University in 2018, both in Electrical and Computer Engineering. Currently, Dr. Al-Habashna is a Postdoctoral Fellow and an Instructor at Carleton University, Ottawa, Canada. He received the fellowship of the School of Graduate Studies at Memorial University of Newfoundland in 2010. Furthermore, Dr. Al-Habashna has won multiple awards including excellence and best-paper awards. He worked as a reviewer for many conferences and journals including the IEEE ICC and Multimedia Tools and Applications journal. Dr. Al-Habashna is also a Technical Program Committee in the International Symposium on Performance Evaluation of Computer and Telecommunication Systems (SPECTS). His current research interests include discrete-event modeling and simulations, signal detection and classification, cognitive radio systems, 5G wireless networks, communication networks architecture and protocols, IoT applications, and multimedia communication over wireless networks.

# Chapter 11
# Using Agent-Based Modeling to Assess Liquidity Mismatch in Open-End Bond Funds

**Donald J. Berndt, David Boogers, Saurav Chakraborty and James McCart**

**Abstract** In this chapter, we present a small-scale heterogeneous agent-based model of the US corporate bond market. The model includes a realistic micro-grounded ecology of investors that trade a set of bonds through dealers. Using the model, we simulate market dynamics that emerge from agent behaviors in response to basic exogenous factors such as interest rate shocks. A first experiment focuses on the liquidity transformation provided by mutual funds and investigates the conditions under which redemption-driven bond sales may trigger market instability. We simulate the effects of increasing mutual fund market shares in the presence of market-wide repricing of risk (in the form of a 100-basis point increase in the expected returns). The simulations highlight robust-yet-fragile aspects of the growing liquidity transformation provided by mutual funds, with an inflection point beyond which redemption-driven negative feedback loops trigger market-wide price instability.

## 11.1 Introduction

The financial crisis of 2008 again highlighted the complex and evolving nature of the financial system and spurred another round of research into the dynamics of financial crises. New regulations aimed to curb risk taking in areas which were at the center of the crisis and limit the potential for contagion to other financial industry segments (with institutions deemed too-big-to-fail being a primary concern). History shows, however, that the financial industry tends to respond to regulation by re-allocating risks across the system. As a result, crises tend to originate in new and unanticipated areas, rarely copying historical patterns.

In the years following the crisis, credit provision to the corporate sector witnessed significant changes. While bank credit contracted, corporate bond markets showed extraordinary growth. Bonds—transferable debt securities—are an

D. J. Berndt (✉) · D. Boogers · S. Chakraborty · J. McCart
Muma College of Business, University of South Florida, Tampa, FL, USA
e-mail: dberndt@usf.edu



J. Sokolowski et al. (eds.), *Summer of Simulation*, Simulation Foundations, Methods and Applications, https://doi.org/10.1007/978-3-030-17164-3_11

important means by which companies fund their business operations and expansion. Well-functioning bond markets—the mechanisms that connect bond issuers with investors and enable trading between investors—are deemed essential for economic activity and growth.

In particular, the US corporate bond market experienced remarkable growth over the past 25 years, with continued expansion following the financial crisis. Analysis of SIFMA data shows the overall US corporate bond market expanding from $5.2 trillion in outstanding nominal (Q4 2007) to over $8.5 trillion (Q3 2016), see Fig. 11.1. Market growth is equally remarkable when measured relative to GDP. Between 2007 and 2015, the investment grade (IG) sector nearly doubled in relative size, with the outstanding nominal relative to GDP increasing from approximately 15–30%.

During the bond market expansion, the risks of investing in bonds increased significantly. With yields at historic lows, bonds offer little compensation for interest rate and credit risks while exhibiting high sensitivity of prices to changes in expected returns. Additionally, concerns around the deterioration of liquidity have taken center stage. The challenges are multi-faceted and include the lack of (pre-trade) price transparency, reduced investor heterogeneity (increased "herding" behavior) and the decline in dealer intermediation capacity.

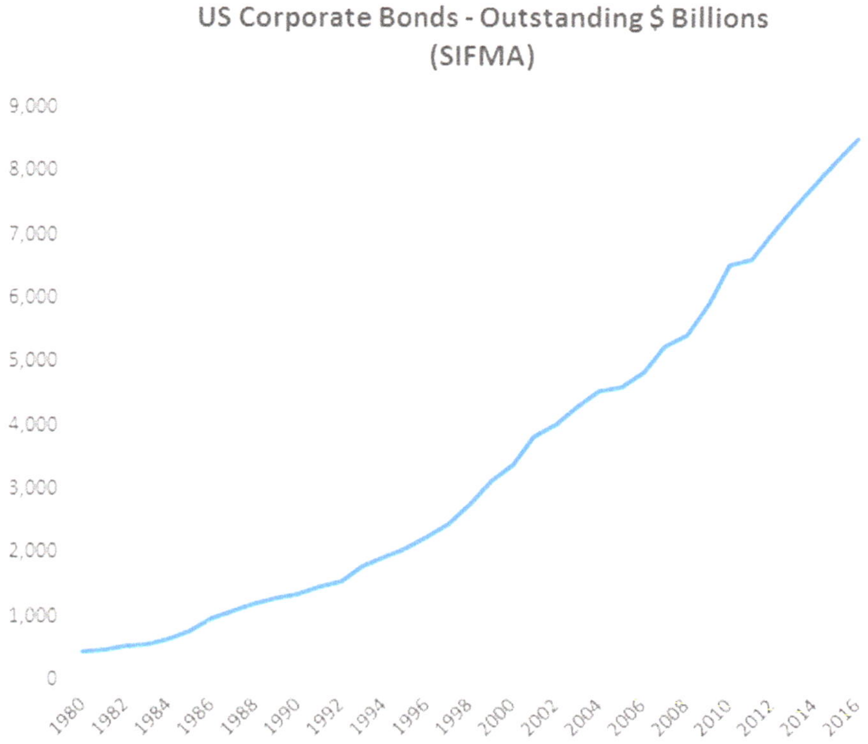


**Fig. 11.1** US corporate bond market size (*Source* SIFMA); data analysis by authors

In light of its current evolution, the bond market ranks highly on the list of potential risks to financial stability, prompting regulators (and industry participants) to question its resilience under stress (Flood 2015). In this chapter, we focus on the potential systemic risk caused by the interaction between impaired market liquidity and the increased reliance on liquidity transformation provided by pooled investment vehicles (such as mutual funds) (Barclays 2015). The two concerns can be summarized as follows.

1. Changes in market micro-structure have led to reduced market liquidity and increased liquidity fragility (the tendency of liquidity to evaporate under conditions of stress).
2. The growing role of mutual funds in bond markets introduces stability risks from rising liquidity transformation in the presence of redeemable shares. Mutual funds contain redemption features that make them susceptible to runs potentially triggering fire sales and price swings similar to those associated with the activities of participants who rely on short-term debt financing.

Over the last few years, both hypotheses have received ample attention from practitioners, regulators and the academic community without reaching consensus on the scope and severity of the issues. The emergence of mutual funds as significant players in bond markets presents some unique risks. When faced with major redemptions, a fund may need to sell its holdings in order to pay out redemptions. Forced selling of bonds, especially in an illiquid market, causes further price drops which in turn leads to more investor redemptions.

Agent-based modeling (ABM) and simulation techniques have the potential to provide critical insight into the dynamics of redemption risk and the effectiveness of regulatory policies or emergency response mechanisms. Given the bottom up approach to modeling systemic risk, ABM is well suited to the analysis of risks that emerge from agent interactions, feedback effects, reaction to stress (and constraints) and changes in regulatory environment. A recent European Central Bank report provides an overview of macroprudential liquidity tools for investment funds. Potential ex-ante tools include liquidity buffers, restrictions on the duration of redemptions and swing pricing. Ex-ante crisis management instruments include suspension of redemptions and the processing of "redemptions in kind" (ECB 2018). Agent-based simulations of fund redemptions offer great potential to assess the efficiency and effectiveness of these liquidity tools.

In this chapter,[1] we seek to use ABM to assess the significance of the price feedback loop associated with mutual fund redemption behavior. In the simulations described here, we analyze redemption-driven feedback loops following market-wide repricing of risk premiums (using an increase in the required yield). The increase in yield leads to drops in the fundamental value of bonds, with the

[1]This chapter is based on our paper from the 2017 *Summer Simulation Multi-Conference* (SummerSim) with the same title (see also a revised version in *Systems* at www.mdpi.com/2079-8954/5/4/54/htm). We are honored to have won both SummerSim-SCSC Best Paper and SummerSim Overall Best Paper, as well as to have our work appear in this retrospective book.

magnitude of the initial price drops determined by the characteristics of each bond's cash flow pattern (duration and convexity). However, the feedback loop caused by investor redemptions and forced liquidations is likely to lead to further price decreases and perhaps panic-driven behavior.

Regulators are charged with developing policies and mechanisms designed to ameliorate these risks. For example, regulators could require mutual funds to hold more cash to meet initial redemption demands. Our goal is to isolate such redemption-driven feedback loop effects and eventually build simulation-based tools to study how proposed regulatory policies could improve stability.

## 11.2 Liquidity Transformation in the Corporate Bond Market

In an environment characterized by persistently low rates and high levels of macro liquidity (driven by structural factors and targeted by active monetary policy), investors have tended to increase their risk appetite and demand for riskier assets.[2] The "reach for yield" response (commonly referred to as the "risk taking channel" in monetary policy discussions[3]) contributed to a significant decrease in corporate bond risk premia, an increase in the risks borne by bond investors and increased participation of non-traditional investors.

Faced with historically low borrowing costs, corporations issued record amounts of debt during the 10 years following the financial crisis. From 2010 to 2017, annual corporate bond issuance in the US averaged $1.4 trillion, a twofold increase from average issuance during the 2000–2007 period. With issuance vastly exceeding maturities, the size of the US corporate bond market ballooned, with the investment grade (IG) segment roughly doubling while the high-yield (HY) segment more than tripled over the same time period.

The amount of risk intermediated through corporate bond markets grew even faster than the stock of debt due to an increase in its risk profile (the amount of risk per average dollar of debt). Market risks grew due to an extension of the maturity profile of outstanding debt, with borrowers aiming to lock in "low rates for longer" through the issuance of longer maturity bonds. Additionally, higher market risks are a direct result of low yield levels given the convex nature of the bond yield-to-price relationship, implying higher price sensitivity at low yields.

---

[2]For a nuanced discussion on the effects of low interest rates on risk taking, please refer to the October 2018 discussion on EconSpark (American Economic Association): "Have low interest rates led to excessive risk taking?"

[3]Borio and Zhu (2008) define the risk-taking channel of monetary policy as "the impact of changes in policy rates on either risk perceptions or risk-tolerance and hence on the degree of risk in the portfolios, on the pricing of assets, and on the price and non-price terms of the extension of funding."

**Table 11.1** IBoxx index credit rating distribution

| Rating | Notional (Euro bn): 2007 | 2017 | Index (%): 2007 | 2017 (%) |
|---|---|---|---|---|
| AAA | 18 | 10 | 3 | 1 |
| AA | 197 | 190 | 27 | 11 |
| A | 319 | 683 | 39 | 40 |
| BBB | 197 | 839 | 27 | 49 |

*Source* Citi Research (2017)

Credit risks have grown steadily as well, as evidenced by the ratings distribution of corporate bonds (ratings are a measure of credit quality), which shows the relative growth of the lowest credit rating in the IG and HY segments of the corporate bond market. For example, Table 11.1 shows the credit distribution of the IBoxx index (an index of European investment grade corporate bonds). In 2007, BBB rated bonds (the lowest rating in the investment grade category) accounted for 27% of the index; by 2017 the share of BBB bonds in the index increased to 49%. Similar observations apply to the US corporate bond sector, where borrowers with lower credit ratings make up for an increasingly large share of the market. Between 2010 and 2017, the Merrill Lynch US investment grade corporate bond index saw the relative share of bonds with BBB ratings increase from 38% to nearly 48%. In addition to lower credit ratings, the issuance of bonds with lower investor protection clauses (so called "covenant-lite bonds") has been on the rise.

While the increase in the amount of risk is significant, it is even more important to understand who is bearing the risk. Richard Bookstaber argues that modeling crisis dynamics requires an understanding of the heterogeneity of the investor ecosystem, including investors' decision cycles (the time frame for deciding what to do in markets) and the specific constraints under which they operate (Bookstaber 2017). Analysis of the Federal Reserve flow of funds data highlights significant changes in the corporate bond investor landscape. Figures 11.2 and 11.3 show the changes in the corporate bond investor ecosystem between 1981 (roughly the start of the secular bond bull market) and 2016. Important changes include the growth in market share of mutual funds and foreign investors at the expense of insurance companies and pension funds.

Brad Delong notes the need to assess how "reaching for yield" impacts investment strategies, especially for organizations whose "business models become unprofitable as a result of low rates and QE and take on risks excessive for them because they have no expertise in judging such risks.[4]" The effect of reaching for yield on the investment decisions of major investor classes is widely documented:

- Natural investors such as insurance companies and pension funds (for whom the investment characteristics of corporate bonds match the nature of their liabilities), have an incentive to move out along the risk spectrum (extending duration and increasing credit risk) when faced with "sticky" rate-of-return targets in a

[4]See https://www.bradford-delong.com/2018/09/is-there-any-reason-to-fear-low-interest-rates.html.

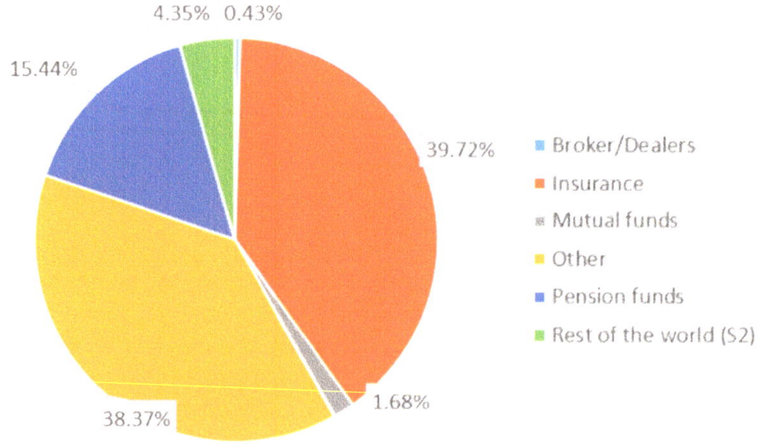


**Fig. 11.2** Investor ecosystem circa 1981 (Q1) by market share (*Source* Federal Reserve flow of funds data); analysis by authors

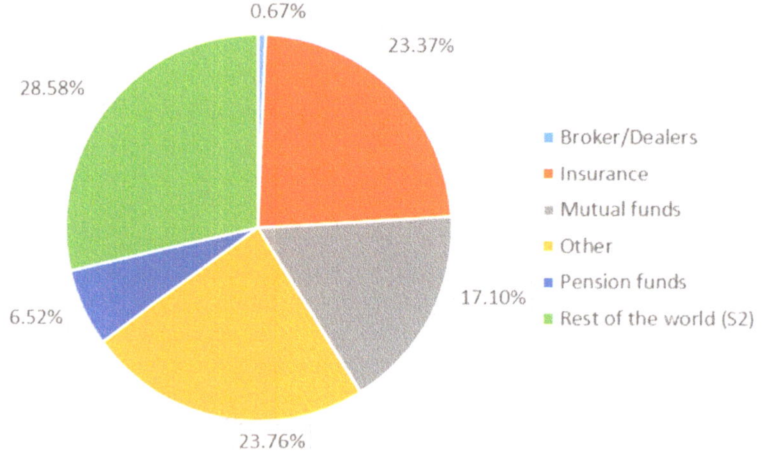


**Fig. 11.3** Investor ecosystem circa 2016 (Q3) by market share (*Source* Federal Reserve flow of funds data); analysis by authors

declining rate environment (Borio and Zhu 2008). Using an empirical model of portfolio allocation decisions of insurance companies and pensions in the UK, researchers at the Bank of England document evidence of the search for yield through a re-allocation of government bond holdings into corporate bonds (Joyce et al. 2017). Becker and Ivashina (2015) focus on the regulatory capital constraints of insurance companies and document a reach for yield incentive driving insurers to seek out riskier bonds (bonds with higher CDS spreads) within a given credit rating category.

- The reach for yield also chased non-natural investors (often referred to as "yield tourists") into corporate bonds.[5] To the extent that yield tourists have less natural ability to take on duration and credit risk (due to more volatile liabilities or a comparatively shorter-term investment horizon), they rely on the capacity of the secondary market to provide liquidity when needed. An environment of high macro liquidity and the assumption of high secondary market liquidity lowers yield tourists' entry barrier into corporate bonds. Analysts at Deutsche Bank characterize the effect of liquidity as "transforming the risk of default into the risk that the securities representing the debt find no purchasers. It replaces responsibility with salability" (Deutsche Bank 2017). The tight link between liquidity (or assumed liquidity) and the risk-taking channel is recognized by regulators as well. "In particular, liquidity and risk-taking are tightly interconnected, and can reinforce each other. For instance, lower perceptions of risk and greater risk tolerance weaken external funding and transferability constraints. In turn, weaker constraints can support higher risk-taking" (Borio and Zhu 2008). In addition to market and credit risks, yield tourists therefore assume significant liquidity risk. The latter derives from the difference between assumed and actual liquidity levels and the fact that liquidity may not be available when most needed as liquidity tends to evaporate in times of stress (a phenomenon known as "liquidity fragility"). In a recent market report, analysts at Gavekal Research note the role of the "illusion of liquidity" in the 2008 crisis dynamic, when numerous fixed income products went from perfectly liquid one day to untradeable the next. They further note "What kills financial market participants time and again is 'the assumption of liquidity', or the idea that, because a certain price is flashing on a screen an asset can be sold at somewhere near this level" (Gavekal Research 2018).

### 11.2.1 The Illusion of Liquidity

The illusion of liquidity in corporate bond markets derives from a number of factors. Bond markets have *limited pre-trade price transparency*. Different from centralized order-driven markets (such as equity markets), bond markets operate through a decentralized quote-driven market. There is no central limit order book (CLOB) to inform participants about latent liquidity nor is consolidated quote data available. Available pre-trade price information consists of indicative prices provided by dealers; getting a firm or tradeable price requires contacting multiple dealers with a "request for quote" (RFQ).

[5]As noted by JP Morgan Asset Management analysts in the *Investment Outlook for 2019*: "A common theme in this cycle has been the hunt for yield, with investors moving into unfamiliar asset classes searching for higher returns to offset the low yields in core bonds."

Transaction-based measures of liquidity are skewed by the *bifurcation of liquidity*. Analysis of secondary market transactions points to a very small core of liquid securities, while the majority of bonds hardly trade. An analysis of 2014 TRACE data by Citi analysts found that—out of 26,000 (publicly registered) corporate bonds—11% didn't see a single trade throughout the entire year and a further 20% only traded on 5 or fewer days. In fact, less than 1% of bonds traded daily (Citi Research 2015). Bifurcation implies that the analysis of market liquidity based on observed transactions (e.g. using metrics such as trading volumes or bid/ask spreads) is skewed by the actively traded bonds in the liquid core.

Structural forces *suppressed demand for secondary market liquidity*. Demand for liquidity decreases as investor heterogeneity declines and participants adopt similar trading strategies. In an environment characterized by a long-term decline in rates and credit spreads, investors tend to concentrate on "long only" strategies. Inflows into bond markets are serviced through primary markets (aided by high levels of new issuance volumes) without drawing heavily on secondary market liquidity.

The *growth and increasing concentration of the asset management sector* further suppressed demand, with large-scale asset managers being able to cross flows across portfolios. This suppresses the need to engage in secondary market transactions as liquidity is provided through internal transactions between portfolios.

The post-2008 evolution in corporate bond markets coincided with the *emergence and continued growth of collective investment vehicles* (mutual funds and ETFs) as significant players. These vehicles shield participants from the operational complexities of bond markets and provide liquidity transformation to their investors. Liquidity transformation by mutual funds contributes to lower transaction volumes in an environment with steady net fund inflows as fund redemptions can be serviced from investor inflows rather than selling bonds. Periods of small net outflows can be serviced from cash balances, limiting the need for bond sales.

### *11.2.2 Redemption Risks in Collective Investment Vehicles*

The emergence of mutual funds as significant players in bond markets presents some unique concerns centered on the pro-cyclical nature of fund flows, and the perceived feedback loop between bond prices and flows. Mutual funds contain redemption features that make them susceptible to withdrawal patterns that resemble classic "bank run" behaviors potentially triggering fire sales and price swings similar to the activities of participants who rely on short-term (runnable) debt financing.

Empirical studies document return-chasing behavior by mutual fund investors and a significant flow-performance relationship (rising prices trigger fund inflows which contribute to rising prices; the opposite logic applies to falling prices) (Chen 2010). Funds with illiquid assets exhibit stronger sensitivity of outflows to past performance than funds with liquid assets. When faced with major redemptions, a fund may need to sell its holdings in order to pay out redemptions. Forced selling of bonds in an illiquid market causes further price drops which in turn leads to more investor redemptions. In a downward market, bond funds can therefore introduce a powerful negative feedback loop which—through the various contagion mechanisms—can spill over to other asset markets and impact the real economy (ECB 2016).

Recent research highlights multiple mechanisms that could lead to disruptive selling by mutual funds, including a first-mover advantage associated with redemption externalities and the dynamic management of cash holdings to buffer against future redemption risk (Morris 2017). When a fund needs to engage in bond sales to pay out redemptions, transaction costs and market impact of sales are borne by the remaining investors, who therefore have an incentive to anticipate redemptions (and be the "first one out the door").

Equally important to the redemption externalities is the "liquidity illusion" (discussed above) which draws yield tourists (some of whom have limited capacity to judge market and credit risks) into mutual funds on the assumption of daily liquidity at the fund's net asset value (NAV). When risks materialize, it triggers a shock to investor risk perceptions, leading the "unsteady hands" to redeem. Synchronized redemptions confronted with limited bond liquidity under stress leads to outsize price moves which trigger more redemptions, a fire sale and the "popping" of the liquidity illusion.

We contribute to the discussion on the risks associated with liquidity transformation in the corporate bond market using a somewhat different approach and explore the conditions under which a run on mutual funds can lead to instability. In this chapter, we seek to assess the significance of the price feedback loop associated with mutual fund redemption behavior. We use agent-based modeling (ABM) and simulation to better understand the endogenous price dynamics following a market-wide reassessment of risk (using a 100-basis point increase in expected bond yields). We analyze the impact of redemption driven feedback loops under increasing mutual fund participation levels (using market shares of 15, 25 and 35%).

## 11.3 Agent-Based Modeling and Simulation

Agent-based modeling and simulation have been applied to financial markets in many creative ways, starting with early work on artificial financial markets (such as the Santa Fe artificial stock market) (LeBaron 1999). For an interesting review of influential early projects, see (LeBaron 2000), and for a wider ranging (and very entertaining) look at agent-based modeling in finance, see (Samanidou 2007). Of course, simulation methods (apart from agent-based approaches) have been used to study economic phenomena for a long time. For example, Stigler's pioneering work that uses Monte Carlo simulations to study trading behaviors in securities markets dates back to 1964 (Stigler 1964). As further noted in Samanidou (2007), Kim and Markowitz developed one of the first "multi-agent" models to investigate the 1987 stock market crash (Kim and Markowitz 1989). Their model has two agent types, "rebalancers" and "portfolio insurers" that trade stocks and hold cash balances. Rebalancers simply aim to keep their portfolios evenly split between the two asset classes. Portfolio insurers use a "constant proportion portfolio insurance" (CPPI) method to maintain the portfolio risk level in relation to an insurance expiration date. The simulation results highlight the destabilizing nature of portfolio insurers and provide at least a partial explanation for crisis dynamics like the 1987 crash. The work of Hommes and in `t Veld provides a much more recent example of modeling booms and busts using heterogeneous agents (Hommes 2017). In this model, two agent types (fundamentalists and chartists) share knowledge of fundamental prices but take different views on how long price trends last. Fundamentalists hold beliefs around mean-reversion of prices, while chartists look to momentum and the belief that trends tend to continue. The agents can gradually switch beliefs based on performance, which leads to amplifications of booms and busts.

The Office of Financial Research (OFR) has published various papers discussing the value of ABM in the analysis of financial systemic risk, for example see (Bookstaber 2012). Furthermore, the European Commission sponsored a major research initiative (CRISIS, the Complexity Research Initiative for Systemic Instabilities), which aims to analyze systemic risks to the financial sector and the wider economy using ABM. Agent-based approaches have not been applied nearly as widely as DSGE (dynamic stochastic general equilibrium) and econometric models (Farmer 2009). While most of the existing literature around the application of ABM to finance is focused on equity markets (with some interesting applications to currency and housing markets), we aim to analyze the liquidity conundrum in the corporate bond market (as outlined above).

Our work draws heavily on the strong foundation of prior agent-based modeling research, including advice on how to pursue this type of research (LeBaron 2001). This bond market model in particular focuses squarely on crisis dynamics following some of the prior work cited. However, our work is marked by several important differences.

1. Most prior agent-based models focus on central limit order book market structures (as in most equity markets).
2. Our corporate bond market model is a decentralized dealer-based quote-driven market, selected since it is considered at risk by many regulatory organizations. Corporate bond markets have been the subject of only one other recent agent-based model (Braun-Munzinger 2016).
3. The agents are drawn from real-world participants based on their investment mandates, constraints and the nature of their liabilities.
4. The agent implementations follow empirical (micro-grounded) financial literature, such as the Treynor model of the dealer function (Treynor 1987).

Traditional approaches to micro-prudential risk management, including stress tests and portfolio risk analytics such as Value-at-Risk (VaR), focus on the resilience of individual firms to specific shocks. They fail to address the broader question of how stress might be transmitted among firms through the dynamics of contagion and fire sales. Agent-based modeling and simulation can capture the second order effects of interactions and feedback loops as it models the reaction functions (or behavior) of individual agents. It is particularly well suited to the analysis of the endogenous nature of liquidity dynamics under stress (Bookstaber 2017).

Numerous studies have attempted to shed light on the changing liquidity conditions. Opinions differ, however, on the significance of the problem, its root causes and potential fixes. Due to limited data availability (or lack of transparency), most liquidity analyses rely on executed transactions and can only paint a partial picture of liquidity (Edwards 2007). Furthermore, historical analyses may not adequately capture the dynamics under stress conditions (which is the fundamental concern); liquidity is typically ample during normal (steady-state or equilibrium) market operations but tends to evaporate under conditions of stress. These are the issues we aim to explore using a heterogeneous agent-based model of corporate bond market.

### *11.3.1 Why Agent-Based Modeling?*

At this point, it is worth considering why agent-based modeling is a reasonable methodology for investigating economic phenomena. In our view, agent-based approaches are particularly well-suited for modeling financial markets, often complex ecosystems of heterogeneous agents. Economic modeling typically proceeds from theory, through mathematical model formulation, toward better understanding of the economic system being studied. This top-down approach employs abstractions that focus more on representative agents and common behaviors expressed in aggregation. Agent-based modeling uses simulation, based on the interactions of autonomous heterogeneous agents, to generate the emergent behaviors of complex systems (and hopefully inform theory). This more bottom-up approach starts with a characterization of the ecosystem, proceeds through the implementation of heterogeneous agent behaviors (and interactions), toward understanding and theory refinement. The record of more traditional economic models in understanding crisis dynamics is not good. Agent-based models may provide some help in understanding economic systems, especially under conditions of stress. For a persuasive and much more expansive case for agent-based modeling, see Richard Bookstaber's recent book *The End of Theory* (Bookstaber 2017). Some of the factors that favor agent-based modeling are outlined below.

- **Heterogeneous agents**: As noted above, agent-based approaches rely on the interactions of heterogeneous agents to generate emergent behaviors of interest. The bond market model described here uses a small (at least to start) set of agent classes to represent a mix of bond investor types with decision-making behaviors grounded in the extensive academic and industry-based finance literature.

- **Emergent phenomena**: Complex system-wide dynamics arise from the interactions between individual agents making simple decisions such as firms trading in a market. Agent-based modeling is based on the actions and interactions between individual autonomous synthetic agents implemented with important behaviors encoded as decision-making algorithms. Basically, the ABM approach is all about emergent system dynamics (simple rules, complex behaviors).
- **Non-ergodic processes**: Individual actions and interactions that take into account prior context make each simulation pathway potentially unique. As Bookstaber suggests: Does history (and context) matter? For roulette, no. For human behavior, typically an emphatic yes.
- **Radical uncertainty**: Emergent phenomena and non-ergodic processes lead to unanticipated uncertainty ("black swans") or "the things we do not know we do not know" (as noted by D. Rumsfeld, US Secretary of Defense). These unknowns are exactly the type of phenomena that characterize crisis dynamics.
- **Computational irreducibility**: Complex systems, based on context-driven interactions, can often be understood only by tracing out the entire ecosystem pathway over time (using simulation rather than mathematical shortcuts). This makes agent-based modeling and simulation a natural fit for studying systems like financial markets.

## 11.4 An Agent-Based Model of the Corporate Bond Market

The bond market agent-based model implements a somewhat stylized investor ecology, with participants trading a limited universe of bonds through dealers who provide transaction immediacy on a principal basis using a request-for-quote (RFQ) protocol. In selecting the set of agents, we aim to model representative corporate bond investor heterogeneity. While there are multiple ways to segment the investor base, we guided our selection of buy-side agents by the nature of their liabilities (leveraged versus non-leveraged, presence of inflows/outflows) and their investment mandate (passive versus active, long only versus long/short). As a result, the market model includes three buy-side agent classes representative of a mutual fund, an insurance company and a hedge fund.

Out of the three types of buy-side agents, two represent "real money" investors (using no leverage), while the third maintains leveraged positions. Real money investors cannot maintain short positions (must be "long only"); these include an insurance company (typical value investor) and a mutual fund (using passive index tracking). The leveraged investor represents an unconstrained participant (such as a hedge fund) who can maintain both long and short positions.

Recent analysis from the Office of Financial Research (OFR) highlights the importance of financial networks in understanding contagion risk and presents a

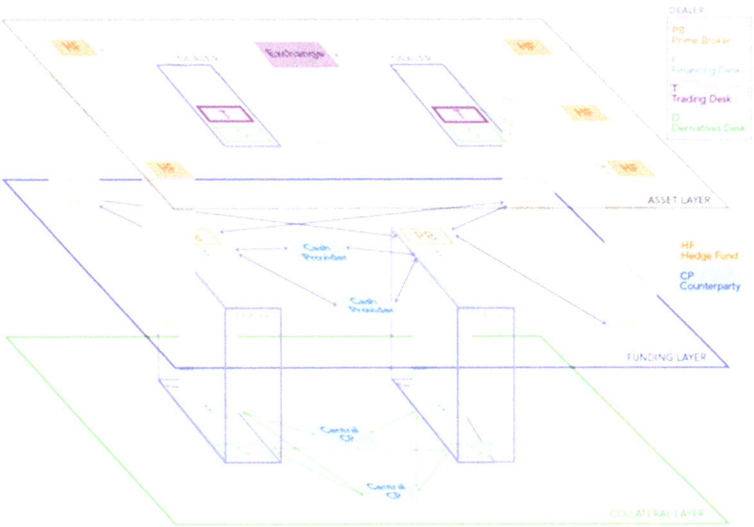


**Fig. 11.4** The financial system as a multilayer network (*Source* Office of Financial Research (Bookstaber 2016)

model of the financial system as a multilayer network (see Fig. 11.4). In line with this multi-layer network view, we modeled interactions between agents using three distinct but interconnected layers representing asset, funding and collateral markets. While our analyses focus on asset price dynamics, there are critical dependencies on the flow of funding (interaction between market liquidity and funding liquidity) and collateral movement. The funding layer focuses on the role of leverage and its impact on fire sales (including feedback loops and cross-asset contagion). It includes critical constraints for leveraged market participants (such as hedge funds and broker-dealers). The collateral layer supports the asset layer (short sales) and funding layer (secured financing) and models the flow of collateral through securities lending arrangements and "repo" transactions.

In addition to the three layers of the bond market, the model includes a few direct and indirect linkages with other asset verticals, including equities and government bonds. Equity markets impact the model through the behavior of one class of buy-side agents (an insurance company) that re-balances positions between equities and bonds based on equity market volatility (as well as absolute yield levels). Government bond markets furthermore provide the risk-free yield curve, which is used as an input to the bond pricing equation.

In summary, we developed heterogeneous agent classes based on two important dimensions that reflect the real ecology of the US corporate bond market.

- **Investment mandate**: The investment mandate or policy is perhaps the most important criterion since it encompasses the goals of the organization. Key factors here include the investment horizon and any performance benchmarks (like the Bloomberg Barclays US Aggregate Bond Index). Is the investment

vehicle passively or actively managed? Can both long and short positions be held or is it long only? These factors map nicely to the asset layer in the multi-layer network view of the financial system (see Fig. 11.4). Basically, the mandate constrains the decision-making options and as such is central to implementing an agent class.
- **Nature of liabilities**: The key factor is leverage with respect to the nature of any liabilities. For example, the hedge fund agent class relies heavily on leverage, while the insurance company does not borrow at all. Are the liabilities more or less runnable? Clearly, an open-ended vehicle such as a mutual fund is subject to investor redemptions at any time, creating inflows and outflows based on performance (or anything else that affects the whims of skittish investors).

In addition to these two dimensions, other taxonomies of bond market investors informed our choices, including buy-side agents based on mutual funds, insurance companies and hedge funds. The first of these taxonomies is drawn from the Federal Reserve flow of funds data referenced in Figs. 11.2 and 11.3. This data is used to develop the approximate market shares reflecting the real investor ecology. A more recent paper from the Bank of England used detailed regulatory bond market data to cluster the behaviors of "dealer banks, non-dealer banks, insurance companies, hedge funds, and asset managers" (Czech 2017). Again, we find support for our buy-side agents based on insurance companies, hedge funds, and mutual funds (as asset managers). Since non-dealer banks are not significant investors in the US corporate bond market, we left that agent class out (at least for now).

### 11.4.1 Mutual Fund Class

The mutual fund acts as a real money investor who aims to replicate the performance of a defined benchmark which includes the full universe of bonds available in the initial model. Any dividends or capital gains distributions are assumed to be re-invested. As noted, the fund is long only and does not leverage positions. The fund also maintains a dynamic cash balance as a buffer against investor redemptions (limiting forced sales) and to minimize transaction costs by parking cash until sizable orders can be made.

### 11.4.2 Insurance Company Class

The insurance company agent implements a long-term value investor with a liability driven investment strategy. This agent manages an investment portfolio across equity and fixed income markets and changes allocation between markets depending on overall market conditions. The insurance company is a "long only" investor with additional constraints limiting the concentration of risk in any specific bond. In the initial model, leveraged positions cannot be established (making this another real money investor) and we further assume there are no external inflows or outflows in the form of premiums or claims.

Trading activity for the insurance company results from changes in portfolio allocations between equity and fixed income markets. Macro allocation decisions are driven by a number of variables, including equity market volatility, as well as the current level and slope of the yield curve. Time series of equity volatility measures such as the VIX are loaded into the model, so different time periods can be used to anchor the simulations in realistic business cycles.

### 11.4.3 Hedge Fund Class

The hedge fund agent acts as a short-term tactical trader who follows a relative value trading strategy. As such, the hedge fund maintains both long and short positions and makes active use of leverage. In the real world, fixed income relative value hedge funds have historically been among the most leveraged market participants.

The hedge fund agent is not subject to external inflows (basically a closed end fund) or redemptions (assume investor lock up); its trading capacity is constrained only by the availability of secured financing (leverage) from broker-dealer agents. We assume the hedge fund finances all positions on margin through prime-brokerage style arrangements with some of the dealers. Broker-dealers limit leverage using security-specific haircuts that can be dynamically adjusted depending on market conditions.

All margining is assumed to occur on an overnight basis. At the start of each trading day (a tick in the agent-based simulation), margin requirements are calculated based on current market prices and security-specific haircuts, as set by the broker-dealer agents. The difference between margin requirements and current wealth determines the trading capacity. If the new margin requirements exceed current wealth, the hedge fund is forced to liquidate positions (de-leverage) to meet margin calls. Any excess wealth is free to be invested.

### 11.4.4 Broker-Dealers and the RFQ Protocol

Dealers respond to requests for quote (RFQ) from the buy-side agents and trade with clients on a principal basis (there is no inter-dealer market in the initial model). Asset owners must trade with the dealer offering the lowest price. Dealers can maintain both long and short positions. Dealer behavior is limited through regulatory constraints and market discipline, the latter expressed through a constraint on value-at-risk (VaR) relative to capital.

## 11.5 Mutual Fund Redemption Experiments: Model Setups

In order to assess the significance of a negative price feedback loop caused by mutual fund redemption behavior, we run agent-based simulations using a simplified model which includes only two agent types: mutual funds and dealers. We model dynamics in the asset layer of the model, without any restrictions in the funding or collateral layers (dealers have unlimited access to financing and there are no restrictions on collateral availability).

### 11.5.1 *Mutual Funds*

Purchases and sales of bonds by mutual funds are driven by inflows (investor fund purchases) and outflows (investor fund redemptions). In line with empirical findings, we model mutual fund flows as a function of historical returns (Goldstein 2015). The flow-performance relationship exhibits a concave shape, which implies that the sensitivity of outflows (to bad performance) is greater than the sensitivity of inflows (to good performance). We specify the daily flow for a given fund at simulation tick (*t*) as a percentage of the fund's prior daily wealth (NAV) as:

$$Flow_t = FlowRatio_t * Wealth_{t-1}$$

$$FlowRatio_t = \alpha + \left(\beta_d * R_{d,t-1}\right) + \left(\beta_{d,neg} * I\left(R_{d,t-1} < 0\right)\right) + \left(\beta_w * R_{w,t-1}\right) + \left(\beta_{w,neg} * I\left(R_{w,t-1} < 0\right)\right)$$

where:

$R_{d,t-1}$ is the fund daily return, lagged by a day (i.e. the daily return at the end of the prior tick)
$R_{w,t-1}$ is the fund weekly return, lagged by a day (i.e. the weekly return at the end of the prior tick)
$I\left(R_{d,t-1} < 0\right)$ is a binary value (indicator) with value 1 *if* $<0$ and value 0 *if* $R_{d,t-1} \geq 0$
$I\left(R_{w,t-1} < 0\right)$ is a binary value (indicator) with value 1 *if* $R_{w,t-1} < 0$ and value 0 *if* $R_{w,t-1} \geq 0$

Using data on corporate bond index returns and fund flows between 1991 and 2014, model parameters were calibrated as follows:

$$\alpha = 0.00017,$$
$$\beta_d = 0.56,$$
$$\beta_{d,neg} = -0.0002,$$
$$\beta_w = 0.6,$$
$$\beta_{w,neg} = -0.0002.$$

Investor inflows and redemptions affect the fund's cash position which is managed to a target cash-to-assets ratio of 5% with a lower bound of 3% and an upper bound of 8%. When the cash position exceeds the upper bound, the fund must put cash to work by buying bonds using the index weights. In the opposite direction, the fund must liquidate a portion of the portfolio if redemptions cause the projected end-of-day cash position to drop below the lower bound. The buying and selling behaviors rank the bonds based on current departures from the index and then try to complete trades bond-by-bond until the free cash is invested (or necessary cash is raised). The behaviors are asymmetrical in that buying a specific bond can be postponed until the next tick if no dealer can be found. However, selling is a bit more aggressive in that the agent attempts to raise the necessary cash using any bonds that can be traded.

### *11.5.2 Dealers*

Dealers are the essential price setters in the initial model. Given the intent to trade, asset owners make a request-for-quote (RFQ) to all dealers. Dealers must respond with "no quote" or a full quote for the requested order size (no partial order fills are allowed for now).

Dealers quote prices using a spread to the bond's fundamental value (in the initial model, the last traded price reflects fundamental value). Spread calculations are based on inventory positioning considerations and follow the logic outlined in Treynor (1987), using the following assumptions (Treynor 1987).

- Dealers are subject to inventory concentration limits which dictate the maximum long or short position a dealer can hold in a given bond. Inventory limits are set to 10% of outstanding nominal for long positions and 7.5% for short positions.
- Within the inventory limits, dealers quote prices with a view on maximizing client flow.
- Similar to Treynor (1987), we assume the existence of a community of value investors who provide the "outside spread" faced by the dealers. The outside spread sets the price at which dealers can unload a position when it approaches inventory limits. Model values for the outside spread(s) are set as follows: outside bid spread = 100 bps, outside ask spread = 125 bps.
- Dealers specialize in specific bonds based on maturity. Dealer specialization coefficients model the depth of the dealer's franchise in a given bond as the percentage of the overall value investor community covered by the dealer's sales force, see Table 11.2. Dealer specialization coefficients are set as follows.
  - An inventory spread is calculated using a step function, with the maximum spread equal to the outside spread applicable to the dealer.
  - The inventory spread is adjusted slightly (increased in case of a bid, lowered in case of an offer) if the dealer can cover any part of the request out of inventory. That is: the inventory spread is increased on a bid quoted against a

**Table 11.2** Dealer specialization across bond maturities

| Dealers | Bonds | | | | |
|---|---|---|---|---|---|
| | MM101 (%) | MM102 (%) | MM103 (%) | MM104 (%) | MM105 (%) |
| Dealer 1 | **90** | **90** | 75 | 50 | 50 |
| Dealer 2 | 50 | 75 | **90** | 75 | 50 |
| Dealer 3 | 50 | 50 | 75 | **90** | **90** |

short position and lowered on an ask quoted against a long position. The amount of the inventory spread adjustment is dependent on dealer specialization and is proportional to the fraction of the quote that can be serviced out of inventory.

- The quoted price in response to an RFQ is equal to the last traded price adjusted for the spread as calculated above.

### 11.5.3 Market Universe and T-Zero Starting Conditions

The market universe consists of five tradable bonds (see Table 11.3 for details). The bonds are identical with respect to structure, form and major covenants including issuer, redemption (bullet redemption at maturity without optionality clauses) and rate provisions (fixed coupon). The bonds differ along only three dimensions.

1. Outstanding nominal amount ranges from 500 M to 2B.
2. Maturities cover major points on the yield curve (1, 2, 5, 10 and 25 years).
3. Coupon rates range from 1.75 to 4.00%.

At any point in time, all asset owners perceive the same fundamental value for a specific bond. That is, all asset owners use the same valuation model and observe the same input prices. The value for the above five bonds is fully reflected in five data points (a par yield curve with five rates) and a simple calculation of a bond's price given its par yield.

**Table 11.3** Characteristics of tradable bonds

| Bonds | MM101 | MM102 | MM103 | MM104 | MM105 |
|---|---|---|---|---|---|
| Nominal | 500 M | 500 M | 1B | 2B | 1B |
| Maturity | 1 year | 2 year | 5 years | 10 years | 25 years |
| Coupon (%) | 1.75 | 2.50 | 2.25 | 2.40 | 4.00 |
| Yield (%) | 1.50 | 1.75 | 2.50 | 2.60 | 4.21 |
| Price | 100.247 | 101.468 | 98.832 | 98.249 | 96.772 |
| Index weight (%) | 10 | 10 | 20 | 40 | 20 |

The starting conditions for the agent-based simulation include the following.

- Bond index composition with weights based on nominal amount (see Table 11.3, Index Weights).
- Initial par yield curve and bond prices (see Table 11.3, Yield and Price).
- Starting bond holdings are allocated to two buy-side agents: a mutual fund and insurance agent (for the minimal model). The mutual fund holds either 15%, 2% or 35% of the outstanding nominal in the market, with the remaining nominal held by the insurance agent. The mutual fund is invested across all bonds in the index based on the index weights. All dealers start with square (zero) inventory positions.

In addition, initial endowments for the buy-side agents are outlined below.

- Mutual fund: In addition to its bond holdings, the fund has an opening cash position reflecting a 5% cash-to-assets ratio.
- Insurance company: Initial portfolio allocation includes a 60/40 split between fixed income and equity markets (assumed to be invested in a broad market index like the S&P 500).

## 11.6 Mutual Fund Redemption Experiments: Model Results

Using the model setup outlined above, we simulate the effect of a 100-basis point rise in expected yields under various mutual fund market shares (using three realistic values: 15, 25, and 35%). The magnitude of the chosen shock (100 basis points) is aligned with market risk measures commonly used by regulators and industry participants. In the latest edition of the Global Financial Stability Report, the IMF estimates losses to fixed income mutual funds following a 100-basis point shock to interest rates (IMF 2017). The IMF analysis puts mutual fund losses around 7% of NAV. In our simulations, the 100 basis point shock triggers instantaneous losses in line with IMF calculations (see further discussion below), with fund losses at tick 50 in the simulation reported as 7.2% of NAV. In a research note focused on the first mover advantage in high-yield bond mutual funds, Barclays analysts investigate a variety of market shocks, with the most adverse case amounting to a 10% market drop, about 3% higher than the market correction used in our simulation.

In all experiments, the interest rate shock causes a decrease in bond prices and hurts fund returns, thereby causing investors to withdraw funds. At a certain point, a mutual fund needs to liquidate part of the bond portfolio to meet redemptions. These forced sales will cause bond prices to decline further, sparking another round of price drops and redemptions. This type of feedback loop is exactly what we are trying to capture using agent-based modeling approaches.

### *11.6.1 Mutual Fund at 15% Market Share*

The first simulations use a market share of 15%, slightly below the real market share, but a reasonable starting point. The simulations are run for 252 ticks, mirroring the typical number of trading days in a year. An economic shock is delivered at tick 50 in the form of a 100-basis point rise in the interest rate. The corresponding price drops across all five bonds are easily seen in Fig. 11.5. Price changes at tick 50 are due only to the interest rate change and are computed for each bond based on its duration and convexity. Bond prices adjust between 1 and 14% with longer maturity bonds showing higher price sensitivity. At the end of tick 50, re-pricing of the mutual fund bond holdings triggers losses around 7.2% of NAV, in line with IMF analysis (see above).

Following market wide repricing at tick 50, a redemption-driven negative price feedback loop drives more gradual secondary price decreases (in the low 1% range) for the next 10 ticks or so. After the market stabilizes, more normal trading activity returns (in the form of sporadic price increases as investor inflows add funds). Quantitative results are presented for both pre-shock and post-shock prices (see Table 11.4), as well as for prices at the bottom of the market (see Table 11.5).

The dynamics of the redemption-driven feedback loop and recovery are outlined in Table 11.6, which includes weekly investor redemptions for four consecutive trading weeks following the market shock (each week consisting of 5 ticks). During week 1 investor redemptions equal roughly $172 million, which equates to 23% of the NAV at the start of the week. The fund engages in significant bond sales in order to pay out redemptions, thereby causing bond prices to decline further while

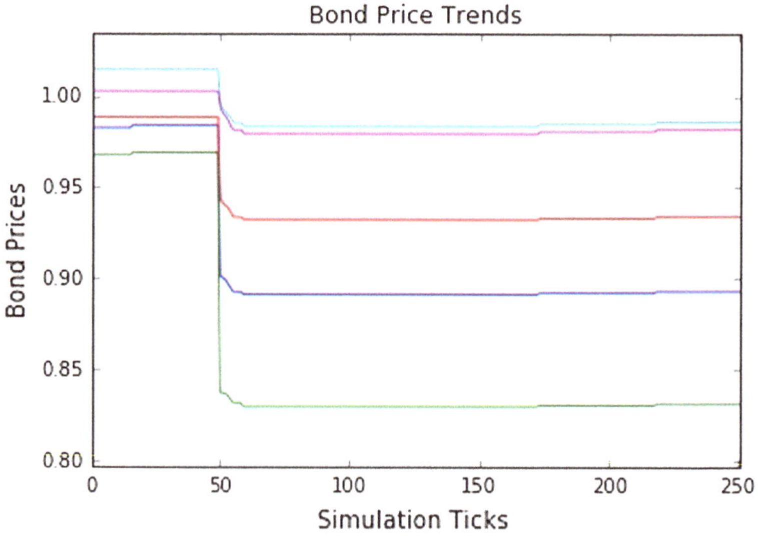


**Fig. 11.5** Bond price trends for a 15% mutual fund market share

**Table 11.4** Initial bond price drops after 100-basis point shock

| Time (tick) | Bond | Price | Price drop (%) |
|---|---|---|---|
| Pre-shock (49) | MM101 | 100.247 | NA |
| Pre-shock (49) | MM102 | 101.468 | NA |
| Pre-shock (49) | MM103 | 98.832 | NA |
| Pre-shock (49) | MM104 | 98.388 | NA |
| Pre-shock (49) | MM105 | 96.911 | NA |
| Post-shock (50) | MM101 | 99.264 | 0.98 |
| Post-shock (50) | MM102 | 99.517 | 1.92 |
| Post-shock (50) | MM103 | 94.312 | 4.57 |
| Post-shock (50) | MM104 | 90.123 | 8.40 |
| Post-shock (50) | MM105 | 83.725 | 13.61 |

**Table 11.5** Redemption-driven bond price drops at a 15% mutual fund market share

| Time (tick) | Bond | Price | Price Drop (%) |
|---|---|---|---|
| Bottom | MM101 | 97.900 | 1.37 |
| Bottom | MM102 | 98.362 | 1.16 |
| Bottom | MM103 | 93.202 | 1.18 |
| Bottom | MM104 | 89.128 | 1.10 |
| Bottom | MM105 | 82.982 | 0.89 |

**Table 11.6** Yield reaction to outflows, measured in bps/percentage flow

| Elapsed time (week since shock) | Investor flows (dollars) | Investor flows (% NAV) | Index yield change (bps) | Yield reaction to flows (bps per percentage flow) |
|---|---|---|---|---|
| Week 1 | −$171,937,047 | −22.9 | 120 | 5.24 |
| Week 2 | −$12,263,176 | −2.1 | 11 | 5.13 |
| Week 3 | −$2,232,899 | −0.4 | 0 | 0 |
| Week 4 | −$475,590 | −0.1 | 0 | 0 |

yields increase. At the end of week 1, the yield on the bond index has increased by about 120 basis points relative to its value just prior to the market shock. Given the 23% decrease in fund NAV, the yield sensitivity to outflows—measured as the yield increase (in basis points) corresponding to fund redemptions at 1% of NAV—is 5.24 basis points. This measure is aligned with the empirical findings reported in a May 2015 research note from Deutsche Bank in which the authors examine the top 10% largest weekly outflows from high yield (HY) and investment grade (IG) mutual funds along with the coincident spread change during those weeks, using a trailing-10-episodes-average spread change per percentage fund outflow (Deutsche Bank 2015). During the period between 2004 and 2014, the authors find the spread/flow sensitivity for IG bonds to vary between 2 and 30 bps, with the average value prior to the 2008 crisis hovering around 5 bps. This value is aligned

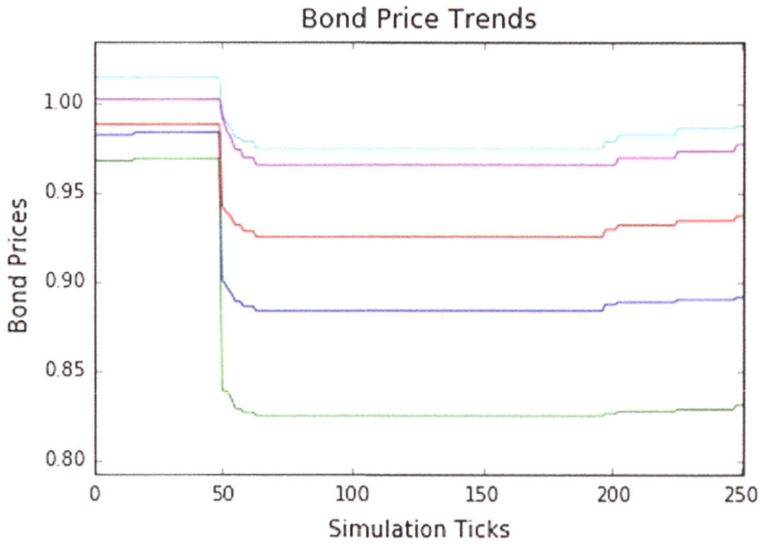


**Fig. 11.6** Bond price trends for a 25% mutual fund market share

with the yield-to-flow sensitivity during our simulation, as reported in the redemption-driven feedback loops during week 1 (5.24 bps) and week 2 (5.13 bps) following the market shock.

## *11.6.2 Mutual Fund at 25% Market Share*

The next simulations use a market share of 25%, a reasonable level somewhat above the current situation. Again, the simulations are run for 252 ticks and a 100-basis point interest rate shock is delivered at tick 50. The interest rate hike causes the expected price drops across all five bonds as before (see Fig. 11.6). These shock-induced drops are followed by further drops in prices for the next 10 to 20 ticks due to the redemption-driven feedback loop. As above, the rate hike impacts increase with maturity, showing price drops in the 1–14% range. The larger mutual fund market share increases the effects of the feedback loop, with secondary price decreases now in the 1.7–2.7% range (see Table 11.7). This is a definite

**Table 11.7** Redemption-driven bond price drops at a 25% mutual fund market share

| Time (tick) | Bond | Price | Price drop (%) |
|---|---|---|---|
| Bottom | MM101 | 96.576 | 2.71 |
| Bottom | MM102 | 97.473 | 2.05 |
| Bottom | MM103 | 92.513 | 1.91 |
| Bottom | MM104 | 88.371 | 1.94 |
| Bottom | MM105 | 82.469 | 1.66 |

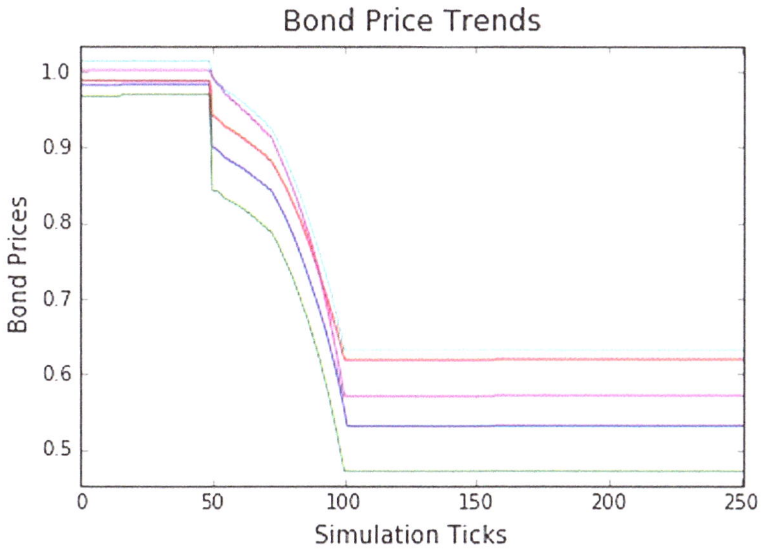


**Fig. 11.7** Bond price trends for a 35% mutual fund market share

increase over the first simulation, but the feedback loop still has much less of an effect as compared to the interest rate shock. Once the market stabilizes and normal trading activity returns, there is a somewhat stronger recovery in prices.

### *11.6.3 Mutual Fund at 35% Market Share*

The final simulations use a market share of 35%, a substantial increase over the current situation. Again, the simulations are run for 252 ticks and a 100-basis point interest rate shock is delivered at tick 50. The interest rate hike causes the expected price drops across all five bonds as before (see Fig. 11.7). However, after that the simulations take a different path. After these shock-induced drops, the prices fall of a cliff. For the next 50 clicks the redemption-driven feedback loop causes a spiral of decreasing prices (see Fig. 11.7). In fact, the concave curve means that the price drops accelerate with correspondingly dramatic wealth destruction. The price drops are more pronounced from the outset but reach an inflection point followed by precipitous price drops until all the dealer capacity is consumed (and the market flatlines). In these simulations, the redemption-driven price drops dwarf the initial interest rate shock effects. The feedback loop causes price drops in the 35–44% range, as compared to the shock-induced drops of roughly 1–14%, see Table 11.8.

In actuality, the instability and all out-market crashes start to occur at approximately a 30% mutual fund market share (see Fig. 11.8). Though notice that the feedback look is not as severe and overall it takes roughly 50 more trading days (or

**Table 11.8** Redemption-driven bond price drops at a 35% mutual fund market share

| Time (tick) | Bond | Price | Price drop (%) |
|---|---|---|---|
| Bottom | MM101 | 57.129 | 42.45 |
| Bottom | MM102 | 63.206 | 36.49 |
| Bottom | MM103 | 61.889 | 34.38 |
| Bottom | MM104 | 53.172 | 41.00 |
| Bottom | MM105 | 47.097 | 44.10 |

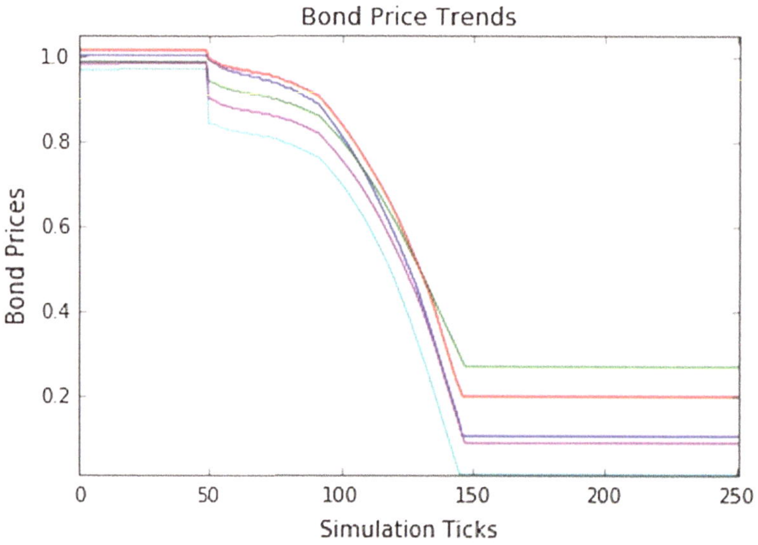


**Fig. 11.8** Bond price trends for a 30% mutual fund market share

ticks) to completely crash. We should not read too much into this exact market share level since we aim to use our model to evaluate regulatory policies or re-create market behaviors, not predict specific thresholds. We are much more interested in the emergent behaviors and relative comparisons that can provide useful guidance.

## 11.7 Alternative Flow of Funds Experiments

One of the challenges in using agent-based modeling is developing the rules that govern agent behaviors. The behaviors exhibited by agents during a simulation are driven by these rules and the specific conditions faced by each individual agent at given point in time, often through the interactions with other agents. These many individual decisions and interactions comprise the simulation and aggregate emergent behaviors of interest.

In the model description above for the investment redemption experiments, the equation (or rule) governing the mutual fund inflow and outflows is a core piece of the mutual fund agent implementation. As noted, this equation is grounded in the literature and mostly based on a handful of key references (including (Goldstein 2015)). The response of each individual mutual fund agent is driven in large part by the size of the inflows or outflows encountered at each tick of the (simulated) clock.

An alternative perspective (and possible basis for a behavioral rule) was recently published in a Bank of America/Merrill Lynch Credit Market Strategist research note (Mikkelsen 2018). In this paper, the authors identify a couple key issues including some observations discussed in this quote.

> One of the key questions right now is why we continue to see large inflows to high grade bond funds and ETFs when total return performance—traditionally the main driver of retail flow—has been flattish over the past five months or so. Our view is that—in addition to retail investors—foreigners increasingly are finding that bond fund and ETFs represent convenient vehicles to gain exposure to US fixed income.

Indeed, the original equation developed for the mutual fund agent class does rely on prior returns, which is called into question (at least in part) by this current research. So clearly some interesting experiments based on swapping flow of funds models are possible. The reference includes a regression model as described below, that has been adapted for use in our simulations.

> … we estimated a linear model for monthly high-grade bond fund and ETF flows using the four explanatory variables and up to three lagged values of them. We employed a general-to-specific principle to reduce this model gradually until all variables were statistically significantly different from zero. Table 11.9 shows the final model using two lags of flows, contemporaneous and one lag of high grade returns as well as contemporaneous equity returns and rates volatility.

So far, this alternative flow model has resulted in similar simulation results. However, this model does include both equity market returns and interest rate volatility, which is certainly an interesting addition. In future simulation experiments these additional factors could produce interesting interactions between the agents. For example, we are working on a set of experiments involving insurance company (and related pension fund) agents that reallocate portfolio positions

**Table 11.9** Band of America linear model for mutual fund flow of funds

| Term | Coefficient | Standard error | t Stat | p-Value |
|---|---|---|---|---|
| Intercept | 10.45 | 4.36 | 2.40 | 0.02 |
| Flow t−1 | 0.42 | 0.09 | 4.48 | 0.00 |
| Flow t−2 | 0.20 | 0.09 | 2.25 | 0.03 |
| HG return t | 226.89 | 61.79 | 3.67 | 0.00 |
| HG return t−1 | 280.02 | 66.47 | 4.21 | 0.00 |
| S&P price return t | 52.91 | 20.10 | 2.63 | 0.01 |
| Interest rate vol t | −0.13 | 0.05 | −2.49 | 0.01 |

between the equity and bond markets. Having the mutual fund inflows and redemptions sensitive to some of the same factors (at least in part) could contribute to interesting emergent behaviors.

## 11.8 Conclusion

In this chapter, we introduce an agent-based model of the corporate bond market and obtain realistic market dynamics with a small-scale heterogeneous model. The model includes three buy-side agent classes that provide a nice breadth of different investment approaches. The sell-side agents are broker-dealers who serve as the price setters in the model. This simple model is being used to refine agent behaviors and assess responses to basic exogenous factors and the introduction of regulatory policies and constraints. Our first experiments focus on redemption-driven price feedback loops under increasing fund market shares. The focus is on the mutual fund class in the experiments described here, with economic shocks delivered as the mutual fund market share is increased. As the market share grows, the simulations exhibit realistic feedback loops. First, the prices drop in direct response to an interest rate shock, with secondary redemption-driven feedback loops causing further declines in price, followed by gradual stabilization and recovery. However, at a 35% market share, the model becomes unstable, with the initial shock-induced price drops followed by an accelerating feedback loop that overwhelms the sell side broker dealers and freezes the market. This is exactly the behavior that concerns both regulators and practitioners. We believe that this agent-based model provides an excellent tool for the evaluation of possible regulatory mechanisms and policies to help minimize future crises.

**Acknowledgement** We gratefully acknowledge the support of the National Science Foundation (NSF Award 1445403) and the Office of Financial Research (OFR) in providing funding for this research. For more details, please see the project website at GSRisk.org (and D. Berndt's research page at dberndt.link/home).

## References

Barclays. (2015). *Mutual funds and credit liquidity, nobody move*. Barclays Credit Research.

Becker, B., & Ivashina, V. (2015). Reaching for yield in the bond market. *Journal of Finance, 70* (5), 1863–1902. American Finance Association.

Bookstaber, R. (2012). *Using agent-based models for analyzing threats to financial stability*. Technical Report 0003, Office of Financial Research.

Bookstaber, R. (2017). *The end of theory*. 41 William Street, Princeton, New Jersey: Princeton University Press.

Bookstaber, R., & Kenett, D. Y. (2016). *Looking deeper, seeing more: A multilayer map of the financial system*. Technical Report 16-06, Office of Financial Research.

Borio, C., & Zhu, H. (2008). *Capital regulation, risk-taking and monetary policy: A missing link in the transmission mechanism*? BIS Working Papers No. 268.

Braun-Munzinger, K., Liu, Z., & Turrell, A. (2016). *An agent-based model of dynamics in corporate bond trading*. Technical Report 592, Bank of England.

Chen, Q., Goldstein, I., & Jiang, W. (2010). Payoff complementarities and financial fragility: Evidence from mutual fund outflows. *Journal of Financial Economics, 97,* 239–262.

Citi Research. (2015). *When agents lose their principals, fixed income liquidity revisited*? European Credit Research.

Citi Research. (2017). *Partying like it's 2007*? European Credit Weekly.

Czech, R., & Roberts-Sklar, M. (2017). *Investor behaviour and reaching for yield: Evidence from the sterling corporate bond market*. Technical Report 685, Bank of England.

Deutsche Bank. (2015). *US credit strategy, signs of liquidity vacuum in unexpected places*. Deutsche Bank Markets Research.

Deutsche Bank. (2017). *Drowned by liquidity*. Deutsche Bank Markets Research.

ECB. (2016). *Shadow banking in the euro area: Risks and vulnerabilities in the investment fund sector*. ECB Occasional Paper 174.

ECB. (2018). *Macroprudential liquidity tools for investment funds–A preliminary discussion*. ECB Macroprudential Bulletin.

Edwards, A. K., Harris, L., & Piwowar, M. S. (2007). Corporate bond market transaction costs and transparency. *Journal of Finance, 62,* 1421–1451.

Farmer, J. D., & Foley, D. (2009). The economy needs agent-based modelling. *Nature, 460,* 685–686.

Flood, M. D., Liechty, J. C., & Piontek, T. (2015). *System wide commonalities in market liquidity*. Technical Report 15-11, Office of Financial Research.

Gavekal Research. (2018). The illusion of liquidity and its consequences.

Goldstein, I., Jiang, H., & Ng, D. T. (2015). Investor flows and fragility in corporate bond funds. Technical Report 2596948, SSRN. https://ssrn.com/abstract=2596948 or http://dx.doi.org/10.2139/ssrn.2596948.

Hommes, C., & Veld, D. (2017). Booms, busts and behavioural heterogeneity in stock prices. *Journal of Economic Dynamics & Control, 80,* 101–124.

IMF. (2017). Global financial stability report. https://www.imf.org/en/Publications/GFSR/Issues/2017/09/27/global-financial-stability-report-october-2017.

Joyce, M., Liu, Z., & Tonks, I. (2017). Institutional investors and the QE portfolio balance channel. *Journal of Money, Credit and Banking, 49*(6).

Kim, G. W., & Markowitz, H. M. (1989). Investment rules, margin and market volatility. *Journal of Portfolio Management, 16,* 45–52.

LeBaron, B. (2000). Agent-based computational finance: Suggested readings and early research. *Journal of Economic Dynamics & Control, 24,* 679–702.

LeBaron, B. (2001). A builder's guide to agent-based financial markets. *Quantitative Finance, 1,* 254–261. https://doi.org/10.1088/1469-7688/1/2/307.

LeBaron, B., Arthur, W. B., & Palmer, R. (1999). Time series properties of an artificial stock market. *Journal of Economic Dynamics & Control, 23,* 1487–1516.

Mikkelsen, H., Seliger, Y., & Zhang, Y. (2018). *Infows to taper*. Bank of America (BoA) Credit Market Strategist, January 28.

Morris, S., Shim, I., & Shin, H. S. (2017). *Redemption risk and cash hoarding by asset managers*. Technical Report 608, Bank for International Settlements.

Samanidou, E., Zschischang, E., Stauffer, D., & Lux, T. (2007). Agent-based models of financial markets. *Reports on Progress in Physics, 70,* 409–450.

Stigler, G. J. (1964). Public regulation of the securities market. *The Journal of Business, 37,* 117–142. JSTOR. http://www.jstor.org/stable/2351027.

Treynor, J. L. (1987). The economics of the dealer function. *Financial Analysts Journal, 62,* 27–34.

# Chapter 12
# A Partially Grounded Agent Based Model on Demonetisation Outcomes in India

**Rishi Bubna, Jayasree Raveendran, Suman Kumar, Mayuri Duggirala and Mukul Malik**

**Abstract** In November 2016 the Indian government announced a policy of demonetising the ₹500 and ₹1000 currency notes in an attempt to move the economy into using cashless modes for financial transactions. Given that India is a cash intensive economy the effects of demonetisation were immediate and far reaching. This chapter examines the likely impact of demonetisation on such a cash intensive economy using an Agent-Based Modeling (ABM) approach. The goals were to understand pathways to transition to a cashless money use behavior and to estimate a population's inconvenience in accessing cash modes, given geographic features, limited resource and readiness towards change. Our model is built on empirical evidence from a survey which studied the systemic and behavioral drivers of money use behavior post-demonetisation. The model is able to illustrate outcomes depicting the pattern of transition to digital payments. We present a framework for optimal remonetisation considering infrastructure density and inconvenience experienced during transition by the population. Simulation analysis found other interesting insights such as service infrastructure acting as a catalyst to noncash transition.

---

R. Bubna (✉) · J. Raveendran · S. Kumar · M. Duggirala · M. Malik
Tata Research Development and Design Center, Pune, Maharashtra, India
e-mail: rishi.bubna@tcs.com

J. Raveendran
e-mail: jayasree.raveendran@tcs.com

S. Kumar
e-mail: suman.kumar4@tcs.com

M. Duggirala
e-mail: mayuri.duggirala@tcs.com

M. Malik
e-mail: mukul.malik2@tcs.com

 
J. Sokolowski et al. (eds.), *Summer of Simulation*, Simulation Foundations, Methods and Applications, https://doi.org/10.1007/978-3-030-17164-3_12

## 12.1 Introduction

The Indian Government announced demonetisation of all ₹500 and ₹1,000 banknotes on November 8, 2016, a fiscal intervention wherein 86% of the total cash in circulation was retrieved with immediate effect (Reserve Bank of India (RBI) 2017b). The decision was guided by the aim of tackling corruption, greater digitalization leading to a cashless society and monetary formalization to tackle the parallel economy. India has traditionally been a cash intensive economy where about 78% of all consumer payments in India are effected in cash (Reserve Bank of India (RBI) 2017c). It was, therefore, obvious that currency squeeze during the demonetisation period would have had some adverse impact on economic activity, although the impact was expected to be transient.

Demonetisation as a tool for systemic restructure measure has been adopted in other economies including the UK (1971), Russia (1991) Australia (1996), the European Union (2002), Zimbabwe (2015) to name a few, with reasons of dealing with hyperinflation, improving durability/security features of bank notes, getting rid of counterfeit notes and corruption. While decoding the consequences of a demonetisation policy may not be straightforward and would need a long-term time frame for assessments, implementation challenges and discomfort that people are prone to in the short-term during the implementation phase has been a prevalent and a significant feature in almost all economies. Likewise, the Indian demonetisation episode was characterized by suddenness (unanticipated announcement with limited time for adjustment), comprehensiveness (nation-wide applicability) and with inadequacy of resources (cash crunch and limited service infrastructure), thus creating a compulsion to use noncash modes for payments leading to possible inconvenience. Apart from macro-economic compulsions that may drive a demonetisation policy, an equally significant element to consider during such policy implementation is human behavior. In achieving a shift towards a cashless phenomenon, two major types of influencers can be acknowledged: one, systemic in nature that deals with the availability of technology and enforcement through policies and secondly, behavioral in nature, that deals with one's inclination to change to adopt digital modes for monetary transactions.

With the demonetization drive in India taking an immediate effect after its announcement, various facets of behavioral manifestations have been observed in the Indian economy. On one hand, people were facing inconvenience due to cash shortage, and faced impact on livelihood and economic activities in some sections. On the other, there was a visible increase in adoption of digital money, which might not have taken place in the absence of the demonetization mandate. While cash shortage impacted the lesser income demographic the most, there were also the emergence of various ways of coping up with this phenomenon—such as use more of debit/credit cards or to use or take help from others in online purchases or transactions. In this setting, the need for a study to understand and analyze the role of behavioral factors that influence money use behavior was conceived and a survey was designed to gather data on behavioral parameters from a lower income demographic segment. In this chapter,[1] the results of behavioral analysis concerning determinants of money use behavior have been used as inputs to Agent based modelling and simulation to understand the manifestation of behavior in a larger societal space. While there exist alternate top-down analytic techniques, it is also interesting and educative to understand the effects of demonetisation using a bottom-up approach like Agent Based Modelling and Simulation. Distinct features of our model are (a) Groundedness—being built from empirical evidence in the context of behavioral drivers for money-use behavior during demonetisation and with additional literature support for model elements (b) Compositeness—money-use behavior being characterized as a composite function of behavioral and systemic factors, and (c) Realism—the feature of convenience/inconvenience faced by an agent in demonstrating preferred Money-Use Behavior (MUB). With these features, the model aims to understand transition to cashless MUB and estimate population's inconvenience levels over the duration. The model outcome on money-use behavior patterns in an economy during demonetisation is analyzed and the implementation of demonetisation is re-assessed to offer suitable suggestions. The rest of the chapter is organized as follows: Sect. 12.2 discusses the literature review and research questions, Sect. 12.3 describes the behavioral and systemic determinants of money-use behavior, Sect. 12.4 elaborates the methodology and computation of the model parameters, Sect. 12.5 mentions the simulation outputs, Sect. 12.6 describes the experimental analysis and Sect. 12.7 concludes and discusses the future work.

## 12.2 Literature Review and Research Questions

Technology adoption by individuals from a behavioral standpoint is explained by technology readiness index, technology acceptance models (Parasuraman 2000; Javadi et al. 2012; Brown et al. 2010), and technology diffusion process (Hall and Khan 2003). While the widespread usage of cash highlights the convenience and transparency it offers in terms of personal cash management, digital payments are also now gaining popularity because of the convenience it offers. Economists have found these vital characteristics in explaining why consumers adopt the payment instruments they use (Schuh and Stavins 2016). Convenience and cost have been found to be especially strong factors affecting payment behavior, although perceived risk, security and privacy concerns in digital payments have also significantly influenced the adoption and/or the use of selected payment instruments as documented in number of studies (Mai 2016; Braga et al. 2013; Jonker 2007; Kupetz 2007; Kosse 2010).

---

[1]This chapter is based on our paper from the 2018 Summer Simulation Multi-Conference (SummerSim) with the same title which won the Best Paper Award. We are honored to have our work appear in this retrospective book.

Major challenges to establishment of a cashless society in the Indian context are continuing financial exclusion, and also an accompanying mistrust of the financial system, infrastructural challenges and a lack of awareness of available financial tools (OCED 2017). In the context of demonetisation in India (Krishnan and Siegel 2017) has studied the awareness levels of noncash forms among select users in a rural area and found that albeit awareness, the adoption levels of digital payments are low.

Extending to a social phenomenon, social cognitive theory (Bandura 1991) explains psychosocial functioning in terms of triadic reciprocal causation, where cognitive and personal factors, behavior and environmental events operate as interacting determinants. In this study, demonetisation policy as an environmental factor can be construed to be a significant influencer of money-use behavior.

Use of several computational techniques have been made to assess an event like demonetization, such as (Chandrakar 2017) where the author uses Bayesian network analysis and recommend better planning and implementation by the Government to curtail unfavorable consequences of demonetisation on population below the poverty line. Bottom-up approaches like agent based modeling (ABM) can be viewed as an potential tool for replicating reality, generating fresh implications, allow for policy implications and predictions as well as providing an alternative view of how an economy would work. Many such attempts have been made where ABM is applied to understand technology adoption (Fagiolo 2005), labor market dynamics (Fagiolo et al. 2004), and firm investment (Dosi et al. 2006). The bank of England has developed ABM models for housing markets to study the impact of macro-prudential policies on key housing market indicators. This work also focuses on the corporate bonds market to analyze feedback effects and the impact of market changes (Braun-Munzinger et al. 2016).

A more recent work based on actor based model (Barat et al. 2017) highlights the need for managing critical situations during demonetisation. Adopting the modelling method recommended by Sargent (2005) a conceptual model has been created from a *problem entity* which largely-considers the interaction between government, banks, citizens and shops. However this model is limited to accounting for of systemic factors in citizen's decision making, e.g. existence of payment patterns which rely on the availability of payment methods and technology.

Summing up, a choice of using digital money forms depends on a number of factors that include: one's openness to adopt a change, confidence to embrace change, perception of security of electronic or digital modes of payment (specifically in adopting new modes of technology), the ease of noncash transactions, level of behavioral stimulation towards adoption of innovative technology and payment modes, awareness and knowledge to use various modes of cashless payment and the regulatory framework existing within the country making the use of cashless transactions safe, secure and easy. The ABM approach we use is founded on research findings of survey based data that identified behavioral factors that impact money-use behavior and further put forth the following research questions for this study:

What are the behavioral manifestations of large scale policy changes? Specifically, how do the individual agent outcomes drive macro level outcomes such as transition of a society into a cashless economy?

How does adoption of digital financial platforms cascade among resource scarce communities?

What are the inputs to policy from a behavioral modeling standpoint that could result in better implementation outcomes for demonetisation?

## 12.3 Behavioral and Structural Determinants of Money Use Behavior

A choice of using digital money forms can depend on a number of factors and a key determinant would be the intrinsic characteristics of an individual such as one's openness to adopt a change, confidence to embrace change, perception of security of electronic or digital modes of payment (specifically in adopting new modes of technology), the ease of noncash transactions, and awareness and knowledge to use various modes of cashless payment. Adopting noncash forms of money use can also be a learned behavior due to sociocultural influences, where factors such as peer influence and presence of enablers such as availability of sufficient infrastructure can influence behavior.

Based on review of literature that has brought out evidences of behavioral and systemic variables at play that influence technology adoption behaviors, the study identified predictor variables, for inferring possible impacts on money use behavior viz. (1) Technology Readiness, (2) Self-efficacy, (3) Peer influence, (4) Mandate effect, (5) Perceived Risk and (6) Service Infrastructure. They were adapted either from existing measures, customized to the context or developed anew. The operational definitions for each are as follows.

### *12.3.1 Technology Readiness*

The technology-readiness construct refers to people's propensity to adopt new technologies For carrying out various activities. It is a multidimensional psychographic construct, offering a way to segment customers based upon their underlying positive and negative technology beliefs. The Technology Readiness Index (TRI) scale developed by Parasuraman (2000) has been adapted to the context of the study.

Technology readiness has four underlying dimensions:

Optimism, a positive view of technology and a belief that it offers people increased control, flexibility, and efficiency

Innovativeness, the tendency to be a technology pioneer and thought leader

Discomfort, perceived lack of control over technology and a feeling of being overwhelmed by it
Insecurity, distrust of technology and skepticism about its ability to work properly.

While optimism and innovativeness are contributors to technology readiness, discomfort and insecurity are inhibitors. It is hypothesized that technology readiness of an individual will play a role in influencing his/her money use behavior, such that those with high levels of technology readiness would be more inclined to use digital forms of money.

### *12.3.2 Self-efficacy*

Self-efficacy theory asserts that personal mastery expectations are the primary determinants of behavioral change (Sherer et al. 1982). Self-efficacy has its roots originated from the Social Cognitive Theory (Bandura 1977), where self-efficacy is defined as one's confidence in his or her abilities to perform a task successfully. Social Cognitive Theory suggests that individuals who have more confidence in their skills and abilities will exert more effort to perform a task, persists longer to overcome any difficulties than those who have less confidence in their abilities (Hasan 2006). Psychological trait of self-efficacy has been found to be a strong determinant of acceptance of computer technology (Parker et al. 2006).

### *12.3.3 Peer Influence*

There is a significant body of evidence in general supporting the viewpoint that social influence plays a critical role in influencing behaviors in a wide variety of domains (Venkatesh and Morris 2000). Subjective norms based on the perceptions of the significant others such as family, peers; media etc. also influence the consumers' behavioral choices and social network impacts significantly on attitudes toward an innovation which, in turn, affects the innovation adoption behavior of employees (Talukder and Quazi 2011). In the context of demonetization, it is likely that individuals could get influenced by the peer network who have adopted non cash modes for monetary transactions. Hence those influenced by peers could resort to a change in money use behavior from cash to non-cash forms.

### *12.3.4 Mandate Effect*

Mandate effect is defined as 'The extent to which demonetization has influenced citizens to change money use behavior towards adoption of digital platforms because of the Government mandate'. The dimensions measured under this were Agency (the mandate is issued by the Government and hence a notion that digital money should be used), Contingency (digital money to be used in the absence of cash), Forfeiture (foregoing routine purchases or payment because of non-availability of cash), Perceived difficulty (hurdles faced due to money shortage) and Perceived impact (the perception of a positive impact of demonetization in the long run).

### 12.3.5 Perceived Risk

Fear of losing money and financial data along with personal details negatively impacts the attitude of people towards using the digital or cashless electronic payment modes. The risk of losing money by disclosing credit card information or other personal details through digital or electronic transactions discourages people from using non cash payment modes (Javadi et al. 2012). Perceived risk in the current context refers to the risk associated with the interaction with a computer or a device with respect to information misuse, the possibility of committing errors in data entry or the notion that one could end up paying more for goods online, in the absence of bargaining possibilities. This is hypothesized to impact money use behavior, such that lower levels of perceived risk will influence higher usage of digital money and higher levels of perceived risk will prompt the use of cash.

### 12.3.6 Service Infrastructure

Service infrastructure as a systemic variable is conceptualized in this study and hypothesized to facilitate adoption of digital money usage. Prior research has acknowledged the role of service infrastructure in enabling technology adoption, and has found that lack of it remains a key bottleneck to growth (Azali 2016). The Availability, accessibility and usability of enabling platforms such as Paytm, BHIM or UPI to use e-money for transactions are assessed to understand their impact on money use behavior.

### 12.3.7 Money Use-Behavior

Money Use behavior is treated as the Dependent Variable in the study and this measures the extent to which demonetization has forced people to change their money use behavior in terms of transition to use of digital money. The variable measures the change in money use behavior that can be attributed to demonetization, which resulted in reduced availability of cash and the impetus for using digital

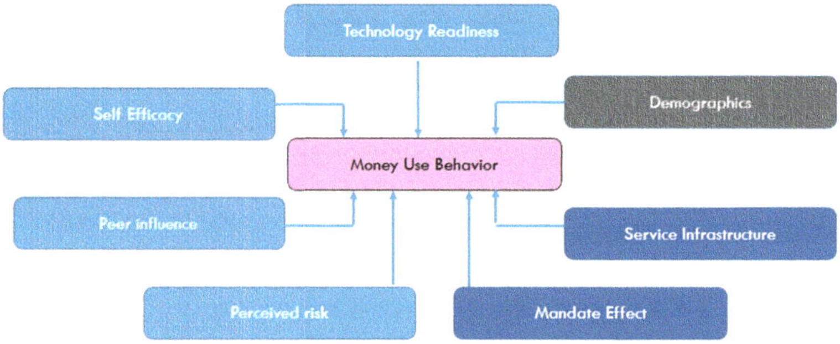


**Fig. 12.1** Conceptual model for money-use behavior

money by the Government. Reduction in cash use, judiciousness in spending, amount of saving post demonetization and the intent to continue to use technology in future are the dimensions studied to evaluate money use behavior. Further, the use of technology post demonetization and despite a scenario with cash availability has also been captured to understand the money use pattern.

The Fig. 12.1 presents the conceptual model for the survey based study of understanding the antecedents of money use behavior.

## 12.4 Approach and Model Description

Agent based modeling (ABM) is well suited to model complex systems, such as the economy, and more particularly those in which different agents' interactions combine to produce unexpected outcomes. ABM offers a way to model social systems that are composed of agents who interact with and influence each other, and adapt their behaviors suited to their environment. Application of such bottom-up approach entails a very wide range of subjects including biology, ecology, social sciences, military planning and physical sciences. In the past, ABM has been identified as useful approach, specifically in macro-economic scenarios, where data maybe scarce and experiments at a larger level can rarely be performed in the real world (Turrell 2016).

The following sections present the overview, description of the model and its implementation.

### *12.4.1 Model Overview*

The model for this study uses empirical evidence from a survey which we conducted to the behavioral drivers of Money-Use Behavior (MUB) in a post-demonetisation economy. The survey showed that a combination of behavioral antecedents (technology readiness (TR), self-efficacy (SE), peer influence (PI), perceived risk (PR)) and systemic attributes (service infrastructure (SI), mandate effect (ME)) influenced the MUB of a person. In this chapter we have attempted to integrate these into a simulate ready decision making model of an agent to study the implications for a population of such agents operating in a landscape with different opportunities for specific cash or noncash modes for financial transactions.

Choice of a mode for making monetary transactions is fundamentally a decision an agent takes under varying situations. We define agent decision making in use of money formats, as a function of behavioral attributes and systemic influencers.

A significant aspect of the ABM in this study is the modelling of inconvenience faced by an agent. When behavioral attribute driven preferences do not match systemic conditions (for example, the behavioral factors lead to a preference for noncash mode but there is absence of service infrastructure for the mode), agent inconvenience is caused, which results in the agent having to move from one location to another to carry out monetary transactions. This behavior is studied in three scenarios; before demonetisation, during demonetisation (when cash availability was restricted and where noncash mode usage was also not ubiquitous) and then in the post-demonetisation scenario (where there was again cash availability).

In our experiments we demonstrate the demonetisation outcomes in the Indian scenario, compare the outcomes in a urban and rural areas and also illustrate a framework for optimal remonetisation in the economy which also takes agent inconvenience into consideration.

### 12.4.2 Model Description

The model elements chosen for this study are based on empirical evidence from a survey which we conducted. The survey studied the behavioral drivers of Money-Use Behavior (MUB) in a post-demonetisation economy. The objective of the survey was to understand behavioral drivers that influenced noncash adoption post demonetisation. Behavioral and systemic variables were identified and data on these variables were gathered among 175 respondents from a population representing a low income demographic offering support services in a technology firm in India (Kummamuru and Raveendran 2017).

The mode for making a monetary transaction is fundamentally a choice available to a decision agent. Understanding the drivers of MUB, i.e., whether ‘cash’ or ‘noncash’ modes is being used for financial transactions in the event of demonetisation has been of research interest in the survey. Behavioral antecedents viz. technology readiness (TR), self-efficacy (SE), peer influence (PI) and perceived risk (PR) identified from literature, along with systemic attributes viz. service infrastructure (SI) and mandate effect (ME) were hypothesized to influence an individual’s MUB.

**Table 12.1** Survey data of intrinsic behavioral parameters

| Parameter—Range [0, 1] | Mean (μ) | Standard deviation (σ) |
|---|---|---|
| Technology readiness (TR) | 0.736 | 0.118 |
| Self-efficacy (SE) | 0.762 | 0.106 |
| Perceived risk (PR) | 0.646 | 0.197 |
| Peer influence (PI) | 0.642 | 0.148 |

A structured questionnaire was administered to collect data on these variables. Multivariate analysis (Multiple Linear Regression and Multiple Analysis of Variance) established statistical significance of the impact of antecedents (TR, SE, PI, ME, PR and SI) on MUB, offering support to the hypotheses of the study, represented in Table 12.1.

Each agent has a behavior model characterized as a function of the variables 'Willingness' and 'Peer Influence', apart from systemic influencers viz. 'Mandate Effect' and 'Service Infrastructure'. The parameters and their computation is discussed below.

#### 12.4.2.1 Willingness (W)

Willingness as a behavioral dimension is assumed in the model as an intrinsic attribute of an agent that represents a state of preparedness to use a money form (cash or noncash) and expressed as a function of Technology readiness (TR), Self-efficacy (SE) and Perceived risk (PR).

The Technology-readiness (TR) construct refers to people's propensity to embrace and use new technologies for accomplishing goals in home life and at work (Parasuraman 2000). It is a multidimensional psychographic construct, offering a way to segment customers based upon their underlying positive and negative technology beliefs.

Self-efficacy (SE), as discussed earlier, is one's confidence in his or her abilities to perform a task successfully. Self-efficacy in this context measures how confident individuals feel about using non-cash forms for performing various transactions.

Perceived risk (PR) in the current context refers to the risk associated with completing any transaction that does not involve the use of physical currency. As explained earlier, the construct measures the fears associated with committing data entry errors or misuse of information supplied by the user.

These characteristics are assigned to every agent based on the distributions we obtained through our survey. Thus taken together TR, SE and PR define an individual's readiness to use noncash means of transaction. The following steps are taken to arrive at an equation of TR, SE and PR that quantifies Willingness:

1. As the components of W, i.e., TR, SE, (-)PR are positively correlated with MUB, we assume that W and MUB are also positively correlated.

2. MUB (noncash) is measured in the survey, and we quantify Willingness as $(W) = k_1(TR) + k_2(SE) - k_3(PR)$, where TR, SE, PR are from survey data and $k_1, k_2, k_3$ are coefficients. Note that the Perceived Risk component is negative indicating that more the sense of PR, less the Willingness.
3. Hence, $corr(W, MUB) = corr(k_1(TR) + k_2(SE) - k_3(PR), MUB)$, which is clearly a function of $k_1, k_2, k_3$.
4. The equation in step 3 serves as the objective function to be maximized using the decision variables $k_1, k_2, k_3$. The optimization technique used was GRG non-linear method to arrive at a plausible co-efficient for TR, SE and PR in the equation of W.
5. It is found that correlation is maximum for this equation of W

$$Willingness(W) = (0.26)TR + (0.48)SE - (0.26)PR \tag{12.1}$$

Once, the aggregate intrinsic behavior of an agent, i.e. Willingness is quantified, a threshold is to be chosen to determine whether the Willingness is towards cash or noncash. If Willingness (W) is greater than the identified threshold, we say that, then the agent's willingness is towards noncash use. The best possible way identified to choose this threshold was to fit the simulation model to the available real-world data. Thus, we take the aid of pre-demonetisation data provided by the Reserve Bank of India (RBI) and match it to the pre-demonetisation model outcomes. The RBI data suggests that 78% of all transactions were in cash pre-demonetisation. The threshold is chosen by varying it in a particular range [0, 1] and observing the %cash users pre-demonetisation for each threshold. The threshold is set to 0.465, where it is observed that close to 78% agents use cash pre-demonetisation.

Evolution of Willingness

The enduring effect of TR on technology acceptance and usage is documented (Lam et al. 2008). Self-efficacy could increase with learning, performance experience, mastery of tasks, vicarious or observational experience (Bandura 1997). According to the consumer decision-making model developed by Blackwell et al. (2001) internal information retrieved from memory (e.g., prior experience, mood, familiarity) could reduce perceived risk. With experience and learning it is possible that there will be changes in the intrinsic characteristics of an individual and thus, it has been modeled such that, with prolonged usage of noncash an agent evolves as its TR and SE improves along with decrease in PR, thus making the agent more willing to use noncash. This acts as a feedback in agent decision making and this agent adaptation technique at an individual level drives emergent behavior of overall economic transition from cash to noncash.

**Table 12.2** Implementation of key model parameters

| Parameter | Parameter implementation | If condition holds true |
|---|---|---|
| Willingness | $(W) = (0.26)TR + (0.48)SE - (0.26)PR > threshold$ | Intrinsic parameters trigger noncash use |
| Willingness evolution | *With noncash use,* $TR_{t+1}, SE_{t+1} = (1.0025)TR_t, SE_t$<br>$PR_{t+1} = (0.9975)PR_t$ | |
| Peer influence | $PI > 1 - (ratio\ of\ noncash\ neighbours)$ | PI triggers noncash use |
| Mandate effect | $p(noncash\ due\ to\ ME) = Demonetisation\ Intensity$ | ME triggers noncash use |
| S. infrastructure | $Distance\ to\ SI(enabled)shop < 1 unit\ distance$ | Enables to use noncash |

For every agent using noncash at a time-step, the intrinsic parameters TR, SE increase by 0.25% and PR decreases at the same rate, mentioned in Table 12.2. This quantification of evolution rate has been achieved through an iterative approach by treating it as a parameter. The evolution rates are varied over a band 0.1–1% over a time span of the scale of a demonetisation (4 months), with increase of 0.05% at every iteration. We observe that the model is able to depict reasonable outcomes when the evolution rate is set to 0.25%. Agent evolution results in increased use of noncash formats of money during the demonetisation period and subsequently shows a decreasing trend post-demonetisation. This subsequent decrease, however, is still higher than noncash money-use in the pre-demonetisation period, which explains the extent of cashless transition in a society. Thus, there is a match with the pattern of noncash use before, during and after demonetisation in the Indian economy as revealed by (Reserve Bank of India (RBI) 2017a) statistics. This approach can be made richer with longitudinal survey of noncash forms by individuals or can be arrived from context specific meta-evidences from literature, which is currently sparse.

#### 12.4.2.2 Peer Influence (PI)

Apart from the construct Willingness (W) computed as a solely intrinsic characteristic of the agent, the model also considers the tendency of an agent to be influenced by peer behavior with regard to use of noncash, denoted as Peer Influence (PI). If individuals observe a model benefitting them, that reinforces the notion that the behavior is acceptable, and the behavior is more likely to be imitated (Brown et al. 2010). This influence is quite likely to be at play in the context of demonetisation to shift to noncash modes of monetary transactions.

The survey dataset captured the tendency of an individual to be influenced towards noncash by its usage by peers as Peer Influence (PI). In the model, the agents observe the ratio of noncash users in their own neighborhood, and tend to adopt noncash modes based on their PI. An individual with higher PI is supposed to have a higher tendency to be influenced towards noncash, thus being easily influenced. In other words, lesser noncash neighbors would trigger noncash influence in an individual with high PI. Using this heuristic, peer influence is modelled such that it triggers noncash use, when the following equation holds true:

$$PI > 1 - (ratio\, of\, noncash\, neighbours) \tag{12.2}$$

As mentioned earlier, each agent has its own PI that is assigned from the survey data and the ratio of noncash neighbors would come from the model scenario the agent is in. In the modelling platform used, Netlogo, each agent has 8 neighbor positions. For instance, if only 2 neighbors use noncash, ratio of noncash neighbors is 2/8 = 0.25. Thus, an agent's PI needs to be greater than (1–0.25), i.e. 0.75 to be influenced towards noncash money-use behavior.

#### 12.4.2.3 Mandate Effect (ME)

Mandate effect denotes the extent to which demonetisation as a policy has influenced citizens to change money-use behavior towards adoption of digital platforms. This behavioral factor may force an agent to use noncash due to a feeling of cash shortage. This notion of cash shortage may trigger noncash use due to ME.

In the model, it is supposed that the notion of cash shortage will depend on the overall cash shortage in the economy, denoted by "Demonetisation Intensity". Demonetisation intensity represents the %cash withdrawn from the economy. Lesser the cash in economy, higher the demonetisation intensity, higher would be the chances of a particular agent to face shortage of cash. Thus, we choose to probabilistically model the trigger toward noncash use due to ME, such that for an agent:

$$\begin{aligned} p(noncash\, use\, due\, to\, ME) &= prob.\, of\, facing\, cash\, shortage \\ &= Demonetisation\, Intensity \end{aligned} \tag{12.3}$$

where, demonetisation intensity gets its value from the %cash withdrawn from the economy i.e. if 85% of cash is withdrawn from the economy, demonetisation intensity is set to be 0.85. Usually, during the phase of demonetisation in an economy, when cash that is withdrawn from the economic system, gradual remonetisation occurs. As remonetisation takes place, cash is restored into the economy and the demonetisation intensity decreases, thus decreasing the probability for agents of facing cash shortage as cash is more readily available (shown later in Fig. 12.8).

#### 12.4.2.4 Service Infrastructure (SI)

Service infrastructure (SI) represents digital services and platforms that enable individuals to perform noncash transactions. SI is conceptualized as digitally equipped shops (for example, with Point of Sale services) in society where agents can use noncash. In a given society, these SI shops are intuitively finite and can be accessed by a limited population. Thus, these SI shops are modelled as specific points situated in the model-world, where agents that are immediately next to these shops can only avail its services. Notably, for an agent to perform noncash use, accessibility to service infrastructure is necessary. Thus, for an agent to have access a particular SI shop to be able to use noncash, this equation should satisfy:

$$\textit{Distance to SI}\ (\textit{enabled})\ \textit{shop} < 1\ \textit{unit distance} \tag{12.4}$$

The number of such SI enabled shops in the model represent the service infrastructure density in the society. Therefore, higher levels of infrastructure density, would indicate more SI shops that allow larger total of individual agents to use noncash.

#### 12.4.2.5 Money-Use Behavior (MUB)

The key parameter of interest for this study is Money-use behavior (MUB), to assess extent of the transition towards a cashless economy. A cashless economy is where majority of financial transactions are on digital platforms rather than through exchange of physical form of money such as banknotes or coins. In the model, MUB is categorized as modes of transactions: cash or noncash. Further, MUB for the overall society is monitored as the % volume of payments made using noncash means.

Each agent, during every transaction decides upon one of these MUB based on behavioral and systemic influencers. The diagram below mentions how behavioral factors together lead to MUB decision making. Individual factors such as W (TR, SE, PR), PI and ME simply act as triggers towards noncash use for an agent, which is fulfilled only if the agent is next to an SI point.

#### 12.4.2.6 Modelling Agent Inconvenience

There is a continuous rise in consideration of convenience in social studies which has been attributed to socioeconomic change, technological progress, competitive business environments, and opportunity costs that have risen with incomes (Berry 1979; Gross 1987). Time and effort saving are two key factors that determine whether a service is perceived to be convenient (Berry et al. 2002). Other measures include appropriateness—fit to specific needs, accessibility—proximity of location, availability, avoidance of unpleasantness (Yale and Venkatesh 1986). Convenience

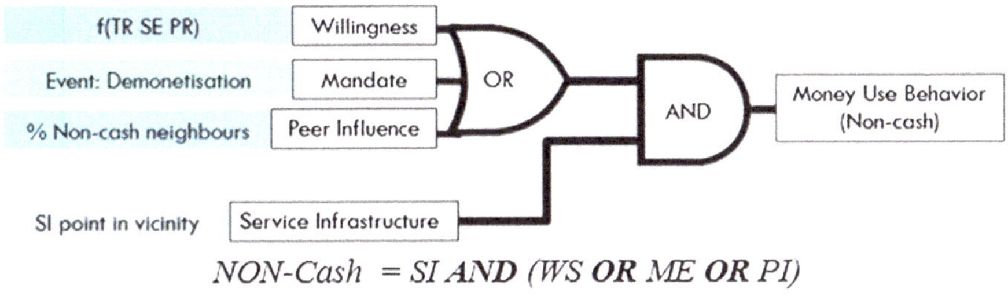


**Fig. 12.2** Logic flow diagram for noncash money-use behavior

is also attributed to the suitable location of service provider (Claessens 2006). Literature also documents the notion of convenience orientation referred to as a person's general preference for convenient services (Berry et al. 2002).

In the context of our study, the notion of convenience orientation is referred as to an individual's general preference to use a particular format of money—cash or noncash. We relate this general preference for use of money format to the behavioural component of 'Willingness (W)' modelled in the study, which is construed as a primary behavioural driver of money-use behaviour. However, actual MUB is not only guided by an individual's preference but also by external factors. Thus, the agent could be exposed to scenarios where its MUB is not as per willingness. Inconvenience is likely to be faced by individuals due to (a) shortage of cash caused by demonetisation (b) lack of service infrastructure or (c) peer pressure of adapting to noncash modes of payment. An agent will also face inconvenience if the inherent preference is to use cash but is forced to use noncash modes for transaction, or similarly vice versa (Fig. 12.2).

In other words, convenience is experienced if an agent is able to use money (cash or noncash) in accordance to the willingness, else, the agent endures inconvenience. In the model, inconvenience is measured as the % population facing inconvenience at a given instant. Further, to evaluate the entire episode of demonetisation, "total-inconvenience" is measured as aggregate inconvenience levels in society over the complete period.

### 12.4.3 Model Implementation

The behavioral antecedents of an agent viz. TR, SE and PR and PI are modeled from the survey data. However, availability levels of material concepts such as cash or service infrastructure is beyond control of an individual. Hence, the model considers for an agent, both presence and absence of (a) demonetisation mandate and (b) adequacy of service infrastructure as parameters of systemic antecedents. The conceptual model (Fig. 12.3) summaries how the model parameters, agent

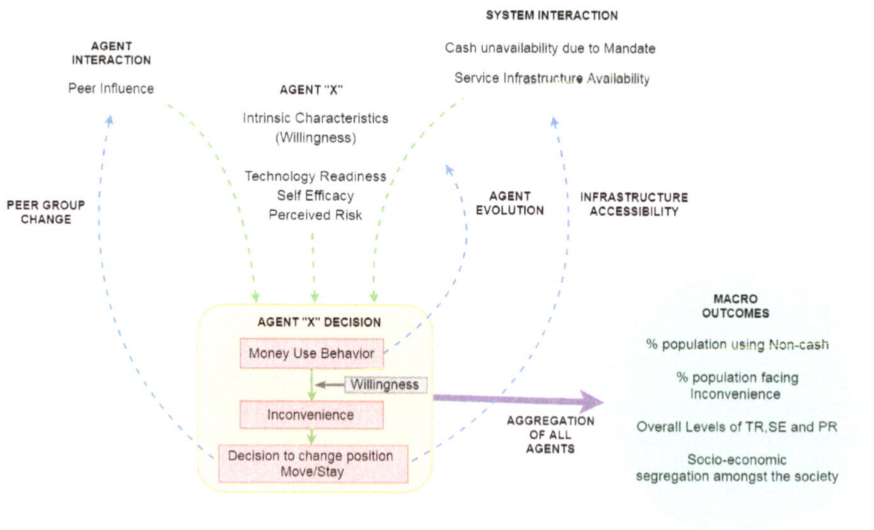


**Fig. 12.3** Conceptual model of the agent based simulation

decision making, agent evolution and macro outcomes of demonetisation fit together in this model.

The modelling platform used for this study is NetLogo, where the agents (turtles) represent common citizens of a society, while specific areas (or patches in Netlogo speak) in the model-world act as shops where essential commodities are readily sold. All shops accept cash as a payment mode, although noncash payment modes are only accepted by limited shops that are digitally equipped (e.g. with Point of Sale (POS) machines). These digitally equipped shops are representative of the service infrastructure in the economy that allows citizens to transact digitally.

A model is initialized with a population of 1000 agents whose intrinsic parameters are assigned through a random-normal distribution ($\mu$, $\sigma$ from Table 12.1) to ensure realistic representation of the population in an economy. These agent-citizens are randomly assigned a position in the model world. The agents can only transact at shops that are immediately next to it. The agents have to necessarily purchase an essential commodity (e.g. food) at every time-step in the model. At every time-step, agent decision on MUB takes place, which is modeled as a function of its own intrinsic behavior, peer group influence, shortage of cash due to demonetisation mandate and availability of infrastructure at the shop (elaborated in Sect. 12.4.3.1). If the agents' choice of payment mode (MUB) is not forced due to systemic or external factors, they experience convenience and continue to hold the position i.e. transact at the same shop using preferred medium for transaction. However, if the agents' choice of payment mode is forced and causes inconvenience, they move to other patches (shops) in the model-world. The

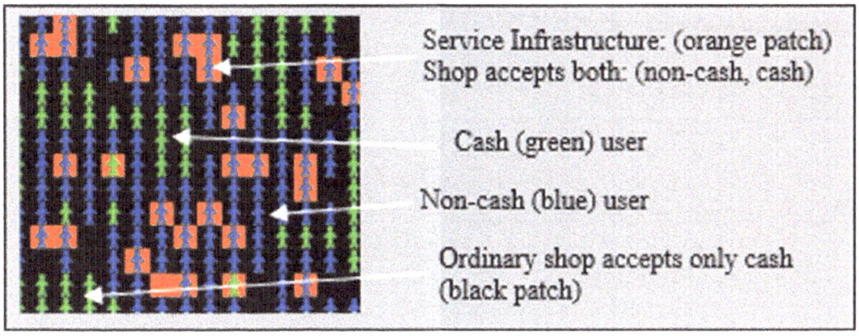


**Fig. 12.4** Model snippet depicting noncash, cash users and infrastructure points

multi-layered feedback driven decision process for every individual agent drives emergent macro-outcome patterns in an economy such as digitalization, population convenience levels and so on (see Fig. 12.4).

#### 12.4.3.1 Agent Decision Making

All agents, at every time-step take a decision to choose a particular MUB—cash or noncash, based on its—Willingness. This MUB decision influences the experience of convenience/inconvenience faced by the agent during the transaction depending on the presence or absence of ME, PI and SI. The pathways of agent decision making is illustrated in Fig. 12.5.

Pathways of agent decision making at every tick in the model is described below: (Fig. 12.6)

1. Perceive: Assess behavioral and systemic factors such as (a) Willingness for noncash (b) Availability of cash (c) Accessibility to infrastructure and (d) Peer behavior.
2. Perform: Agent decision on mode of transaction and associated (in)convenience.

    a. Willingness towards noncash

        i. If there is access to service infrastructure; perform noncash transaction, with convenience.
        ii. If service infrastructure is unavailable; perform cash transaction, with inconvenience.

    b. No Willingness towards noncash—i.e. preference to use cash.

        i. If there is shortage of cash, and/or peer influence towards noncash but service infrastructure exists, inconveniently attempt to use of noncash.
        ii. In absence of Mandate effect and Peer influence, use cash with convenience.

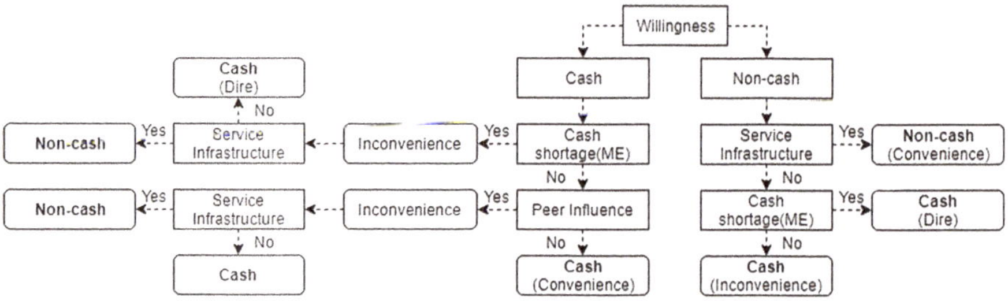


**Fig. 12.5** Pathways of agent decision making on MUB and associated convenience

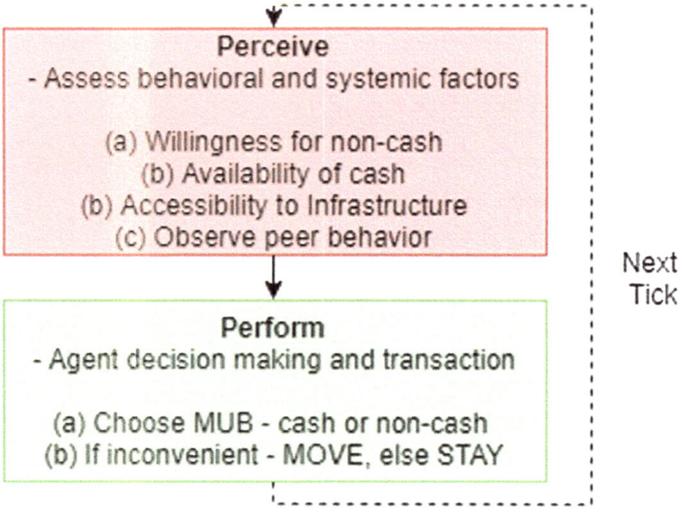


**Fig. 12.6** Process description for agents at every tick

Service infrastructure is indispensable for conducting any noncash operation. Absence of service infrastructure, coupled with cash shortage would lead an agent to a dire situation which indicates a situation of distress or extreme anxiety. Also, whenever an agent faces inconvenience, it would move to change its position to find service infrastructure or to change peer group, in order to transact with convenience. The model considers these set of decision making capabilities for an agent, however it could be hypothesized that prolonged exposure to large spells of dire situation might lead to an agent's exit from the economy through permanent relocation outside the economy or sadly, even death (Worstall 2016).

## 12.5 Simulation Output

After implementing the model as a Logo algorithm on the NetLogo platform, the simulation interface of the platform was customized. The model elements such as Service Infrastructure and Remonetisation rate were chosen as parameters, to provide controls to define the simulation while setup. By varying the levels of Service Infrastructure, the number of digitally-enabled shops in the economy could be regulated, while by varying the Remonetisation rate, the period of remonetisation has been specified before running the simulation. Along with these parameter controls, a toggle is provided to trigger the event of Demonetisation during the simulation run.

As the simulation progresses, the model outputs are recorded and displayed on the interface to track the simulation. The following parameters are captured in the model to enable further analyses: (1) A model-world displays the location of digitally-enabled shops, the position of each agent and their payment mode, the total number of agents using noncash, total number of agents facing inconvenience and the number of agents in a dire state are also plotted. (2) Average values of Technology Readiness, Self-Efficacy and Perceived Risk are plotted to observe the trend of these intrinsic parameters at the society-level. (3) Current 'demonetisation intensity' in the model-economy. These displays and plots are recalculated and updated on every tick, which represents 1-day, in the simulation.

The Fig. 12.7 provides snapshots of the various model outputs generated during the simulation. Figure 12.7a depicts the trend of noncash usage in the economy during the demonetisation period. The noncash usage rises sharply on introduction of demonetisation, due to a high 'demonetisation intensity' in the model-economy. The shortage of cash forces agents to perform noncash transactions. In the next stage, a slowly decreasing trend in noncash usage is observed, influenced by an interplay between gradually decreasing 'demonetisation intensity' that promotes cash usage, and also evolving intrinsic parameters (TR, SE and PR) of agents—towards noncash usage, However, the post-demonetisation noncash usage stabilizes above the noncash usage observed during the pre-demonetisation phase, thus marking an increase in digital payments due to the event of demonetisation.

Figure 12.7b depicts the trend of the number of agents facing inconvenience during the demonetisation period. The proportion of population facing inconvenience rises sharply, on introduction of demonetisation since agents are forced to use payments modes that may be different from their preferred payment mode. The proportion of population facing inconvenience shows a decreasing trend as the 'demonetisation intensity' decreases with gradual remonetisation as well as due to increased preference for noncash caused by 'Evolution of Willingness'. A similar trend is observed for the number of agents in 'dire' state, as depicted in Fig. 12.7c.

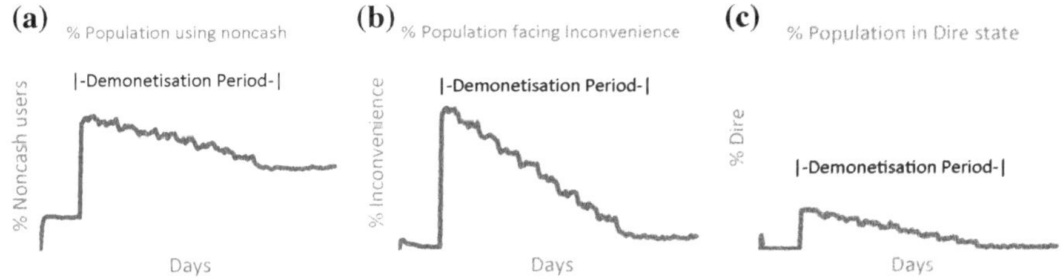


**Fig. 12.7** Model output graphs depicting: **a**—non-cash usage (left), **b**—population inconvenience (center) and, **c**—population in dire state (right)

## 12.6 Experiments and Analysis

The experimentations on our behavior focused ground-up model are: (a) simulating the Indian episode of demonetisation, (b) comparing demonetisation implementation in urban and rural areas, and (c) offering a framework for policy makers to effectively plan demonetisation based on existing service infrastructure and choice of remonetisation. The controls of the model in these experiments are infrastructure density and remonetisation rate. Although, population density and behavior characteristics of agents can be adjusted as per the economy under analysis, they are kept constant for all our experiments.

### *12.6.1 Indian Episode of Demonetisation*

The experiment displays that the model simulation of the case of demonetisation in India, which acts as a base to study the outcomes of demonetisation. During the Indian demonetisation, almost 86% of the cash was retrieved with immediate effect and remonetisation took place gradually over a period of 4 months. The new currency was released into the economy in highly regulated manner by having multiple constraints on withdrawal limits. This resulted in a remonetisation rate of roughly 7–8% every 10 days, which is taken as an input for this experiment (Fig. 12.8).

This is implemented by replicating the Indian scenario in the model-world. The model is initiated with 1000 agents with behavioral characteristics as per the survey parameters and data thereof. The infrastructure density is set to 30% (digital accessibility for India as per the World Bank Group). Further, the remonetisation rate for the model-world is set to be same as it was implemented by the Indian government.

The model run has three phases, pre-demonetisation—no mandate effect (no cash shortage), demonetisation (mandate implemented with gradual remonetisation) and post-demonetisation (cash availability restored). Noncash usage is monitored for each of these phases in the model and depicted in Fig. 12.9.

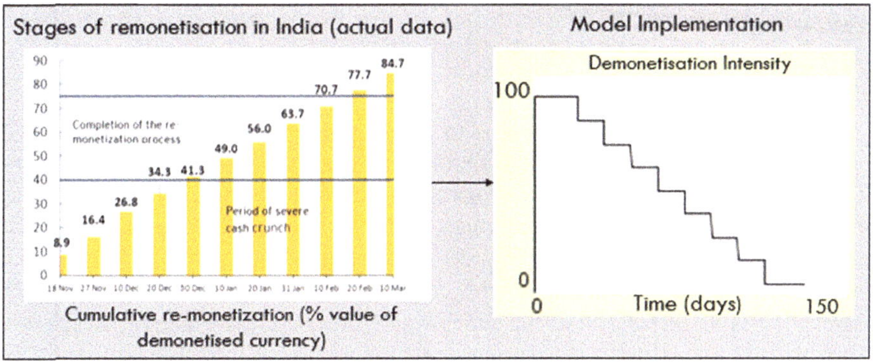


Fig. 12.8 Remonetisation in the Indian economy

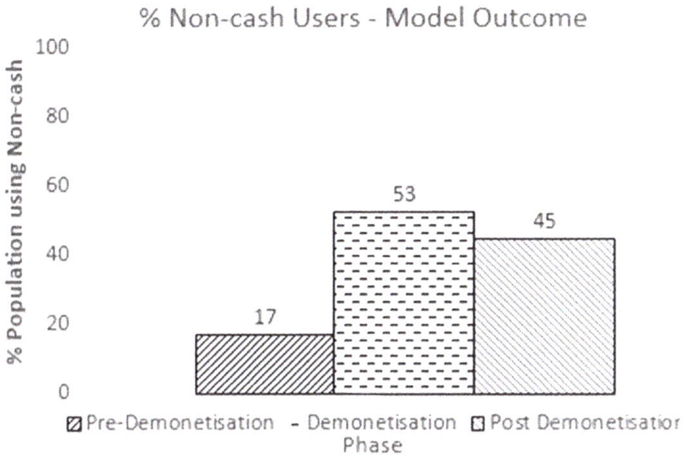


Fig. 12.9 Model output on noncash use during phases of demonetisation

As anticipated, noncash transactions through cards, cheques etc. increases sharply on introduction of demonetisation explained by cash shortage in the economy during the period. Post demonetisation, with restoration of cash in the economy, noncash transactions dropped but remain higher than that in the pre-demonetisation period, which depicts the extent of transition to cashless society in India.

### *12.6.2 Comparison of Demonetisation Outcomes in Urban and Rural Areas*

The goal of the second experiment is to compare the demonetisation outcomes in urban and rural areas. India being a country with a massive urban-rural divide, 70% of the Indian population stay in rural areas (as per 2011 Census), where availability of service infrastructure is scarcer than the rest of India. This scenario is also reflected in the 2017 Internet & Mobile Association of India (IAMAI) report that presents e-infrastructure index for the country considering the extent of digitalization across India. We take the index vales of two representative states—one urban and the other rural, to examine the significance of infrastructure levels by comparing the two demonetisation outcomes.

As per the IAMAI report, Karnataka (urban economy) has an e-Infrastructure index of 50.17, whereas Bihar (rural economy) has an index of 25.44. With these as inputs representing infrastructure density, and other controls same as the earlier experiment, two scenarios are simulated to compare the effects of demonetisation in both urban and rural geographies.

The experiment reveals that population using noncash is significantly higher in urban area coupled with lesser inconvenience and a lesser proportion of the population in a dire state (Fig. 12.10). This reinforces the claim that effects of demonetisation are largely dependent on its infrastructure levels in an area. This experiment result suggests that in the real world, infrastructure equipped urban areas would have higher and sustained noncash usage in the economy alongside lower inconvenience levels, compared to a resource scarce rural area would observe much lower noncash usage together with higher inconvenience, which is in fact also observed by (RBI 2017c; Sood and Baruah 2018). Further, during demonetisation, citizens in rural areas have a substantial chance of facing dire situation due to unavailability of both, cash and infrastructure. Thus, making evident that adequate infrastructure (a) upholds greater adoption towards digital MUB, (b) promotes lower levels of inconvenience and (c) ensures rare emergence of dire situations.

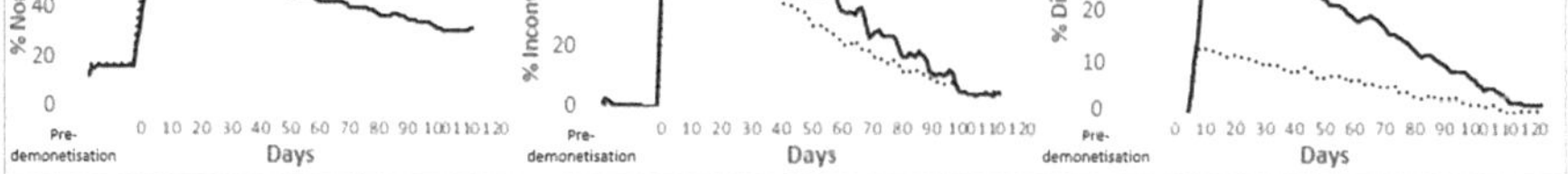


**Fig. 12.10** Model outcomes for urban and rural areas

### 12.6.3 Framework for Future Demonetisation Implementations

This experiment aims at illustrating the outcomes of remonetisation at varying levels of infrastructure density. The effectiveness of policies would depend on how well regulators anticipate stakeholder behaviors, which involves taking a holistic view of the incentives and contexts of citizens—what people experience and perceive, what they want, and what they know about their environments. In an event of demonetisation, a policy maker should plan its implementation in such a way that it ensures (a) sustained digitalization in the economy and (b) minimum total-inconvenience in the society. This implementation planning would include arriving at an appropriate remonetisation period—time duration in which currency is reintroduced into the economy with the objective of maximizing sustained digitalization of payments along with minimizing citizen inconvenience.

The decision of an optimal remonetisation period could be dilemmatic for any country across the globe. Figure 12.11 depicts an increasing trend in sustained noncash transition and total-Inconvenience with longer remonetisation periods. This illustrates, longer remonetisation periods would force adaptation of the majority towards digital payment modes, in return of higher total-inconvenience. Likewise, shorter remonetisation periods would ensure lower total-inconvenience alongside poorly sustained digitalization in the economy. This presents a dilemma between longer remonetisation period—for higher sustained noncash transition and shorter remonetisation periods—for lower inconvenience for the population. The dilemma indicates that the choice of remonetisation period cannot be solely outcome-based, and should be guided by consideration of relevant behavioral and systemic factors.

We devise a construct, "Transition-Inconvenience ratio" to assess the efficacy of a demonetisation plan to maximize transition to noncash usage along with minimum total-inconvenience. This construct is a measure of the sustained noncash transition upon the inconvenience faced by the society. Higher the Transition-Inconvenience ratio, better is the sustained digitalization of the economy alongside lesser inconvenience faced by the society during the transition.

$$Transition\,to\,Inconvenience\,Ratio = \frac{Sustained\,Noncash\,Transition}{Total\,Inconvenience} \tag{12.5}$$

In the experiment, sustained noncash transition was measured as number of noncash users even after the end of the demonetisation phase and total-inconvenience is measured as the aggregate inconvenience over the complete period. To illustrate the use of this construct for choosing the optimal remonetisation period, the periods used for experimentation were: (a) short remonetisation periods: 2 months; 3 months (b) Indian remonetisation period: 4 months, and (c) long remonetisation periods: 5 months; 6 months. These periods were controlled in the model through the rate at which demonetisation intensity decreases.

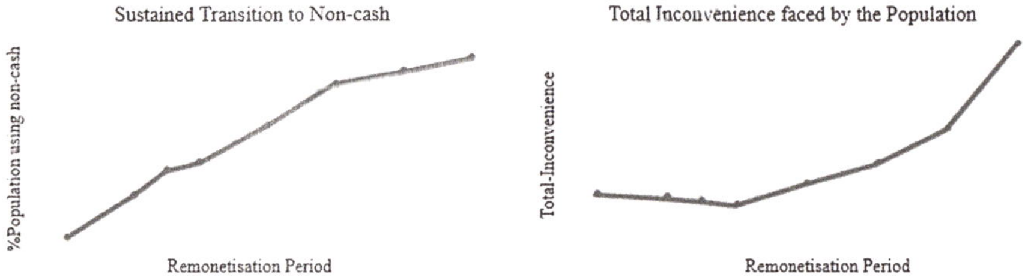


**Fig. 12.11** Model output for non-cash usage and inconvenience with remonetisation period

These 5 remonetisation periods were implemented to model-worlds with service infrastructure density varying from 10 to 100%, to aid the assessment of the choice of remonetisation rate for an economy with a particular infrastructure density through the Transition-Inconvenience ratio.

Figure 12.12 illustrates the "Transition-Inconvenience ratio" for demonetisation periods for every infrastructure density which potentially serves as a tool for policy makers to choose appropriate re-monetisation rates based on an ideal trade-off between noncash transition and convenience. For instance, a country like Singapore (digital internet accessibility—roughly 80% as per The World Bank, 2015) is

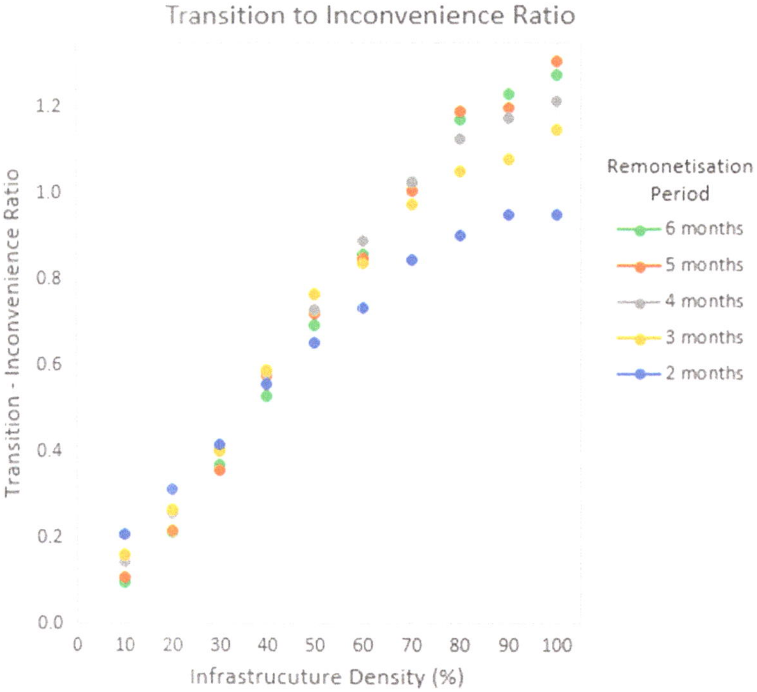


**Fig. 12.12** Transition-Inconvenience ratio for Infrastructure Density and Remonetisation period

optimally suggested a longer remonetisation period of around 5 months. In scenarios with poor infrastructure, the construct "Transition-Inconvenience ratio" favors a shorter remonetisation period, whereas in scenarios with adequate infrastructure a longer remonetisation period is suggested. These findings highlight the optimal remonetisation period for an economy based on its infrastructure.

## 12.7 Conclusion and Future Work

This chapter uses ABM approach to depict money-use behavior considering behavioral antecedents that can influence the adoption of noncash modes for conducting financial transactions by agents. The study contributes to literature by adopting a methodology that is both grounded (in behavioral theories) and composite (with consideration of behavioral and systemic elements). The behavior focused approach helps in modelling inconvenience faced by agents during an societal transition to a cashless economy. The study thus helps in understanding the macro-level implications of demonetisation through a behavioral lens.

While the model is able to depict the demonetisation outcomes in India, it also suggests ways in which other countries could better implement demonetisation, particularly to ensure greater digitalization of the economy by the end of demonetisation, with minimum inconvenience faced by the society. This study could also aid governments and policy makers to formulate feasible interventions towards smoother transitions. Pre-demonetisation drills to understand readiness for a cashless society, presence of adequate service infrastructure in all areas to be affected, etc. can be tested through this model before implementation to appease an otherwise turbulent progression.

Participant numbers from the survey can be augmented to create the larger population of agents, with retention of behavioral parameters. A plausible method is generation of synthetic population which is a microscopic representation of the actual population. The inputs for this method of generating synthetic data are fragments of micro-data and corresponding macro-data of the given demographic. In our study, the survey dataset can serve as the micro-data and relevant macro-data can be acquired from sources such as census data and other public large-scale datasets that represent the demographic of interest. Using this approach of synthetic population, larger demographic societies such as states and countries can be modelled with greater statistical relevance.

Formation of groups based on money use behavior could also be of research interest to further explore the associated societal segregation. This society-level segregation can emerge due to the fact that individuals prefer to surround themselves (and transact) within groups with similar money-use preference. The segregation could be hypothesized to be shaped by the agent's decision to move or stay and could be further analyzed.

The model can be extended by considering different intensity levels of (in)-convenience experienced by agents. Future work could further look at the nitty-gritties of interaction effects of variables such as the moderating role of peer influence or perceived risk grounded with field data.

## Appendix

See Table 12.3 and Fig. 12.13.

**Table 12.3** Operational definitions of key variables

| Variables | Definition |
|---|---|
| Technology readiness (TR) | Individual's propensity to embrace and use new technologies for accomplishing goals in home life and at work |
| Self-efficacy to adapt (SE) | One's judgment about one's capability to perform particular tasks, including the ability to carry out tasks more effectively, persist at tasks, cope more effectively with change, choose more difficult goals, and adopt more efficient task strategies |
| Peer influence (PI) | The extent to which participants feel influenced by their peers to participate in adopting e-transactions |
| Mandate effect (ME) | The extent to which demonetization has enforced citizens to change money use behavior towards adoption of digital platforms because of the Government mandate causing cash unavailability |
| Perceived risk (PR) | Risk perceived in using cashless modes of transactions |
| Service infrastructure (SI) | Enabling platforms such as e-wallets and UPI that enables citizens to use e-money for transactions |
| Money use behavior (MUB) | The extent to which demonetization has forced people to change their money use behavior in terms of transition to use of e-money |
| Willingness | An intrinsic attribute of an agent that represents a state of preparedness to use a money form (cash or non-cash) and expressed function of TR, SE and PI |
| Convenience | A state that the agent experiences when the attribute of willingness to use a particular form of money use matches actual money use behavior |

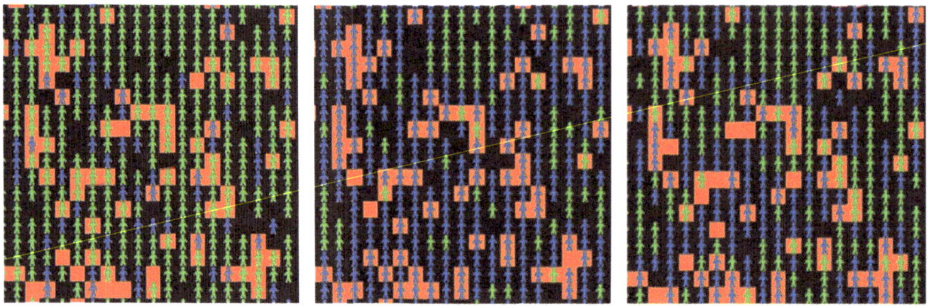

**Fig. 12.13** Model-world output for the three phases of the simulation—Pre-demonetisation (left), During demonetisation (center), Post-demonetisation (right)

## References

Azali, K. (2016). Cashless in Indonesia: Gelling mobile e-frictions? *Journal of Southeast Asian Economies (JSEAE), 33*(3), 364–386.

Bandura, A. (1977). Self efficacy: Toward a unifying theory of behavioral change. *Psychological Review, 84*(2), 194–215.

Bandura, A. (1991). Social cognitive theory of self regulation. *Organizational Behavior and Human Decision Processes, 50,* 248–287.

Bandura, A. (1997). *Self-efficacy: The exercise of control.* New York: W. H. Freeman and Company.

Barat, S., Kulkarni, V., Clark, T., & Barn, B. (2017). An actor-model based bottom-up simulation—An experiment on Indian demonetisation initiative. In *Simulation Conference (WSC), 2017 Winter* (pp. 860–871). IEEE (December).

Berry, L. L. (1979). The time-buying consumer. *Journal of Retailing, 55*(4), 58–69.

Berry, L. L., Seiders, K., & Grewal, D. (2002). Understanding service convenience. *Journal of Marketing, 66*(3), 1–17.

Blackwell, R. D., Miniard, P. W., &. Engel, J. F. (2001). *Consumer behavior.* Troy, MO: Harcourt College Publishers.

Braga, F., Isabella, G., & Mazzon, J. (2013). Do digital wallets as a payment method influence consumer in their buying behavior. *Encontro Da Anpad–Enanpad., 35.*

Braun-Munzinger, K., Liu, Z., & Turrell, A. (2016). *An agent-based model of dynamics in corporate bond trading.* Bank of England Staff Working Paper No. 592.

Brown S. A., Dennis, A. R. &. Venkatesh, V. (2010). Predicting collaboration technology use: Integrating technology adoption and collaboration research. *Journal of Management Information Systems, 27*(2), 9–54.

Chandrakar, I. (2017). Analysis of the effects of currency demonetization in India using Bayesian prediction. In *2017 2nd International Conference on Computing and Communications Technologies (ICCCT)* (pp. 117–122). IEEE (February).

Claessens, S. (2006). Access to financial services: A review of the issues and public policy objectives. *The World Bank Research Observer, 21*(2), 207–240.

Dosi, G., Fagiolo, G., & Roventini, A. (2006). An evolutionary model of endogenous business cycles. *Computational Economics, 27*(1), 3–34.

Fagiolo, G. (2005). Endogenous neighborhood formation in a local coordination model with negative network externalities. *Journal of Economic Dynamics and Control, 29,* 297–319.

Fagiolo, G., Dosi, G., & Gabriele, R. (2004). Matching, bargaining, and wage setting in an evolutionary model of labor market and output dynamics. *Advances in Complex Systems, 14,* 237–273.

Gross, B. L. (1987). Time scarcity: Interdisciplinary perspectives and implications for consumer behavior. *Research in consumer behavior* (pp. 1–54). Greenwich: JAI Press.

Hall, B., & Khan, B. (2003). Adoption of new technology. In D. C. Jones (Ed.), *New economy handbook.* San Diego: Academic Press.

Hasan, B. (2006). Delineating the effects of general and system-specific computer self-efficacy beliefs on IS acceptance. *Information & Management, 43*(5), 565–571.

Javadi, M. H., et al. (2012). An analysis of factors affecting on online shopping behavior of consumers. *International Journal of Marketing Studies, 4*(5), 81–98.

Jonker, N. (2007). Payment instruments as perceived by consumers–Results from a household survey. *De Economist, 155*(3), 271–303.

Kosse, A. (2010). The safety of cash and debit cards: A study on the perception and behaviour of Dutch consumers.

Krishnan, D., & Siegel, S. (2017). Survey of the effects of demonetisation on 28 slum neighbourhoods in Mumbai. *52*(3).

Kummamuru, S., & Raveendran, J. (2017). Behavioural drivers towards digital money usage post demonetisation: Insights from a study of pink collar workers. In *5th Biennial Indian Academy of Management Conference*.

Kupetz, H. (2007). The cashless future. *The Futurist.* May-June.

Lam, S. Y., Chiang, J., & Parasuraman, A. (2008). The effects of the dimensions of technology readiness on technology acceptance: An empirical analysis. *Journal of Interactive Marketing, 22*(4), 19–39.

Mai, H. (2016). *Cash, freedom and crime use and impact of cash in a world going digital.* Deutsche Bank Research (s.l.).

OCED. (2017). *Ensuring financial education and consumer protection for all in the digital age.* G20/OECD INFE report on digitalisation, consumer protection and financial literacy.

Parasuraman, A. (2000). Technology readiness index (TRI) a multiple-item scale to measure readiness to embrace new technologies. *Journal of Service Research, 2*(4), 307–320.

Parker, K., Williams, M., & Turner, N. (2006). Modeling the antecedents of proactive behavior at work. *Journal of Applied Psychology, 91*(3), 636–652.

Reserve Bank of India (RBI). (2017a). *From cash to non-cash and cheque to digital: The unfolding revolution in India's payment systems.*

Reserve Bank of India (RBI). (2017b). *Impact of demonetisation on the financial sector.* https://rbidocs.rbi.org.in/rdocs/Bulletin/PDFs/IDFS5EBBDCCB9C274F0E921997DA8EC93CCA.PDF.

Reserve Bank of India (RBI). (2017c). Macroeconomic impact of demonetisation-A preliminary assessment.

Sargent, R. (2005). Verification and validation of simulation models. In *Proceedings of the 37th Conference on Winter Simulation, FL* (pp. 130–143).

Schuh, S., & Stavins, J. (2016). How do speed and security influence consumers' payment behavior? *Contemporary Economic Policy, 34*(4), 595–613.

Sherer, M., Maddux, J. E., Mercandante, B., Prentice-Dunn, S., Jacobs, B., & Rogers, R. W. (1982). The self-efficacy scale: Construction and validation. *Psychological Reports, 51*(2), 663–671.

Sood, A., & Baruah, A. (2018). The new moral economy. In *Note-bandi: Demonetisation and India's elusive chase for black money.*

Talukder, M., & Quazi, A. (2011). The impact of social influence on individuals' adoption of innovation. *Journal of Organizational Computing and Electronic Commerce, 21*(2), 111–135.

Turrell, A. (2016). Agent-based models: understanding the economy from the bottom up.

Venkatesh, V., & Morris, G. (2000). Why don't men ever stop to ask for directions? Gender, social influence, and their role in technology acceptance and usage behavior. *MIS Quarterly, 24*(1), 115–139.

Worstall, T. (2016*). India's demonetisation kills 100 people apparently-this is not an important number*. Forbes Mumbai Forbes India. Retrieved December 8.

Yale, L., & Venkatesh, A. (1986). Toward the construct of convenience in consumer research. *ACR North American Advances.*

**Rishi Bubna** Mechanical Engineering and Management minor from IIT Gandhinagar, works as a researcher at Behavior Business and Social Sciences (BBS).

**Jayasree Raveendran** is a Ph.D. from IIT Madras and works in the field of behavioural economics at BBS.

**Suman Kumar** is a researcher at BBS working in modelling & simulation and its real world applications.

**Mayuri Duggirala** Ph.D. (Management) from IIT Madras, works in behavioral science, particularly in organizational behavior research at BBS.

**Mukul Malik** is a researcher at BBS working in the areas of Machine Learning and Data Mining. He did his Masters in Statistics from IIT Kanpur.

# Index





J. Sokolowski et al. (eds.), *Summer of Simulation*, Simulation Foundations, Methods and Applications, https://doi.org/10.1007/978-3-030-17164-3

Printed by Printforce, the Netherlands